Lecture Notes in Intelligent Transportation and Infrastructure

Series editor

Janusz Kacprzyk, Systems Research Institute, Polish Academy of Sciences, Warszawa, Poland

The series "Lecture Notes in Intelligent Transportation and Infrastructure" (LNITI) publishes new developments and advances in the various areas of intelligent transportation and infrastructure. The intent is to cover the theory, applications, and perspectives on the state-of-the-art and future developments relevant to topics such as intelligent transportation systems, smart mobility, urban logistics, smart grids, critical infrastructure, smart architecture, smart citizens, intelligent governance, smart architecture and construction design, as well as green and sustainable urban structures. The series contains monographs, conference proceedings, edited volumes, lecture notes and textbooks. Of particular value to both the contributors and the readership are the short publication timeframe and the world-wide distribution, which enable wide and rapid dissemination of high-quality research output.

More information about this series at http://www.springer.com/series/15991

Aboul Ella Hassanien · Mohamed Elhoseny
Syed Hassan Ahmed · Amit Kumar Singh
Editors

Security in Smart Cities: Models, Applications, and Challenges

 Springer

Editors
Aboul Ella Hassanien
Faculty of Computers and Information
Cairo University
Giza, Egypt

Mohamed Elhoseny
Faculty of Computers and Information
Mansoura University
Dakahlia, Egypt

Syed Hassan Ahmed
Department of Computer Science
Georgia Southern University,
  Statesboro Campus
Statesboro, GA, USA

Amit Kumar Singh
Department of Computer Science
  and Engineering
National Institute of Technology Patna,
  Patna University Campus
Patna, Bihar, India

ISSN 2523-3440   ISSN 2523-3459 (electronic)
Lecture Notes in Intelligent Transportation and Infrastructure
ISBN 978-3-030-13185-2   ISBN 978-3-030-01560-2 (eBook)
https://doi.org/10.1007/978-3-030-01560-2

This Springer imprint is published by the registered company Springer Nature Switzerland AG
The registered company address is: Gewerbestrasse 11, 6330 Cham, Switzerland

Preface

Due to the increase in Internet of Things (IoT) application size, complexity, and the number of components, it is no longer practical to anticipate and secure all possible interactions and data processing in these applications using the traditional information security models. The emerging of new engineering research areas is a clear evidence of the emergence of new demands and requirements of IoT applications to be more secure. The IoT offers advanced connectivity of devices, systems, and services that go beyond machine-to-machine communications and covers a variety of domains and applications. The interconnection of embedded devices is expected in many fields including Mobility and Intelligent Transport. It is being promoted by the software engineering community to use such systems as the adequate solution to handle the current requirements of complex big data processing problems that demanding distribution, flexibility, and robustness. However, information security of such systems is still on the fly a big challenge. The main objective of this book is to provide an exhaustive review on the challenges of information security in smart and intelligent applications, especially those are working in big data environments. In addition, it aims to provide a collection of high-quality research works that address broad challenges in both theoretical and application aspects of intelligent and smart real-life applications. This book contains a set of chapters that stimulates the continuing effort on the application of the big data processing that leads to solve the problem of big data processing in smart cities applications.

In this book, we present the concepts associated with secure IoT applications in three distinct parts.

1. **Part I** discusses the challenges surrounding the IoT-based healthcare applications. This part aims to provide a clear view about the fundamental concepts for building a secure data transmission for healthcare service applications. It contains five chapters.
2. **Part II** discusses different security models for IoT applications in smart cities. This part includes five chapters.

3. **Part III** discusses the applications of secure network communications in smart IoT applications. Different topics are covered in that part and are well presented in five chapters to ensure the context is simple for understanding.

Giza, Egypt   Aboul Ella Hassanien
Dakahlia, Egypt   Mohamed Elhoseny
Statesboro, USA   Syed Hassan Ahmed
Patna, India   Amit Kumar Singh
Springer 2018

Contents

Part I
Challenges of Smart Healthcare Applications

Big Data Challenges and Opportunities in Healthcare Informatics and Smart Hospitals

Mohammed K. Hassan, Ali I. El Desouky, Sally M. Elghamrawy
and Amany M. Sarhan

Abstract Healthcare informatics is undergoing a revolution because of the availability of safe, wearable sensors at low cost. Smart hospitals have exploited the development of the Internet of Things (IoT) sensors to create Remote Patients monitoring (RPM) models that observe patients at their homes. RPM is one of the Ambient Assisted Living (AAL) applications. The long-term monitoring of patients using the AALs generates big data. Therefore, AALs must adopt cloud-based architectures to store, process and analyze big data. The usage of big data analytics for handling and analyzing the massive amount of big medical data will make a big shift in the healthcare field. Advanced software frameworks such as Hadoop will promote the success of medical assistive applications because it allows the storage of data in its native form not only in the form of electronic medical records that can be stored in data warehouses. Also, Spark and its machine learning libraries accelerate the analysis of big medical data ten times faster than MapReduce. The advanced cloud technologies that are capable of handling big data give great hope for developing smart healthcare systems that can provide innovative medical services. Building smart Remote patient monitoring models using cloud-based technologies will preserve the lives of patients, especially the elderly who live alone. A case study for monitoring patients suffering from chronic diseases (blood pressure disorders) for 24 h with a reading every

M. K. Hassan (✉) · A. I. El Desouky
Department of Computer Engineering and Systems, Faculty of Engineering, Mansoura
University, Mansoura, Egypt
e-mail: m_kamal1976@hotmail.com

A. I. El Desouky
e-mail: prof.dr.ali.eldosouky@gmail.com

S. M. Elghamrawy
Department of Computer Engineering, MISR Higher Institute for Engineering and Technology,
Mansoura, Egypt
e-mail: sally_elghamrawy@ieee.org; sally@mans.edu.eg

A. M. Sarhan
Department of Computer Engineering and Automatic Control, Faculty of Engineering, Tanta
University, Tanta, Egypt
e-mail: prof.dr.amany.sarhan@gmail.com

© Springer Nature Switzerland AG 2019
A. E. Hassanien et al. (eds.), *Security in Smart Cities: Models, Applications,
and Challenges*, Lecture Notes in Intelligent Transportation and Infrastructure,
https://doi.org/10.1007/978-3-030-01560-2_1

15 min using a cloud-based monitoring model shows its effectiveness in predicting the health status of the patients.

Keywords Smart hospitals · Big data · Cloud computing
Smart remote patient monitoring (RPM) · Ambient assisted living (AAL)
Hadoop · Spark · Healthcare informatics · Elderly · Chronic diseases

1 Introduction

In the digital universe era, there is an explosion in the amount of data generated every day. The flood of data is increasing exponentially because of the diversity of data generation sources such as Internet of Things (IoT) sensors, Radio-Frequency Identification (RFID), social media, Global Positioning System (GPS), log files, images, videos, texts and the computerization of all life aspects. The generated data from these sources has a large volume, high velocity and a significant degree of variety, which is referred to as Big Data. The term Big Data become a buzzword because the size of data is doubling every two years and will reach to around 45 ZB ($45*10^{21}$ bytes) by 2020. Therefore, there are many worries and challenges to keep pace with this huge increase and how to benefit from it [1].

Modern organizations found that there is an imminent need to collect, store, analyze this data to exploit these precious assets. The expected contributions of big data in the field of health informatics include developing innovative smart services, treating illnesses, maximizing the utilization of the health budgets, maximizing the utilization of the limited medical crews, enhancing the quality of patient life by preventing emergencies, and predicting epidemics to deal with them quickly [2].

There is an exponential increase in big electronic health datasets, which cannot be managed using the conventional techniques. The optimal utilization of big data sets in the healthcare informatics faces many challenges includes the volume of data, the speed of generated data, the variety of data types, the veracity of data, and the privacy of patient's medical information [3]. Healthcare informatics field is expected to benefit from the rapid development of big data analytics tools in solving critical problems such as knowledge representation, diagnosis of diseases, clinical decision support [4].

By the end of 2017, the number of electronic health records will be more than ten billion for tens of millions of patients, which poses many needs and challenges

- The need to develop infrastructures that are capable of processing data in parallel.
- The need to provide safe storages for the huge amount of unstructured data sets.
- The need to provide a fault-tolerant mechanism with high availability.

Hadoop technology has succeeded in dealing with the most of above challenges that face the healthcare industry. The HDFS and MapReduce engine is capable of processing terabytes of data in an innovative manner using commodity servers [5].

Spark is capable of using big medical data generated from multiple sources to gain insights and knowledge using various Machine Learning library (MLlib). In conventional medicine, doctors use their knowledge in diagnosing the illness and describing the treatments, but in the coming years, a significant shift toward evidence-based medicine will take place. The development of smart healthcare applications will provide accurate treatment decisions by analyzing the clinical data of the patients using advanced Machine Learning (ML) techniques.

Aggregating patient's datasets into big datasets according to the illnesses' types will enable ML algorithms to provide robust evidence about the predicted health status of the patients and comprehend nuances between patients, which will make a significant shift in the development of smart healthcare applications [6].

The rest of this chapter is organized as follows. Section 2 summarizes big data definitions and attributes. A brief about cloud computing is presented in Sect. 3. Section 4 illustrates big data analytics such as Hadoop and Spark. Section 5 provides a literature review about data mining in health informatics such as AAL and RPM models. Section 6 presents a case study for monitoring patients suffering from chronic diseases such as blood pressure disorders, and Sect. 7 concludes our work.

2 Big Data

2.1 Big Data Definition

In the current decade, Big Data became the buzzword of the computer science field. In the past, this term has been formulated to reference the big volume of data. This term is loose because what is considered big data today will be considered very small in the future because of the continuous development of the storage methods. Nowadays, Big data is used to describe data with volume in Petabyte (10^3 TB), Exabyte (10^3 PB), or zettabyte (10^3 EB). Also, the term big data is used to express the technology required for handling a large amount of data using tools and processes.

The most practical definition of Big Data is datasets with characteristics beyond the ability of commonly used hardware and software to deal within a reasonable time. One of the most comprehensive definitions is *"Big Data is a term that is used to describe data that is high volume, high velocity, and/or high variety; requires new technologies and techniques to capture, store, and analyze it; and is used to enhance decision making, provide insight and discovery, and support and optimize processes"* [7, 8].

Big Data is handled by splitting big files into smaller data chunks; then they will are replicated and distributed among different clusters according to the WORM (write once read many times) methodology. The processing of big data is accomplished using a distributed parallel framework that processes data and logic in parallel on nodes where data is located, unlike the conventional method where data is collected from different sites to be processed in a central node.

Fig. 1 The three V's model
of big data

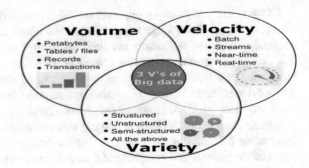

2.2 Big Data Attributes

As shown in Fig. 1, Gartner's analyst Doug Laney introduced the first model that use "Vs" to describe the characteristics of big data in 2001. This model had selected volume, velocity, and variety as the main dimensions that describe Big Data [9].

2.2.1 Volume

Volume refers to the quantity of the generated or stored data. The size of the data indicates whether to be considered as big data or not, the term big data is used nowadays for data above hundreds of terabytes. In big data era, the volume of data became insanely large as Facebook alone has more than 300 billion images. The volume of the digital universe today is about 2.7 ZB, and it is likely to increase 50% every year. This massive amount of data need to be analyzed using advanced analytical techniques to gain knowledge and provide helpful insights to the decision makers [10].

2.2.2 Velocity

Velocity refers to the speed of data flow and the elapsed time in collecting, processing and taking the appropriate action, which required to be performed in real-time in many applications. For instance, Online store monitors the login of the users and what they are searching about to suggest some goods based on their historical transactions. This technique increases the sales and the profit of the site taking into account that all these operations must be performed at a super speed because they are done in real time. One of the challenging applications that need high speed is the fraud detection in the Electronic Commerce (e-Commerce) and online banking fields.

2.2.3 Variety

Variety refers to the diversity of data structures and types. Data is categorized to structured data such as databases, semi-structured data such as XML, and unstructured data such as texts. Nowadays, the most generated data is unstructured data such as images, texts, videos, and interactions on social networks.

2.2.4 Veracity

International Business Machines (IBM) defined Big Data using "The 4 V's of Big Data" by adding a new characteristic called "Veracity" to the previous V's. Veracity is the operation that checks the accuracy and trustworthy of data that will be used in the decision-making operation [11].

2.2.5 Value

As shown in Fig. 2, in 2014 a new characteristic called "Value" was added to expand the model to 5 V's. This characteristic is interested in determining the value of analyzing Big Data. Big organizations use this characteristic to evaluate the potential benefits from any analytical operation to avoid sinking in a flood of Big Data and complex analytic operations without tangible results [12].

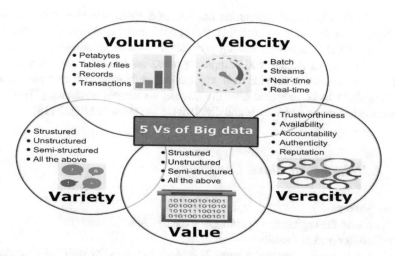

Fig. 2 The Five V's model of big data

2.2.6   Variability

It seems that the idea of describing Big Data using characteristics that begin with 'V' letter has attracted researchers. They have expanded the model of Big Data by adding more 'V' characteristics. In 2013, some researchers proposed the six V's model of Big Data by adding variability, which is different from variety. This Characteristic is concerned with the effect of the continuous data change on the data homogeneity [13]. Some researchers have described the six V's model of Big Data by adding validity and Volatility to IBM 4 V's instead of variability [14].

2.2.7   Visualization

In 2013, the seven V's model of Big Data had been proposed by adding the character-istic of visualization to the previous models. The Visualization of complex Big Data using graphs and charts is much more informative than text reports or spreadsheets [15]. The passion of V's didn't stop at this point, in 2014 models with 8 and 10 V's have been proposed [10, 11]. In 2017, the biggest Big Data model that contains 42 V's has been proposed [16].

3   Cloud Computing

In the past decade, cloud computing has received much attention and became a buzzword in the Information Technology (IT) field, and it is always associated with the Big Data field. Cloud computing provides a scalable, elastic infrastructure at low cost by sharing storage and computing resources to save time and cost. The innovative architectures provide data, application, architecture or even platform as a service that can be executed over the internet on remote machines [17]. Cloud computing has many advantages and disadvantage as shown in Fig. 3 [18].

3.1   Advantages of Cloud Computing

- **Cost Savings**
  Minimizing the capital and operational expenses of infrastructures.
- **Availability and reliability**
  Reliance on service provider ensures 24-h service availability with fail-over mech-anisms.
- **Integration and manageability**
  Using a simple web-based interface for accessing services, software, and applica-tions without paying attention to installation, maintenance or software integration.

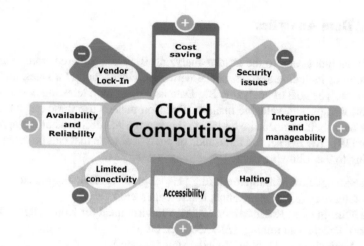

Fig. 3 Cloud computing advantages and disadvantages

- **Accessibility**
  Data stored in the cloud can be accessed easily anywhere and anytime.
- **Quick deployment**
  Easily and quick deployment with focusing on the objectives and activities and forgetting about technology.

3.2 Disadvantages of Cloud Computing

- **Security issues**
  The sensitive information of the company will be vulnerable to be hacked, as it is transferred to a third-party cloud service provider.
- **Halting**
  The system is prone to be wholly stopped if the internet connection is interrupted or when the cloud service provider suffers from technical glitches.
- **Vendor Lock-In**
  Switching cloud service provider is difficult because of the need to reintegrate current cloud applications with another platform (e.g., deploying applications that are running on windows platform to Linux platform instead).
- **Limited connectivity**
  The customer has no administrative roles on the infrastructure, but he can only manage services, applications, and data.

4 Big Data Analytics

Big data analytics refer to the task of analyzing Big Data gathered from many different sources including sensors, social networks, videos, digital images, and sales transactions. The goal of analyzing Big Data is to discover interesting patterns, hidden relationships, and valuable insights. Decision makers use these results to gain an edge over their competitors by taking the appropriate decisions at the appropriate time [19]. The market has thousands of analytical tools that can be categorized according to the following tasks:

- **Data Storage and Management** (e.g., Hadoop, MongoDB, Talend, etc.)
- **Data Cleaning** (e.g., OpenRefine, DataCleaner, etc.)
- **Data Mining** (e.g., Rapid Miner, Waikato Environment for Knowledge Analysis (Weka), Oracle data mining, IBM Modeler, etc.)
- **Data Analysis** (e.g., Qubole, BigML, Statwing, etc.)
- **Data Visualization** (e.g., Tableau, Silk, Charito, Plot.ly. etc.)
- **Data Integration** (Blockspring, Pentaho, etc.)
- **Data Languages** (R, Python, etc.)
- **Data Collection** (Import.io, etc.)

4.1 Apache Hadoop

Hadoop is an open source software developed by Apache™ Hadoop® foundation for scalable, distributed and reliable computing.

4.1.1 Hadoop Definition

Hadoop is a software framework that distributes the processing of big datasets across sets of computers' clusters using agile programming models. It offers a high-level of scalability by providing hundreds or even thousands of machines; each machine is offering local storages and computations rather than depending on centralized servers. Failures are handled in the application layer by the library itself to ensure the availability of services with this cluster of computers [5]. Doug cutting inspired the idea of Hadoop framework from Google File System (GFS) and gave this architecture the name of his son's elephant toy, and since then it becomes the standard for storing, processing and analyzing the massive amount of data [20]. Hadoop is used in many applications such as health care, web applications, telecommunications, e-commerce, finance, genomics, politics, energy, and travel.

4.1.2 Hadoop Characteristics

- **Distributed**

  Hadoop relies on Hadoop Distributed File System (HDFS) to provide reliable shared storage and on MapReduce to provide distributed data analysis. Data is written once and can be read several times.
- **Scalable**

  Hadoop shows a high level of scalability where any node can be added or removed easily. The number of servers contained in a single cluster can range from tens to hundreds or even thousands.
- **Fault-tolerant**

  Hadoop adopts a fault-tolerant technique by replicating data across multiple nodes. When a node fails, data will be available on other nodes having the same data.
- **Cost-effective**

  Besides that Hadoop is an open source, its operation depends on a large number of inexpensive commodity servers distributed on different clusters.
- **Flexibility**

  Structured, semi-structured and unstructured data can be processed using Hadoop.

4.1.3 Hadoop Components

Hadoop framework consists of four main components (Hadoop Common, Hadoop Yarn, Hadoop Distributed File System (HDFS), and Hadoop MapReduce). The HDFS and MapReduce are the most important components in Hadoop's architecture.

- **Hadoop Common**

  Hadoop Common has Supporting utilities for Hadoop modules.
- **Hadoop YARN**

  Hadoop Yarn is responsible for managing cluster's resources and job scheduling.
- **HDFS**

  A distributed file system facilitates storage and access to application data. As shown in Fig. 4, HDFS is a distributed file system written in Java and has a master-slave architecture that provides redundant storage for a massive amount of data may be in a terabyte or even petabyte [21].

Steps for storing the file in HDFS

- Files are split into data chunks (blocks) with size 64 or 128 MB for each block.
- Data chunks are distributed over different data nodes (machines).
- Data chunks are replicated over different data nodes (machines).
- The name node (master node) holds metadata about the chunks of each file and the location of each chunk.

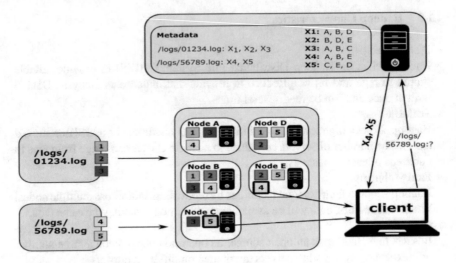

Fig. 4 Storing and retrieving files using HDFS architecture

– **Hadoop MapReduce**
   As illustrated in Fig. 5, MapReduce is a software framework that facilitates developing applications for processing big datasets (terabyte or petabyte data-sets) in-parallel on a large cluster (number of nodes) of commodity servers in a fault-tolerant, reliable manner. The main components of MapReduce function are:

• **Mapper function**
   Mapper function parses raw source data and maps values to unique keys in the form of <key, value> pairs. Mapper is described as the operation of extracting and organizing information. The Map task usually works on one HDFS block and runs on the node where this block is stored.
• **Shuffle and Sort function**
   It is an intermediate function that works after the completion of all Mappers and before the beginning of Reducers to put the output of mappers in a suitable form for the execution of the reducers.
• **Reducer**
   Reducer function operates on shuffled and sorted intermediate data generated from the mapper functions to produce the final output [22].

4.1.4 Hadoop Clients

Many big enterprises rely on Hadoop in its operation such as:

• **Google**
   Google is the inventor of this technology and uses it for reverse search indices, page ranking, web crawling, etc.

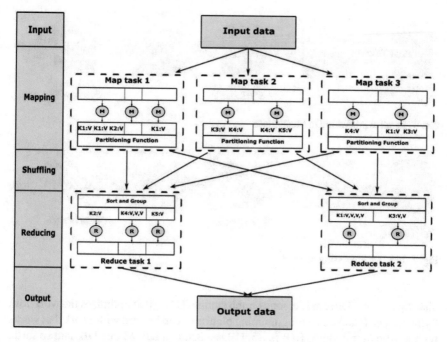

Fig. 5 MapReduce functions

- **Yahoo**
  Yahoo is an active contributor to the Hadoop community, and they are confirming that they are running more than 100,000 Central Processing Units (CPUs) for search and research applications.
- **eBay**
  eBay is using it for search and research optimization work.
- **Amazon**
  Amazon has hundreds of millions of items in its catalogs and hundreds of millions of customers that need to be managed, so they are using Hadoop for their product search indices, various analytical jobs, and log analysis.
- **Facebook**
  Also, Facebook is a very active member in the Hadoop community, as they have contributed in the development Hive. Facebook has the biggest Hadoop cluster that has been used in ML and reporting analytics.

4.2 Spark

Spark is a fast general engine which can be used in the processing of big datasets. Spark is one hundred times faster than MapReduce in memory and ten times on

Fig. 6 Iterative processing in MapReduce

Fig. 7 Iterative processing in Spark

disk. Spark has a Directed Acyclic Graph engine (DAG) that optimizes the workflow. Python, Java, R and Scala programming platforms can be used with Spark that works with Elastic MapReduce, HDFS, S3, HBase, Sequoia DB, MongoDB, and so forth. The Resilient Distributed Dataset (RDD) is the primary concept that supports the way spark works. The main components of Spark are Spark core, Spark Structured Query Language (SQL), MLlib, Spark Streaming, and GraphX.

4.2.1 Resilient Distributed Dataset (RDD)

The weak point of MapReduce is the iterative distributed computations that are needed for logistic regression; k means clustering and many other algorithms. These algorithms need to reuse or share data between different jobs or need to perform multiple queries with the shared datasets. As illustrated in Fig. 6, MapReduce needs to store the intermediate data in distributed repositories such as HDFS. This operation needs many I/O operations and replications which affect the overall speed. As illustrated in Fig. 7, Spark solved this problem by doing computations in memory, not in HDFS or Amazon S3 and this is why Spark is superior to MapReduce regarding the speed of processing.

4.2.2 Spark Machine Learning Library (MLlib)

Spark MLlib is the main reason for Spark's success as it is offering a library of algorithms that can be used in the most of problems related to machine learning. Spark MLlib has algorithms for classification, regression, clustering, recommendation, topic modeling, frequent item set, etc. Spark MLlib is simple, scalable and compatible with the most of data science tools like R, Python, Weka, etc.

5 Big Data and Data Mining in Biomedical and Health Informatics

Healthcare Informatics is an interdisciplinary field that connects medical sciences, computer sciences, and social sciences. Biomedical informatics, clinical informatics, medical informatics, nursing informatics and healthcare informatics are different names for the same domain. Healthcare informatics use innovative information technologies to optimize the acquisition, storage, retrieval, and apply information in biomedicine and healthcare. The continuous increase in the generated data of healthcare domain is the primary challenge in this field because this huge amount of data needs advanced data analytics to extract knowledge. The big data analytical tools extract knowledge from patients' care data to provide innovative healthcare services to the patients to save their lives and to develop innovative techniques to diagnose and treat various diseases [23].

5.1 Health Informatics Era

In the last decade, clinical and healthcare informatics have gained increasing capabilities because of the continuous development of hardware and software such as RFID, Wireless Sensor Network (WSN), IoT sensors and cloud computing [24–26]. Health Care Assistants (HCA) such as Remote Patients Monitoring (RPM) are generating a huge amount of data in real time using IoT medical sensors and ambient sensors. Cloud repositories are used to store patient's medical data to enable applying different analytical techniques to extract the medical knowledge such as detecting patients' health status, innovating methods for the diagnosis of different diseases, and how to treat them. This development puts biomedical Informatics on the cusp of a new era where these technologies are used to deal with big data and get unprecedented knowledge in the medical field. All these technologies participate in the development of models that monitor, help, and cure patients [27].

Smart hospitals use these models to monitor patients remotely at their homes especially elderly patients suffering from chronic diseases and live alone. These models aim to comprehend the real health status of each patient and the different nuances between patients to recommend a suitable treatment for this patient according to his health status which is called personalized treatment [28, 29].

5.2 Ambient Assisted Living (AAL) Models

AAL is a type of RPMs that enable caregivers to observe patients remotely using smart IoT sensors that record physiological data, ambient data, behavioral data, and associated activities at the same time. Then, data is transferred electronically to the smart hospital for doing assessments and taking preventive actions [30, 31]. The reliance on AALs in monitoring elderly patients will minimize the number of sudden deaths and improve the quality of healthcare services provided to the elderly patient by putting him under the continuous monitoring to make an early intervention in the emergency cases. AALs must be smart enough to understand personal differences between patients and sense variations in physiological signals to give early alerts in the case of health deterioration. AALs should use the context-awareness techniques to take accurate decisions and prevent false alarms [32–35].

Traditional AAL is developed for a specific patient suffering from a particular disease by deploying a local application that works on a handheld peripheral [36]. These applications are non-generalizable and always fail when used to monitor other patients or other diseases because it depends on its operation on the general medical rules. Also, domain experts criticize this kind of healthcare assistive applications because of the need of patient's participation in the monitoring operation, the absence of interaction with a central healthcare provider, and inability to comprehend the real health situations of the patient during his activities. Therefore, Traditional AALs had increased the cost of the healthcare procedures without providing tangible benefits.

These obstacles are tackled by inserting IoT sensors in patient's clothes for measuring his physiological signals and then transferred to different locations using different communication media. Moreover, embedding ambient sensors in the surrounding environment of the patient help monitor the changes in his ambient such as temperature, humidity, and lighting, in addition, to studying the effect of these changes on the vital signs of the patient. Also, registering Activities of Daily Living (ADLs) such as relaxing, watching TV, exercising, sleeping or eating [29, 37].

Wearable sensors enable the elderly patient with chronic diseases or disabilities to do his regular daily routine while monitoring his vital signs and activities without limiting his movement [38]. The monitoring technique that takes into consideration all factors affecting the physiological signals of the patient is called context-aware AAL which is capable of explaining the variations in patient's vital signs [39–44].

In 2014, the Center Medicaid Services (CMS) used AAL for monitoring elderly patients suffering from chronic diseases to minimize the health care expenditures [45]. AAL is a multidisciplinary field benefit from the advancement in many areas such as continuous care [41], activity monitoring [46], cloud technologies to achieve a tangible shift in this area [47–49].

In the last decade, this idea has attracted many universities and research centers to propose many context-aware RPM such as Code Blue and Care Predict that has been proposed by University of Harvard [50, 51]. Also, Massachusetts Institute of Technology (MIT) had presented a system for residential monitoring and assisted living called AlarmNet [52]. CareMerge is a virtual toolbox that presents quality care for the elderly patients [53]. Guardian Angel had been developed to monitor the health status of the elderly patients and alert the medical service provider in emergency cases using a service developed for the users of Symbian mobile phones [54]. GetMyRx is a pharmaceutical service that enables patients to automatically receive the prescribed medicine by the doctor without any extra charge [55].

Earlier trials for developing context-aware AAL have some disadvantages such as its inability to handle big data due to its local architecture and the limited storage capacity. Moreover, each AAL is designed for a specific disease, and supporting a limited number of context-aware services [56, 57]. Many research had been proposed to solve these shortcomings by developing flexible Context-aware cloud-based models to extract knowledge from continuous monitoring data of the patients [58, 59].

6   Case Study

This case study has been conducted to monitor three patients suffering from different categories of blood pressure disorders (hypertensive patient P1, hypotensive patient P2, and normotensive patient P3).

6.1   Datasets

Three elderly patients (older than 65 years) have been selected from PhysioBank MIMIC-II database that contains continuous measurements of patients' vital signs for more than 24 h such as Systolic Blood Pressure (SBP), Diastolic Blood Pressure DBP, Mean Blood Pressure (MBP), Heart Rate (HR), Respiratory Rate (RR), and Oxygen saturation in blood (SPO2) [60]. The other attributes of the dataset are generated synthetically using MATLAB taking into account the correlation between activities, symptoms and vital signs [61]. The structure of the generated dataset is listed in Table 1. Table 2 as well as Figs. 8, 9 and 10 show statistics about the generated datasets for the three patients at 24 h (a sample every 15 min).

6.2   Tools and Techniques

- All experiments are performed on the PC with following configuration (Intel ® Core ™ I5 3317U/1.7 GHz/6 GB RAM/Windows 10 64 bit).
- MATLAB R2016b (9.1) 64 bit used for datasets generation.
- WEKA 3.8.1 used to compare classifiers accuracy and elapsed time.
- Plugins of DistributedWekaBase and DistributedWekaSpark.

Table 1 A sample of monitoring dataset

Timestamp	HR	SBP	DBP	MBP	RR	SPO₂	Room Temp.	Activity	Last Activity	Medication	Symptoms	Class
07-01-17 0:00	75	157	92	105	18	99	1	3	1	0	4	Warning
07-01-17 2:45	104	142	62	112	11	100	1	2	3	0	16	Alert
07-01-17 3:00	65	144	80	94	17	98	0	4	3	1	0	Normal
07-01-17 4:15	84	141	66	102	16	100	0	2	4	1	0	Normal
07-01-17 22:30	61	183	94	103	22	100	2	1	1	0	2	Emergency

Table 2 The daily monitoring datasets of the three patients

Patient	No. of contexts	IHCAM-PUSH			
		Normal	Warning	Alert	Emergency
P1	96	18	66	11	1
P2	96	46	44	5	1
P3	96	31	61	3	1

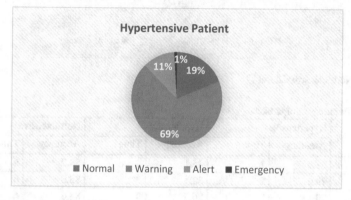

Fig. 8 Hypertensive dataset

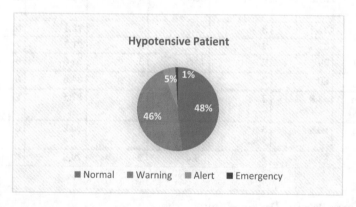

Fig. 9 Hypotensive dataset

- As illustrated in Fig. 11, an Intelligent Hybrid Context-Aware Model for Patients Under-Supervision at Homes (IHCAF-PUSH) has been used to classify the health status of the patients to one of the following categories (Normal-Warning-Alert-Emergency) [28]. The IHCAF-PUSH is an example of modern hybrid models that exploit the advantages of local and cloud architectures and avoid their disadvantages [62, 63].

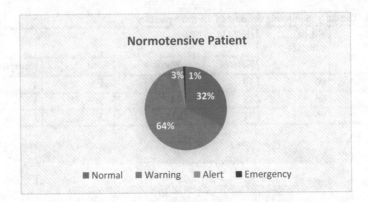

Fig. 10 Normotensive dataset

Table 3 Accuracy comparison

	Hypertensive		Hypotensive		Normotensive	
	Accuracy	Time	Accuracy	Time	Accuracy	Time
ZeroR	62.6	2	45.1	3	58.7	6
OneR	79.0	2	82.4	3	85.6	6
JRip	80	2	76.5	3	85.6	6
Naive Bayes	87.6	2	82.4	3	96.2	6
SVM	71.4	3	62.7	3	72.1	6
MLP	87.6	2	76.5	3	91.3	7
Boosting	79	2	74.5	3	84.6	6
Bagging	80	2	74.5	3	84.6	6
Decision tree (J48)	83.8	2	85.3	3	87.5	7
Random Forest	90.5	3	91.2	3	96.2	6
IBK	81	5	79.4	3	79.8	6

6.3 Results and Discussions

Waikato environment for knowledge analysis (Weka) and its plugins (distributed Weka and distributed spark) have been used to run the learning and evaluation phases of the IHCAF-PUSH in parallel using the five cores of the PC to prove the concept of handling big data using Spark.

As listed in Table 3 as well as Fig. 12, the case study compares the accuracy of 10 classifiers from different classification families in addition to ZeroR classifier as a baseline for the classification performance. Figure 13 shows the comparison of time elapsed in the learning and evaluating the same classifiers.

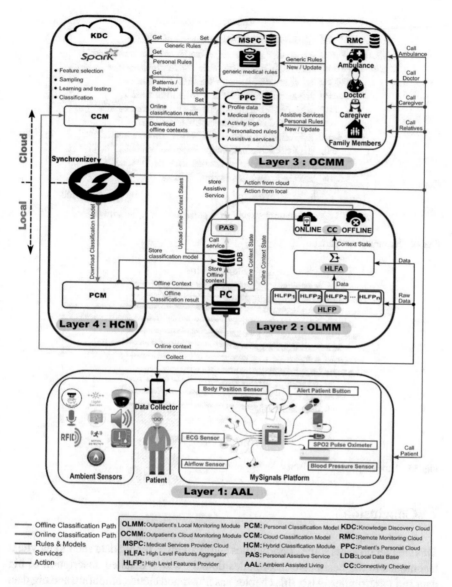

Fig. 11 Intelligent hybrid context-aware model for patients under-supervision at homes (IHCAF-PUSH)

As listed in Table 3 and as illustrated in Fig. 12, Random Forest, decision trees, and Naïve Bayes are the best classifiers that can detect the health status of patients as they have the highest accuracy while the worst is SVM. As depicted in Fig. 13, the classification elapsed time is nearly the same for all classifiers except IBK and MLP that consume more time.

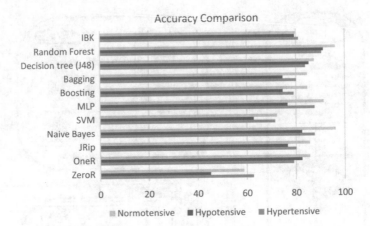

Fig. 12 Accuracy comparison

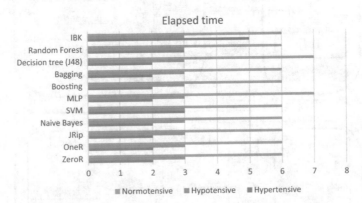

Fig. 13 Time elapsed comparison

7 Conclusion

This chapter presents some opportunities and challenges for big data and its analytical techniques in the healthcare informatics domain with a detailed description of big data and its attributes. Also, this chapter has illustrated cloud computing and big data analytics such as Hadoop and Spark. Moreover, a literature review for data mining and big data in biomedical informatics is presented in this chapter. A case study for monitoring patients suffering from blood pressure disorders has been presented in this chapter to show the efficiency of using AALs such as IHCAF-PUSH and big data analytics such as Spark in detecting the health status of the patients. The experimental results conclude that J48 and Naïve Bayer show good performance with the monitoring datasets when Spark is used to distribute storage and processing over the different clusters.

References

1. Erevelles S, Fukawa N, Swayne L (2016) Big data consumer analytics and the transformation of marketing. J Bus Res 69:897–904. https://doi.org/10.1016/j.jbusres.2015.07.001
2. Yang CC, Veltri P (2015) Intelligent healthcare informatics in big data era. Artif Intell Med 65:75–77. https://doi.org/10.1016/j.artmed.2015.08.002
3. Elhoseny M, Ramirez-Gonzalez G, Abu-Elnasr OM, Shawkat SA, N A, Farouk A (2018) Secure medical data transmission model for IoT-based healthcare systems. IEEE Access pp 1–1. https://doi.org/10.1109/access.2018.2817615
4. Herland M, Khoshgoftaar TM, Wald R, Access O (2014) A review of data mining using big data in health informatics. J Big Data 1:2. https://doi.org/10.1186/2196-1115-1-2
5. Apache Hadoop (2014) Welcome to Apache™ Hadoop®! 2014. http://hadoop.apache.org/index.html. Accessed on 15 Dec 2017
6. Kumar S (2016) HealthCare Use Case With Apache Spark 2016. https://acadgild.com/blog/healthcare-use-case-apache-spark/. Accessed on 18 Jan 2018
7. Big Data Commission (2012) Demystifying big data: a practical guide to transforming the business of government. Transp Sci 35:61–79
8. Sicular S (2013) Gartner's big data definition consists of three parts, not to be confused with three "V" s. http://www.ForbesCom/Sites/Gartnergroup/2013/03/27/Gartners-Big-Data-Definition-Consists-of-Three-Parts-Not-to-Be-Confused-with-Three-Vs/. vol 3. Accessed on 15 Dec 2017
9. Laney D (2001) 3 D data management: controlling data volume, velocity and variety. https://doi.org/10.1016/j.infsof.2008.09.005
10. Mulcahy M (2017) Big Data-interesting statistics, Facts & Figures 2017. https://www.waterfordtechnologies.com/big-data-interesting-facts/. Accessed on 19 Jan 2018
11. IBM (2015) 4-Vs-of-big-data. IBM. http://www.ibmbigdatahub.com/tag/587/. Accessed on 13 Dec 2017
12. Gandomi A, Haider M (2015) Beyond the hype: big data concepts, methods, and analytics. Int J Inf Manage 35:137–144. https://doi.org/10.1016/j.ijinfomgt.2014.10.007
13. Eapen B (2017) The 6 V's of big data. https://community.mis.temple.edu/mis520817/2017/04/07/the-6-vs-of-big-data/. Accessed on 13 Dec 2017
14. Normandeau K (2013) Beyond Volume. Variety and velocity is the issue of big data veracity, Insid Big Data
15. Devan A (2016) The 7 V's of big dataImpact radius 2016. https://www.impactradius.com/blog/7-vs-big-data/. Accessed on 13 Dec 2017
16. Shafer T The 42 V's of big data and data science. https://www.kdnuggets.com/2017/04/42-vs-big-data-data-science.html. Accessed on 30 Dec 2017
17. CS Odessa (2017) Cloud computing architecture. http://www.conceptdraw.com/How-To-Guide/cloud-computing-architecture. Accessed on 14 Dec 2017
18. LevelCloud (2017) Advantages and disadvantages of cloud computing|LevelCloud 2017. http://www.levelcloud.net/why-levelcloud/cloud-educationcenter/advantages-and-disadvantages-of-cloud-computing/. Accessed on 15 Dec 2017
19. Watson HJ (2014) Tutorial: big data analytics: concepts, technologies, and applications. Commun Assoc Inf Syst 34:1247–1268
20. Vibhavari C, Phursule RN (2014) Survey paper on big data. Int J Comput Sci Inf Technol 5:7932–7939
21. Shvachko K, Kuang H, Radia S, Chansler R (2010) The hadoop distributed file system. In: 2010 IEEE 26th Symposium Mass Storage Systems and Technologies MSST2010. https://doi.org/10.1109/msst.2010.5496972
22. de Kruijf M, Sankaralingam K (2009) MapReduce online. IBM J Res Dev 53:10:1–10:12. https://doi.org/10.1147/jrd.2009.5429076
23. O'Donoghue J, Herbert J (2012) Data management within mhealth environments: patient sensors, mobile devices, and databases. J Data Inf Qual 4:5:1–5:20. https://doi.org/10.1145/2378016.2378021

24. Elhoseny M, Farouk A, Zhou N, Wang M-M, Abdalla S, Batle J (2017) Dynamic multi-hop clustering in a wireless sensor network: performance improvement. Wirel Pers Commun 95:3733–3753. https://doi.org/10.1007/s11277-017-4023-8

25. Elsayed W, Elhoseny M, Sabbeh S, Riad A (2017) Self-maintenance model for wireless sensor networks. Comput Electr Eng. https://doi.org/10.1016/j.compeleceng.2017.12.022

26. Elhoseny M, Yuan X, Yu Z, Mao C, El-Minir HK, Riad AM (2015) Balancing energy consumption in heterogeneous wireless sensor networks using genetic algorithm. IEEE Commun Lett 19:2194–2197. https://doi.org/10.1109/LCOMM.2014.2381226

27. Yuan X, Elhoseny M, El-Minir HK, Riad AM (2017) A genetic algorithm-based, dynamic clustering method towards improved WSN longevity. J Netw Syst Manag 25:21–46. https://doi.org/10.1007/s10922-016-9379-7

28. Hassan MK, El Desouky AI, Elghamrawy SM, Sarhan AM (2018) Intelligent hybrid remote patient-monitoring model with cloud-based framework for knowledge discovery. Comput Electr Eng. https://doi.org/10.1016/j.compeleceng.2018.02.032

29. Hassan MK, El Desouky AI, Badawy MM, Sarhan AM, Elhoseny M (2018) Gunasekaran. EoT driven Hybrid Ambient Assisted Living Framework with Naïve Bayes-Firefly Algorithm, Neural Comput Applic. https://doi.org/10.1007/s00521-018-3533-y

30. Miami University (2007) Telehealth. http://telehealth.med.miami.edu/what-is-telehealth. Accessed On 19 Sep 2017

31. Himss U (2008) Defining key health information technology terms. Heal San Fr. http://www.himss.org/defining-key-health-information-technology-terms-onc-nahit. Accessed on 5 Oct 2017

32. Kleinberger T, Becker M, Ras E, Holzinger A, Müller P (2007) Ambient intelligence in assisted living: enable elderly people to handle future interfaces. Univers Access Human-Computer Interact Ambient Interact pp 103–12. https://doi.org/10.1007/978-3-540-73281-5_11

33. Belbachir AN, Drobics M, Marschitz W (2010) Ambient assisted living for ageing well—an overview. Elektrotechnik Und Informationstechnik 127:200–205. https://doi.org/10.1007/s00502-010-0747-9

34. Costin H, Rotariu C, Adochiei F, Ciobotariu R, Andruseac G, Corciova F (2011) Telemonitoring of vital signs—an effective tool for ambient assisted living. Processing International Conference on Advanced Medical Health. Care through Technology. vol 29. Springer, Cluj-Napoca, Rom, pp 60–65

35. European Commission (2007) CORDIS programmes. Ambient assisted living (AAL) in the ageing society. http://cordis.europa.eu/programme/rcn/9273_en.html. Accessed on 15 Jan 2018

36. Oresko JJ, Jin Z, Cheng J, Huang S, Sun Y, Duschl H et al (2010) A wearable smartphone-based platform for real-time cardiovascular disease detection via electrocardiogram processing. IEEE Trans Inf Technol Biomed 14:734–740. https://doi.org/10.1109/TITB.2010.2047865

37. Elhoseny M, Shehab A, Yuan X (2017) Optimizing robot path in dynamic environments using genetic algorithm and bezier curve. J Intell Fuzzy Syst 33:2305–2316

38. Kern SE, Jaron D (2002) Healthcare technology, economics, and policy: an evolving balance. IEEE Eng Med Biol Mag 22:16–19. https://doi.org/10.1109/MEMB.2003.1191444

39. Zhou F, Jiao J, Chen S, Zhang D (2011) A case-driven ambient intelligence system for elderly in-home assistance applications. IEEE Trans Syst Man Cybern Part C Appl Rev 41:179–189. https://doi.org/10.1109/TSMCC.2010.2052456

40. Taleb T, Bottazzi D, Guizani M, Nait-Charif H (2009) Angelah: a framework for assisting elders at home. IEEE J Sel Areas Commun 27:480–494. https://doi.org/10.1109/JSAC.2009.090511

41. Paganelli F, Spinicci E, Giuli D (2008) ERMHAN: a context-aware service platform to support continuous care networks for home-based assistance. Int J Telemed. https://doi.org/10.1155/2008/867639

42. Cho K, Hwang I, Kang S, Kim B, Lee J, Lee SJ et al (2008) HiCon: A hierarchical context monitoring and composition framework for next-generation context-aware services. IEEE Netw 22:34–42. https://doi.org/10.1109/MNET.2008.4579769

43. Hong JY, Suh EH, Kim SJ (2009) Context-aware systems: a literature review and classification. Expert Syst Appl 36:8509–8522. https://doi.org/10.1016/j.eswa.2008.10.071
44. Gu T, Pung HK, Zhang DQ (2004) Toward an OSGi-based infrastructure for context-aware applications. IEEE Pervasive Comput 3:66–74. https://doi.org/10.1109/MPRV.2004.19
45. Jeste DV (2011) Promoting successful ageing through integrated care. BMJ 343:1076. https://doi.org/10.1136/bmj.d6808
46. Lymberopoulos D, Bamis A, Savvides A (2011) Extracting spatiotemporal human activity patterns in assisted living using a home sensor network. Univers Access Inf Soc 10:125–138. https://doi.org/10.1007/s10209-010-0197-5
47. Forkan A, Khalil I, Tari Z (2014) CoCaMAAL: A cloud-oriented context-aware middleware in ambient assisted living. Futur Gener Comput Syst 35:114–127. https://doi.org/10.1016/j.future.2013.07.009
48. Forkan A, Khalil I, Ibaida A, Tari Z (2015) BDCaM: big data for context-aware monitoring—a Personalized knowledge discovery framework for assisted healthcare. IEEE Trans Cloud Comput pp 1–1. https://doi.org/10.1109/tcc.2015.2440269
49. Hoang DB, Chen L (2010) Mobile Cloud for Assistive Healthcare (MoCAsH). Proceedings - 2010 IEEE Asia-Pacific Services Computing Conference APSCC 2010, pp 325–332. https://doi.org/10.1109/apscc.2010.102
50. Klenk J, Kerse N, Rapp K, Nikolaus T, Becker C, Rothenbacher D, et al (2015) Physical activity and different concepts of fall risk estimation in older people-results of the ActiFE-Ulm study. PLoS One 10. https://doi.org/10.1371/journal.pone.0129098
51. Malan D, Fulford-Jones T, Welsh M, Moulton S (2004) Codeblue: an ad hoc sensor network infrastructure for emergency medical care. Implant Body Sens 12–4
52. Wood AD, Stankovic JA, Virone G, Selavo L, He Z, Cao Q et al (2008) Context-aware wireless sensor networks for assisted living and residential monitoring. IEEE Netw 22:26–33. https://doi.org/10.1109/MNET.2008.4579768
53. Caremerge. Care coordination and communication software for senior care n.d. http://www.caremerge.com/web/. Accessed on 16 Jan 2018
54. Panou M, Touliou K (2013) Mobile phone application to support the elderly. Int J Cyber Soc Educ 6:51–56. https://doi.org/10.7903/ijcse.1047
55. GetMyRx (2016) GetMyRx delivered free today. https://www.getmyrx.com/. Accessed on 16 Jan 2018)
56. Haghighi PD, Zaslavsky A, Krishnaswamy S, Gaber MM (2009) Mobile data mining for intelligent healthcare support. In: Proceedings of the 42nd annual hawaii international conference on system *sciences* HICSS. https://doi.org/10.1109/hicss.2009.309
57. Panagiotakopoulos TC, Lyras DP, Livaditis M, Sgarbas KN, Anastassopoulos GC, Lymberopoulos DK (2010) A contextual data mining approach toward assisting the treatment of anxiety disorders. IEEE Trans Inf Technol Biomed 14:567–581. https://doi.org/10.1109/TITB.2009.2038905
58. Ekonomou E, Fan L, Buchanan W, Thüemmler C (2011) An integrated cloud-based healthcare infrastructure. In: Proceedings - 2011 3rd IEEE International Conference on Cloud Computing Technology and Science. CloudCom 2011, pp 532–536. https://doi.org/10.1109/cloudcom.2011.80
59. Forkan A, Khalil I, Tari Z (2013) Context-aware cardiac monitoring for early detection of heart diseases. Comput Cardiol 2013(40):277–280
60. Saeed M, Villarroel M, Reisner AT, Clifford G, Lehman L-W, Moody G et al (2011) Multiparameter intelligent monitoring in intensive care II: a public-access intensive care unit database. Crit Care Med 39:952–960. https://doi.org/10.1097/CCM.0b013e31820a92c6
61. Sharma G, Martin J (2009) MATLAB®: a language for parallel computing. Int J Parallel Program 37:3–36

62. Elhoseny M, Abdelaziz A, Salama AS, Riad AM, Muhammad K, Sangaiah AK (2018) A hybrid model of Internet of things and cloud computing to manage big data in health services applications. Futur Gener Comput Syst (In press)
63. Darwish A, Hassanien AE, Elhoseny M, Sangaiah AK, Muhammad K (2017) The impact of the hybrid platform of internet of things and cloud computing on healthcare systems: opportunities, challenges, and open problems. J Ambient Intell Humaniz Comput 1–16. https://doi.org/10.1007/s12652-017-0659-1

Healthcare Analysis in Smart Big Data Analytics: Reviews, Challenges and Recommendations

Ahmed Ismail, Abdulaziz Shehab and I. M. El-Henawy

Abstract Increasing demand and costs for healthcare is a challenge because of the high populations and the difficulty to cover all patients by the available doctors. The healthcare data processing and management became a challenge because the problems with the data itself like irregularity high-dimensionality, and sparsity. A number of researchers worked on these problems and provided some efficient and scalable healthcare solutions. we present the algorithms and systems for healthcare analytics and applications and some related solutions. The solution what we propose is depending on adding a new layer as middleware between the sources of heterogeneous data and the Map reduce Hadoop cluster. The solution solved the common problems of dealing with heterogeneous data effectively.

Keywords Healthcare system · Hadoop map reduce · Data analytics · Big data Machine learning · IoT

1 Introduction

The high volume of heterogeneous medical data has become available in various healthcare organizations and sensors (e.g. wearable devices). The Electronic Health Record (EHR) is any record which supports medical practice or supports healthcare aspects. The benefits may include earlier disease detection, more accurate prognosis, and faster clinical research advance and better management for the patients. The main problems to get value from dealing with big data are complexity, heterogeneity, timeliness, noise and incompleteness. Big Healthcare Analytics is no different in general. There are some steps must be done on the HER information such as collection,

A. Ismail · A. Shehab (✉)
Information Systems Department, Faculty of Computers and Information,
Mansoura University, Mansoura, Egypt
e-mail: Abdulaziz_shehab@mans.edu.eg

I. M. El-Henawy
Faculty of Computer Science and Information Systems, Zagazig University, Zagazig, Egypt

© Springer Nature Switzerland AG 2019
A. E. Hassanien et al. (eds.), *Security in Smart Cities: Models, Applications, and Challenges*, Lecture Notes in Intelligent Transportation and Infrastructure,
https://doi.org/10.1007/978-3-030-01560-2_2

integration, cleaning, storing, analysis and interpreted in an optimal manner. The whole process of analyzing the data pipeline where different algorithms or systems focus on different specific targets and are coupled together to deliver an end-to-end solution.

The view can be such as a software stack where in each phase there are multiple solutions and the actual choice based on the data type [1]. The EHR is one of two types, sensor data electronic medical records (EMR). Only one of two directions of sensor data and EMR data can be chosen. One direction is to understanding the basic EMR from hospitals. The second direction is to use sensors technologies such as wearable devices, and smart phones by getting more medical related data sources [2]. EMR data is usually collected from hospitals and then analyzed to give valuable information. EMR data is timestamp data which collects patients' data. EMR data is defined with heterogeneous features of medical historical data of patients such as diagnoses, medications, lab tests, unstructured text data (i.e. doctors notes), images (i.e., magnetic resonance imaging (MRI) data). The EMR data can be used as a useful tool for diseases classification, modelling of disease progression, phenotyping [3].

Although the EMR data is a useful support for healthcare application, it suffers from a great challenge. The first challenge, EMR data is a high dimensional because it contains a large number of medical features. Second issue, EMR data is often dirty or incomplete due to the collection process is collected over a disconnected time and long period of time [4]. Third issue, EMR data is irregular because the patients usually visit the hospitals only when it is necessary only. The sensors data is very important also to monitor the health of patients and normal people by analyzing the signals of those sensors. From a big data perspective, such sensor signals exhibit some unique characteristics [5].

The signals come from millions of mobile devices, users and sensors form different sources and different environments will produce heterogeneous data streams in real time. In order to Monitor and analyze those streams will be an effective solution to understand the physical, psychological and physiological health conditions of patients. For examples, using cameras, pressure sensors installed inside a patient room can monitor the activities of elderly people remotely and can support the any emergency case; ECG and EEG sensors are used for the emotions and the stress level control and depression usually of the people [6]. The oxygen saturation and pH of the human body can be monitored by the Carbon Nano-tube sensors, which are used to be employed against cancer tissues to support patients [7].

The paper gives a picture of the current problems and challenges in big healthcare data in Sect. 2. In Sect. 3 it describes several key steps for processing healthcare data before doing data analytics. Section 4 presents various healthcare applications and services that can be supported by data analytics, and various healthcare systems. In Sect. 5. It discusses the proposed solution. In Sect. 6, it presents the conclusions and the future work.

2 The Healthcare System Challenges

The mean challenges in mining EMR data are because of the following reasons: High-dimensionality, irregularity, missing data, noise, sparsity, and bias.

2.1 High-Dimensionality

The challenge of high-dimensional is a complex issue because the model contains more parameters and those parameters associated with noise and sparsity issues. There are two categories, feature selection and feature extraction are used in order to give a solution for the high-dimensionality.

2.1.1 Feature Selection

The main process of feature selection method is to select a set of relevant predictive features for model construction. There are some feature selection solutions such as wrapper methods, filter methods, and embedded methods [8]. The Different from filter methods, wrapper methods take the relationships between features into consideration. A predictive model will be built to evaluate the combinations of features and a score will be assigned to each set of feature combinations based on the model accuracy. Wrapper methods take a much longer time since they need to search a large number of combinations of features. Also, if the data is not enough, this method will have over-fitting problem. Embedded feature selection methods shift the process of feature selection into the building process of the model. The embedded method is robust and fast to over-fitting and considers relationships between features. Unfortunately, these methods are not generic as they are designed for specific tasks with certain underlying assumptions [1].

2.1.2 Feature Extraction

The feature extraction embeds original features in a lower dimensional space, in this method each dimension will correspond to a combination of original features. There are mainly two types, linear or non-linear. The linear transforming method discovers complex non-linear relationships between the original features in the second method, the non-linear transforming method. In [9], the Gaussian regression cam be an effective solution to show the densities probability for uric acid sequences. Using this transforming step, an auto-encoder is then used to make a sense of features from the transformed probability densities.

2.2 *Irregularity*

The EMR data suffers usually from irregularity problem. Because the EMR data is organized into a matrix where a dimension represents a time and another dimension for patient medical features, and the EMR records will be scattered within uneven-spaced time spans. Also, the different patients do not visit the hospital in regular time plan so every record is varied with differ times. The irregularity issue can be solved according to some categories which should be followed [10].

2.2.1 Use of Baseline Features

The method which lead to utilize patients "baseline" features (i.e., the recorded data of the first hospital visit) for EMR data analytics tasks. MRI scans used baseline feature [11] to provide the user with the probability of Alzheimer's disease development. In this work, a relevance vector regression, a novel sparse kernel method in a Bayesian network, is employed. The baseline MRI features with baseline of Mini Mental State Examination (MMSE) features, and demographic features [12] to predict the changes in the MMSE feature in one year.

In [13], a risk score based on patients' baseline features and demographic features is proposed to predict the probability of developing Type 2 diabetes. The performance of the prediction of the baseline method

Based on a selected category is not always effective because the patients' health conditions change all over the time. Another issue with this method when it is used in multi task learning as it can act only with linear relationships among features. However, in the medical area, relationships between medical features, relationships between medical features and labels can be quite complicated and may not be described easily by simple linear relationships.

2.2.2 Data Transformation

Some existing solutions are used to organize EMR data of the patients along with time by dividing such longitudinal data into window. In [14], used Markov jump method with a probabilistic disease foe modeling the transition of disease states for Chronic Obstructive Pulmonary Disease (COPD) patients. They segmented the time dimension into non-overlapping windows to process the EMR data over 90 days, and the regularly resulted data of this method can be used for better modelling and analysis.

In [15], two kinds of features are used: daily recorded features and static features. The models of deep learning such as (Long Short-Term Memory (LSTM)) which is based on those two features to employ Gradient Boosting Trees. In [16], an hourly rate was used to process the training data in each hourly window. The diagnoses of

128 diseases are classified by using an LSTM model to detect the dynamic patterns in the input features [17].

In [18], the dynamic changing trends are captured using an alternative approach. The prior domain knowledge with the Multi-Layer Perceptron (MLP) are used to gather to the predict set of defined diseases. In this method, the irregular data is transformed firstly into regular time series to use some efficient methods directly like linear algebra. There are some limitations with this method, the resampling process may lead to missing data and the sparsity problems because of the possibility of no observations during certain time windows. Another issue, the model may be less sensitive to capture short time feature patterns when the longitudinal data is divided into windows.

2.2.3 Using Irregular Data Directly

In [19], the LSTM model used between follow of medical features through combining the time periods to handle the irregularity. They used this method for disease progression modeling, the necessary recommendations and the future risks prediction. This category of methods demonstrates the possibility of fully utilizing available data. The main limitation of this method is in in cooperation the time with historical medical features, may cause over-parameterization or under parameterization.

2.3 Corrupted and Missing Data

Missing EMR data can be caused by either inefficient data collection or lack of documentation. In data collection problem, patients are not checked specifically for a certain medical feature. In documentation problem, patients are checked for a certain feature, but either their outcomes are negative, which means that they are not needed to be documented, or the outcomes are positive but are not recorded due to human errors [20]. There are some common solutions used to solve corrupted and missing data. In [15], missing temporal features problem is solved using a simple imputation strategy based on binary values, the majority value is used for filling; and for features with numerical values, the mean values are used for imputation. In [16], the missing data is filled by employing the forward-filling and back-filling method during the resampling. For a feature that is totally missing, the clinically normal value suggested by medical experts is used instead.

In [21] the time duration and masking method is used to solve missing data problem inside the deep learning model structure. The method is based on capturing the missing EMR data. the sparsity problem is solved by forward-filling and back-filling, mean imputation, are also commonly used to get a dense patient matrix [22]. The matrix completion method is based on recovering unknown data from a few observed entries to fill the gabs by multiplying these two derived matrices to densify the raw patient matrix.

2.4 Noise

EMR data suffers from a noisy problem because of inconsistent naming, coding inaccuracies, conventions, etc. There are many researchers tend to use machine learning algorithms to learn latent representations to derive more robust representations to solve the noisy problem. The common methods such as topic models, inconsistent naming or factorization-based methods. In [9], the raw noisy data of uric acid sequences (UCA) is transformed into continuous probability density function by a Gaussian process. Each UCA is simulated as noisy samples which came from the source function. After that, an auto-encoder proceeds the mean vector of the learned Gaussian distribution to assume the hidden representations. In [23], the raw EMR data is translated to meaningful medical concept by performing predictive tasks.

2.5 Bias

The meaning of Bias is that the sampling rate is not based only on states of patients but also based on judgment of doctors on patients. The samples of patients are always available when they are ill but the samples are less when they are healthy [24]. Other sources of bias include (i) the same patient may visit different healthcare organizations for medical help and different organizations do not share information between each other; (ii) patients fail to follow up in the whole medical examination process; (iii) the recorded data in one specific healthcare organization is incomplete [25]. In [26], the EMR data is derived by analyzing the relationships between solid lab test results and time distribution between nonstop tests and utilizing the lab test time examples to give extra information. The main problem of using this method is that the bias is dependent on coarse-grained patterns, and the intra-pattern biases are still unsolved. The study found that the sequence time method could perform the best among EMR data parameterizing methods methods, perhaps due to clinicians' tendency to change sampling rate according to patients' severity.

3 Data Analysis

The process of understanding the medical information to make decisions and export reports. In [27], the structural smoothness is incorporated into the RBM model via a regularization term. Its basic underlying assumption is that two diseases which share the same parent in the taxonomy are likely to possess similar characteristics.

There are some steps should be processed on the data before using EHR as an input into the model to be analyzed. The EHR data, first of all should be stored, accessed and acquired. After that, the raw EHR data from different resources (i.e. EMR data, Sensors data) is obtained. The cleansing of data is used to remove errors and inconsistencies, and then the annotation of data with medical experts to ensure effectiveness and efficiency of this whole process from acquisition to extraction and finally cleansing. Third step is integration of the data to combine various sources of data for the same patient. The last step is processing EHR data to be modelled and analyzed, and then analytics results are interpreted and visualized [28].

3.1 Data Annotation

The EHR data is incomplete because there are many missing variables of the learning model. The HER data is uncertain, fortunately, this problem can be solved by machine learning algorithms [14]. However, the healthcare problems in most cases are too complex to be solved by using limited EHR data. Usually annotating EHR data by medical experts or doctors is the best choice to help the machine to interpret EHR data in correct manner. However, in current status, all these methods have limitations in real healthcare applications. Due to the difficulty of quantifying the supervised information by a human.

3.2 Data Cleansing

EMR data is usually noisy because of some factors such as erroneous inputs, coding inaccuracies, etc. The cleansing techniques should be developed before EMR data is used. So it is very important to understand the healthcare background to define the dirty EMR data for better cleansing performance. Data cleansing is quite challenging when we consider sensor data. The data from sensors and mobile devices is uncertain due to lack of precisions, battery life, communication problems etc. The data cleansing composes of some basic steps (i) identifying and removing inaccurate, incomplete, redundant and irrelevant data from collected data and (ii) interpolate or replace incorrect and missing records with reasonably assigned values. These processes are expected to improve data quality assessed by its accuracy, validity and integrity, which lead to reliable analytics results.

3.3 The Data Integration

The heterogeneous data which come from multiple different sources needs to processed to be integrated to provide users with a unified view [29]. The EHR data should be organized by integration techniques of the data. The HER data came from multiple sources such as mobile devices and wearable sensors at different sampling rates. The second challenge due to the heterogeneous data while integrating data streams because of a tradeoff between the data processing speed and the quality of data analytics [30]. Although the high degree of multi-modality provides more reliability of analytics results, it needs a longer time for data processing. The efficient data integration helps reduce the size of data to be analyzed without dropping the analytics performance (e.g., accuracy).

3.4 The Data Visualization

The medical features representation by using the data visualization of the EHR data information provides better data analytics performance. The HER data Regularization is used to transform the EHR data into a multivariate time series, solve the problems of missing data, irregularity, data sparsity and bias. The result data is an unbiased, regularly sampled EHR time series. the medical features representation is used to present feature-time relationships. To be specific, we learn for each medical feature whether this feature has influence after a certain time period and which features it poses influence on.

After EHR data regularizing into more suitable formats for analytics and features representation to discover underlying relationships, the next step is rewriting medical features to enhance the analysis performance. Medical Knowledge Support, in this phase, we propose to instill medical knowledge into typical machine learning and deep learning models for better analytics performance. This will involve finding the best structures to represent existing medical knowledge (i.e., domain knowledge) and developing the model training scheme using such knowledge.

4 Healthcare Systems Techniques

There are a lot of researches and application which tried to use all possible parameters in healthcare area to provide a system which can provide the users or the doctors with valuable information.

4.1 Clustering

Clustering can help to detect similar patients or diseases from the HER data. The researchers usually use two kinds of approaches to derive meaningful clusters because the HER data is not clean. The first method depends on learning robust latent representations, then clustering method. In the other hand the second method develops a probabilistic clustering models to deal with raw healthcare data effectively. In [31], firstly, the diseases are separated into 200-dimension using a modified RBM model, eNRBM model. These hidden dimensions' vectors are then projected into 2D space

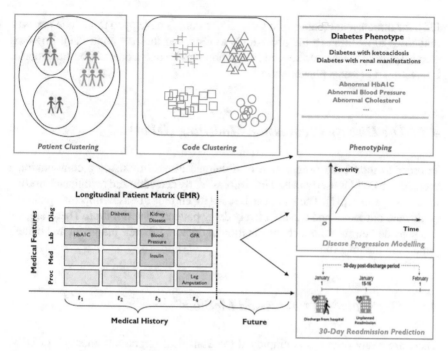

Fig. 1 The clustering from historical EHR data [33]

[32]. The 2D presentation provides many meaningful groups consisted of related diagnoses and procedures as shown in Fig. 1.

Experiments show some groups of patients are closely related to a specific disease condition (say Type I diabetes). identifies multivariate patterns of perceptions using cluster analysis. In [34], the probabilistic clustering method used for multidimensional, sparse physiological time series data. Moreover, this clustering model can be used to construct high-quality predictive models.

4.2 Phenotyping

In Phenotyping, the most diseases are defined by medical experts [35]. At first, the domain experts choose some features; then using some statistical approaches such as chi-square test to identify the significant features for developing acute kidney injury during hospital admissions. PheKB5 is a phenotype knowledge base that shows many rules for different diseases and medical conditions. Traditional methods using statistical models are easier to be implemented and interpreted, but they may require a large amount of human intervention. There are many researchers work on Phenotypes method to discover latent factors which impact on common diseases. In

[36], the dimensional tensor with three dimensions is developed to include patients, diagnoses in addition to the procedures to represent the raw input data. The next step is splitting tensor into several interaction tensors and a bias tensor. Each tensor interaction represents phenotyping.

4.3   The Disease Progression Modelling (DPM)

In order to model the progression of a defined disease by using a computational method, the DPM is used as a tool for a disease early detection and for enhance the disease management [37]. The DPM can be used for chronic diseases, improves patients' healthcare outcomes and can effectively delay patients' deterioration. Therefore, we can provide helpful reference information to doctors for their judgment and benefit patients in the long run.

4.4   The Statistical Regression Methods

There are many researchers employed the Statistical Regression method for DPM [38]. Then, via such correlation, we can have access to the progression of patients with patients' features. In [13], A risk score model is used to predict the diabetic disease through five years. Another line of research focuses on "survival analysis", which is to link patients' disease progression to the time before a certain outcome. For instance, in [39], a disease progression model is proposed to predict liver transplant patients' long-term survival. The statistical methods are efficient because of their simple models and computation; we should note that this is accomplished with an underlying assumption that the disease progression follows a certain distribution. However, for real-life applications, this assumption may not be true, and the performance of statistical regression methods would suffer.

4.5   The Machine Learning Method

There are many researchers used machine learning algorithms to solve DPM problem with different techniques like neural network or Markov models [40]. In [14], a Markov jump process is employed to model COPD patients' transition behavior between disease stages. In [40], the proposed method as developed to predict the change between different stages for abdominal aortic aneurysm patients considering by using a multi-state Markov model. The advantage of this method that it provides good causality and interpretability. However, medical experts need to be involved to determine the causal relationships during model construction. In [41], they formalized developed a multi-task learning system to predict the future of patients in multiple time points and select informative features of progression. The use of the multi-task learning methodology is limited by employing the medical features of the patients' visits to the doctors instead of time-related features. Also, they can deal only with linear relationships in the model.

4.6 The Deep Learning Method

In [42], the prediction of the breast cancer recurrence after surgery was analyzed by using the artificial neural network methodology. With a deeper neural network than this, deep learning models become more widely applicable with its great power in representation and abstraction due to its non-linear activation function. In [19], in order to track the diabetes progression, they employed the LSTM method. They used "Precision at K" as the metric to evaluate the performance of models. The limitation of this method is due to the lack of interpretability. Furthermore, more training data is of vital importance in order to improve deep learning models' performance.

4.7 The Image Data Analysis

MRI scans are commonly useful because they use radio waves, magnetic fields and field gradients to export images. Then these images are Analyzed and used for various diagnoses and used in some studies for data classification or segmentation methods. This method depends on the image recognition algorithm efficiency.

4.8 Mobile Healthcare

The user activities can be tracked anytime and anywhere by the smart phones; it is called mobile healthcare. The Human activity recognition (HAR) is another solution to use smart phones and wearable sensors to retrieve features in terms of activities of the patient. There are some solutions are developed by defining patterns, and those patterns are used for modelling and classifications with Hidden Markov Modelling (HMM) [43]. There are many researchers worked on machine learning and high-performance computing, to provide personalized healthcare services such as diagnoses, medications and treatments remotely. The HAR method can be a video-based or a wearable sensors-based. The video-based approach continuously tracks human activities through cameras deployed in rooms; however, it raises privacy issues and requires the targeted person to allow the camera [44]. Moreover, the feature extraction from the captured video/images requires complex computation for further analytics [44]. Because of these limitations, there has been a shift towards the use of wearable sensors requiring less data processing.

4.9 The Environment Monitoring

The chemical sensors are used recently in applications for healthcare aspects by detecting specific molecules in the test area. For example, we can collect nitrogen dioxide (NO_2), carbon monoxide (CO) and ozone (O_3), from individual users and construct a fine-grained pollution map together with images and location information. The environmental monitoring for haze, sewage water and smog emission etc. has become a significant worldwide problem. Combined with the cloud computing technology, a large number of smart mobile devices make a distributed data collection infrastructure possible [45]. A major contribution of their work is that this platform can be used for various targets such as traffic condition measuring, environmental pollution monitoring, and vehicle emission estimating.

4.10 Disease Detection

Biochemical-sensors deployed in/on the body can detect particular volatile organic compounds (VOCs) [46]. Developments of Nano-sensor arrays and micro electro mechanical systems have enabled artificial olfactory sensors, called electronic noses [47], as tiny, energy efficient, portable devices. In [47] discusses the essential cause of obesity from overeating and an intake of high-calorie food and presents the way to compute energy expenditure from exhaled breath. In [48], Nano-enabling electrochemical sensing technology is introduced, which rapidly detects beta-amyloid peptides, potential bio-markers to diagnose Alzheimer's disease, and a tool is developed to facilitate fast personalized healthcare for AD monitoring.

5 The Proposed System

The proposed healthcare system is aiming to integrate various types of EHR data, sensory data and user input data all over the time to predict and pre-empt illnesses, improve patient care and treatment, and ease the burden of clinicians by providing timely and assistive recommendations. The system provides structured data, unstructured data and graph data to predict the diseases according to the integration of the different data.

5.1 Data Collection

The system collects data from different resources using decision level fusion technique to make a decision or prediction depending on the information from different sources. The decision level fusion technique [48] is mainly a technique for fusion of heterogeneous data by extracting the information from each source as a separated system and collect that information to make a decision. The first source in the proposed system is a wearable device from collection of sensors such as temperature sensors, voice-sensors microphone, video sensors, image sensors, IR (infrared) sensors, optical sensors, ultrasonic sensors, piezoelectric sensors, blood pressure, heart rate, and accelerometer.

The wearable device can provide the system information about movements, heart rate, sleep rate, temperature, respiration, oxygen consumption, skin, attention, food ingestion, blood levels, among others. Several diseases studied by epidemiology are known to be caused not only by genetic or biological factors but also by environmental and behavioral ones. Thus, the need to understand the interplay among the various factors that have an effect on disease onset, development, and care affecting diverse populations. The second source of the data is by collecting information from the user individually or by the smartphone, about the current treatments, drugs, habits, daily meals, weight, height, previous diseases, current diseases, genetics problems, family history with diseases, the current living area, the most common disease in the living area, and the life system routine.

The system proposes a different solution for the data integration problem by adding a new middleware layer before the streaming the data. This layer will make the map reduce process more efficient and more accurate. As shown in Fig. 2, the patient sends have a profile on the proposed mobile application on his smartphone to register his profile such as his living area, weight, height, health information, etc. the wearable device is connected via Bluetooth with the smartphone to stream the changes in the sensors readings. When the patient goes to a clinic, the clinician can register all possible information about the patient on the system. The data then be transferred from the Hadoop cluster to the Knowledge discovery system to predict the patient's possible diseases to take care.

5.2   The Healthcare Knowledge Based System

The proposed system provides a solution which depends on Hadoop map reduce to solve the streaming data problems like irregularity, inaccuracy, errors, and missing data. The system developed a new middle ware layer as a supervisor layer to get the data and transform it to valuable information before transmitting it to the Hadoop Cluster as shown in Fig. 3. The Hadoop MapReduce framework is applied to get every information from sources for every parameter as information with value without any errors or missing data. The proposed system used alarm method technique which based on MQTT (Message Queue Telemetry Transport) [49] to send only the data which is critical according to the threshold for every parameter. So the system for example sends only the heart rate only when the rate is not between 60 and 100 beats per a minute. So the system makes the load on the bandwidth of the network is very light.

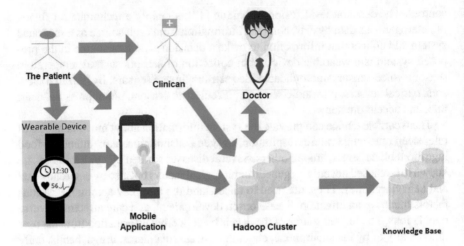

Fig. 2 The data collection structure

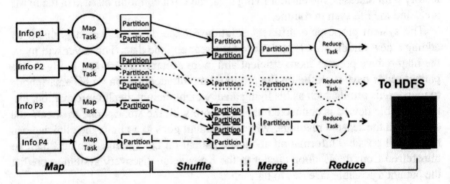

Fig. 3 The proposed hadoop cluster map reduce

There are 4 Parallel Map tasks and 3 parallel Reduce tasks, respectively. Each of the partition consists of certain key-value data pairs, whose keys can be classified into one group. The simplest classification method is a hash function. Within such a hash capability, data partition belonging to the same group are shuffled across all the compute nodes and merged together locally. There are three data partitions shown in the figure. These merged data partitions, as indicated by three different framed rectangular boxes, are consumed by three Reduce tasks separately. The output data generated by all the Reduce tasks are written back to the HDFS.

Each Map task resides in a Map slot of a compute node. Usually Two Map slots reside in one compute node. The slot number per node can be adjusted in the configuration file. The total number of the Map slots determines the degree of parallelism that indicates the total number of Map tasks that can be concurrently launched.

The rational for the Reduce task and Reduce slot is the same. The whole Hadoop MapReduce workflow is controlled in a Job Tracker located in the main computer node, or what is called the Name Node. The Map and Reduce tasks are launched at the Task Nodes, with each task corresponding to one Task Tracker to communicate with the Job Tracker. The communication includes heart-beat message to report the progress and status of the current task. If detecting a task failure or task straggler, the Job Tracker will reschedule the Task Tracker on another Task slot.

The system gets the information from the Hadoop cluster to knowledge discovery base for every parameter with rates to help to make reports which can be very useful for doctors. The system solved the problems of data streaming load and data streaming integration by using the filtration on every data source to transmit the data into valuable information. The system could predict the possible diseases under supervision of a doctor to collect all possible information about the health information for the abnormal readings from sensor and HER.

6 Conclusion and Future Work

In this paper, we summarize the challenges of Big Healthcare Analytics and their solutions to relevant applications from both EMR data and sensor data. However, specific algorithms which are adopted must be adjusted by modelling the unique characteristics of medical features. The proposed system is done by using Hadoop cluster buy we added a new middleware layer to solve the data integration problems. The future work, we will use the system on real sample of patients to define the impact of the system and the real time recommendations. The possibility to integrate the system with deep learning modules will add the ability to predict the possible diseases for the patients. In the future work, our proposed model will be applied to different applications using a variety of optimization techniques to be reduce the processing time [50–68]

References

1. Saiod AK van Greunen D, Veldsman A (2017) Electronic health records: benefits and challenges for data quality. Springer International Publishing AG, pp 3–319
2. Darwish A, Hassanien AE, Elhoseny M, Sangaiah AK, Muhammad K (2017) The impact of the hybrid platform of internet of things and cloud computing on healthcare systems: opportunities, challenges, and open problems. J Ambient Intell Humanized Computing. https://doi.org/10.1007/s12652-017-0659-1
3. Kuang Z, Thomson J, Caldwell M, et al (2016) Computational drug repositioning using continuous self-controlled case series, arXiv preprint arXiv: 1604.05976
4. Avendi MR, Kheradvar A, Jafarkhani H (2016) A combined deep-learning and deformable model approach to fully automatic segmentation of the left ventricle in cardiac mri. Med Image Anal 30:108–119
5. Cort R, Bonnaire X, Marin O (2015) Stream processing of healthcare sensor data: studying user traces to identify challenges from a big data perspective. In: Proceedings of the 4th international workshop on body area sensor networks

6. Sioni R, Chittaro L (2015) Stress detection using physiological sensors. IEEE Computer 48(10):26–33
7. Kumar S, Willander M, Sharma JG (2015) A solution processed carbon nanotube modified conducting paper sensor for cancer detection. J Mater Chem B 3:9305–9314
8. Guyon I, Elisseeff A (2003) An introduction to variable and feature selection. J Machine Learning Res 3:1157–1182
9. Lasko TA, Denny JC, Levy MA (2013) Computational phenotype discovery using unsupervised feature learning over noisy, sparse, and irregular clinical data. PLoS ONE 8(6):1–13
10. Sajjad M, Nasir M, Muhammad K, Khan S, Jan Z, Sangaiah AK, Elhoseny M, Baik SW (2018) Raspberry Pi assisted face recognition framework for enhanced law-enforcement services in smart cities. Future Generation Comput Syst. https://doi.org/10.1016/j.future.2017.11.013
11. Yuan X, Li D, Mohapatra D, Elhoseny M (2017) Automatic removal of complex shadows from indoor videos using transfer learning and dynamic thresholding. Computers and electrical engineering, in press. Available online. https://doi.org/10.1016/j.compeleceng.2017.12.026
12. Duchesne S, Caroli A, Geroldi C (2009) Relating one-year cognitive change in mild cognitive impairment to baseline MRI features. Neuroimage 47(4):1363–1370
13. Schulze MB, Hoffmann K, Boeing H (2007) An accurate risk score based on anthropometric, dietary, and lifestyle factors to predict the development of type 2 diabetes. Diabetes Care 30(3):510–515
14. Wang X, Sontag D, Wang F (2014) Unsupervised learning of disease progression models. In: Proceedings of the 20th ACM SIGKDD international conference on knowledge discovery and data mining, pp 85–94
15. Che Z, Purushotham S, Khemani R (2015) Distilling knowledge from deep networks with applications to healthcare domain. arXiv preprint arXiv: 1512.03542
16. Lipton ZC, Kale DC, Elkan C (2015) Learning to diagnose with lstm recurrent neural networks. arXiv preprint arXiv: 1511.03677
17. Hochreiter S, Schmidhuber J (1997) Long short-term memory. Neural Comput 9(8):1735–1780
18. Che Z, Kale DC, Li W (2015) Deep computational phenotyping. In: Proceedings of the 21th ACM SIGKDD international conference on knowledge discovery and data mining, pp 507–516
19. Pham T, Tran T, Phung D (2016) Deepcare: a deep dynamic memory model for predictive medicine. In: Pacific-Asia conference on knowledge discovery and data mining, pp 30–41
20. Wells BJ, Nowacki AS, Chagin K (2013) Strategies for handling missing data in electronic health record derived data. eGEMs (Generating Evidence & Methods to improve patient outcomes), 1(3):7
21. Che Z, Purushotham S, Cho K (2016) Recurrent neural networks for multivariate time series with missing values. arXiv preprint arXiv: 1606.01865
22. Zhou J, Wang F, Hu J (2014) From micro to macro: data driven phenotyping by densification of longitudinal electronic medical records. In: Proceedings of the 20th ACM SIGKDD international conference on knowledge discovery and data mining, pp 135–144
23. Ho JC, Ghosh J, Sun J (2014) Marble: high-throughput phenotyping from electronic health records via sparse nonnegative tensor factorization. In: Proceedings of the 20th ACM SIGKDD international conference on knowledge discovery and data mining, pp 115–124
24. Hripcsak G, Albers DJ, Perotte A (2015) Parameterizing time in electronic health record studies. J Am Med Inform Assoc 22(4):794–804
25. Hersh WR, Weiner MG, Embi PJ (2013) Caveats for the use of operational electronic health record data in comparative effectiveness research. Med Care 51:S30–S37
26. Pivovarov R, Albers DJ, Sepulveda JL (2014) Identifying and mitigating biases in her laboratory tests. J Biomed Inform 51:24–34
27. Tran T, Nguyen TD, Phung D (2015) Learning vector representation of medical objects via emr-driven nonnegative restricted boltzmann machines (eNRBM). J Biomed Informatics, pp 96–105
28. Cui B, Mei H, Ooim BC (2014) Big data: the driver for innovation in databases. National Sci Rev 1(1):27–30

29. Dong XL, Srivastava D (2013) Big data integration. In: 2013 IEEE 29th international conference on data engineering (ICDE), pp 1245–1248
30. Doan A, Halevy A, Ives Z (2012) Principles of data integration. Elsevier
31. Torres-Huitzil C, lvarez-Landero A (2015) Accelerometer-based human activity recognition in smartphones for healthcare services. Springer, pp 147–169
32. Nguyen TD, Tran T, Phung D (2013) Latent patient profile modelling and applications with mixed-variate restricted boltzmann machine. In: Pacific-Asia conference on knowledge discovery and data mining, pp 123–135
33. Krumholz H, Normand S-L, Keenan P (2008) 30-day heart failure readmission measure methodology. Technical report, Yale University/Yale-New Haven Hospital Center for Outcomes Research and Evaluation (YNHH-CORE)
34. Marlin BM, Kale DC, Khemani RG (2012) Unsupervised pattern discovery in electronic health care data using probabilistic clustering models. In: Proceedings of the 2nd ACMSIGHIT International Health Informatics Symposium, pp 389–398
35. Matheny ME, Miller RA, Ikizler TA (2010) Development of inpatient risk stratification models of acute kidney injury for use in electronic health records. Med Decis Making 30(6):639–650
36. Hirsch JS, Tanenbaum JS, Lipsky Gorman S (2014) Harvest, a longitudinal patient record summarizer. J Am Med Inform Assoc 22(2):263–274
37. Mould D (2012) Models for disease progression: new approaches and uses. Clin Pharmacol Ther 92(1):125–131
38. Evani AS, Sreenivasan B, Sudesh JS (2013) Activity recognition using wearable sensors for healthcare. In: Proceedings of the 7th international conference on sensor technologies and applications
39. Pearson RK, Kingan RJ, Hochberg A (2005) Disease progression modeling from historical clinical databases. In: Proceedings of the eleventh ACM SIGKDD international conference on knowledge discovery in data mining, pp 788–793
40. Jackson CH, Sharples LD, Thompson SG (2003) Multistate markov models for disease progression with classification error. J Royal Stat Soc: Series D (The Statistician) 52(2):193–209
41. Zhou J, Liu J, Narayan VA (2012) Modeling disease progression via fused sparse group lasso. In: Proceedings of the 18th ACM SIGKDD international conference on knowledge discovery and data mining, pp 1095–1103
42. Street N (1998) A neural network model for prognostic prediction. In: Proceedings of the 15th international conference on machine learning, pp 540–546
43. Korchiyne R, Farssi SM, Sbihi A, Touahni R, Alaoui MT (2014) A combined method of fractal and GLCM features for MRI and CT scan images classification, arXiv preprint arXiv: 1409.4559
44. Margarito J, Helaoui R, Bianchi AM (2016) User-independent recognition of sports activities from a single wrist-worn accelerometer: a template-matching-based approach. IEEE Trans Biomed Eng 63(4):788–796
45. Bulling A, Blanke U, Schiele B (2014) A tutorial on human activity recognition using bodyworn inertial sensors. ACM Comput Survey 46(3):1–33
46. Elhoseny M, Shehab A, Osman L (2018) An empirical analysis of user behavior for P2P IPTV workloads, AMLTA 2018: the international conference on advanced machine learning technologies and applications (AMLTA2018) pp 405–414, Cairo Egypt, Springer. https://doi.org/10.1007/978-3-319-74690-6_25
47. Filipe L, Fdez-Riverola F, Costa N (2015) Wireless body area networks for healthcare applications: Protocol stack review. Int J Distributed Sens Networks 2015:1:1–1:1
48. Bandodkar AJ, Jeerapan I, Wang J (2016) Wearable chemical sensors: present challenges and future prospects. ACS Sensors 1:464–482
49. Lockhart JW, Weiss GM, Xue JC (2011) Design considerations for the wisdm smart phone-based sensor mining architecture. In: Proceedings of the 5th international workshop on knowledge discovery from sensor data, pp 25–33
50. Tharwat A, Mahdi H, Elhoseny M, Hassanien AE (2018) Recognizing human activity in mobile crowdsensing environment using optimized k-NN algorithm, expert systems with applications. Available online 12 April 2018. https://doi.org/10.1016/j.eswa.2018.04.017

51. Hosseinabadi AAR, Vahidi J, Saemi B, Sangaiah AK, Elhoseny M (2018) Extended genetic algorithm for solving open-shop scheduling problem. Soft Comput. https://doi.org/10.1007/s00500-018-3177-y

52. Tharwat A, Elhoseny M, Hassanien AE, Gabel T, Arun kumar N (2018) Intelligent beziér curve-based path planning model using chaotic particle swarm optimization algorithm, cluster computing, Springer, pp 1–22 https://doi.org/10.1007/s10586-018-2360-3

53. Abdelaziza A, Elhoseny M, Salama AS, Riad AM (2018) A machine learning model for improving healthcare services on cloud computing environment. Measurement, 119:117–128 https://doi.org/10.1016/j.measurement.2018.01.022

54. Abd El Aziz M, Hemdan AM, Ewees AA, Elhoseny M, Shehab A, Hassanien AE, Xiong S (2017) Prediction of biochar yield using adaptive neuro-fuzzy inference system with particle swarm optimization, In: 2017 IEEE PES Power Africa Conference, June 27–30, Accra-Ghana, IEEE, pp 115–120. https://doi.org/10.1109/powerafrica.2017.7991209

55. Ewees AA, Abd El Aziz M, Elhoseny M (2017) Social-Spider Optimization Algorithm for Improving ANFIS to Predict Biochar Yield. In: 8th international conference on computing, communication and networking technologies (8ICCCNT), July 3–5, Delhi-India, IEEE

56. Elhoseny M, Tharwat A, Yuan X, Hassanien AE (2018) Optimizing K-coverage of mobile WSNs". Expert Syst Appl 92:142–153. https://doi.org/10.1016/j.eswa.2017.09.008

57. Elhoseny M, Abdelaziz A, Salama A, Riad AM, Sangaiah AK, Muhammad K (2018) A hybrid model of internet of things and cloud computing to manage big data in health services applications. Future generation computer systems, Elsevier, Accepted March 2018, In Press

58. Sarvaghad-Moghaddam M, Orouji AA, Ramezani Z, Elhoseny M, Farouk A, Arun kumar N (2018) Modelling the Spice parameters of SOI MOSFET using a combinational algorithm. Cluster computing, Springer, March 2018, In Press. https://doi.org/10.1007/s10586-018-2289-6

59. Rizk-Allah RM, Hassanien AE, Elhoseny M (2018) A multi-objective transportation model under neutrosophic environment. Computers and electrical engineering, Elsevier, in Press, 2018. https://doi.org/10.1016/j.compeleceng.2018.02.024

60. Batle J, Naseri M, Ghoranneviss M, Farouk A, Alkhambashi M, Elhoseny M (2017) Shareability of correlations in multiqubit states: optimization of nonlocal monogamy inequalities. Phys Rev A 95(3):032123. https://doi.org/10.1103/PhysRevA.95.032123

61. Elhoseny M, Nabil A, Hassanien AE, Oliva D (2018) Hybrid rough neural network model for signature recognition. In: Hassanien A, Oliva D (eds) Advances in soft computing and machine learning in image processing. Studies in computational intelligence, vol 730. Springer, Cham. https://doi.org/10.1007/978-3-319-63754-9_14

62. Elhoseny M, Tharwat A, Farouk A, Hassanien AE (2017) K-coverage model based on genetic algorithm to extend WSN lifetime. IEEE sensors letters, 1(4):1–4, IEEE. https://doi.org/10.1109/lsens.2017.2724846

63. Yuan X, Elhoseny M, El-Minir HK, Riad AM (2017) A genetic algorithm-based, dynamic clustering method towards improved wsn longevity. J Network Syst Manage, Springer US, 25(1):21–46 https://doi.org/10.1007/s10922-016-9379-7

64. Elhoseny M, Nabil A, Hassanien AE, Oliva D (2018) Hybrid rough neural network model for signature recognition. In: Hassanien A, Oliva D (eds) Advances in soft computing and machine

65. Elhoseny M, Shehab A, Yuan X (2017) Optimizing robot path in dynamic environments using genetic algorithm and bezier curve. J Intell Fuzzy Syst 33(4):2305–2316, IOS-Press. https://doi.org/10.3233/jifs-17348

66. Elhoseny M, Tharwat A, Hassanien AE (2017) Bezier curve based path planning in a dynamic field using modified genetic algorithm. J Comput Sci Elsevier. https://doi.org/10.1016/j.jocs.2017.08.004

67. Metawaa N, Hassana MK, Elhoseny M (2017) Genetic algorithm based model for optimizing bank lending decisions. Expert Syst with Appl, Elsevier, 80:75–82. https://doi.org/10.1016/j.eswa.2017.03.021

68. Metawa N, Elhoseny M, Hassan MK Hassanien AE (2016) Loan portfolio optimization using genetic algorithm: a case of credit constraints. In: Proceedings of 12th international computer engineering conference (ICENCO), IEEE, 59–64. https://doi.org/10.1109/icenco.2016.7856446

Realization of an Adaptive Data Hiding System for Electronic Patient Record, Embedding in Medical Images

Shabir A. Parah, Ifshan Ahmad, Nazir A. Loan, Khan Muhammad, Javaid A. Sheikh and G. M. Bhat

Abstract An adaptive Electronic Patient Record (EPR) hiding system utilizing medical images as cover media has been presented in this chapter. The proposed system is based on embedding the EPR information in Discrete Cosine Transform (DCT) coefficients. Cover medical image is divided into 8×8 non-overlapping blocks and DCT is applied on each block. The two predefined coefficients of each 8×8 DCT block are used for embedding the EPR information. The proposed system uses an adaptive embedding factor based on mean of a selected block unlike the conventional way of predefining an embedding factor. In order to add an additional layer of security to the EPR, it has been encrypted before embedding it. The proposed algorithms have been evaluated using various image quality matrices like Peak Signal to Noise Ratio (PSNR), Structural Similarity Index (SSIM), and Normalized Cross Co-relation (NCC). The experimental investigations reveal that the proposed scheme is capable of providing better quality watermarked medical images compared to the conventional embedding scheme while maintaining good robustness thus it can be a better option for utilization in a typical e-health care communication system.

Keywords Discrete cosine transform · Embedding · Medical image watermarking Electronic health record

S. A. Parah (✉) · I. Ahmad · N. A. Loan · J. A. Sheikh
Department of Electronics and Instrumentation Technology,
University of Kashmir, Srinagar 190006, Jammu and Kashmir, India
e-mail: shabireltr@gmail.com

K. Muhammad
Digital Contents Research Institute, Sejong University, Seoul, Republic of Korea

G. M. Bhat
Department of Electronics Engineering, Institute of Technology,
Zakoora College of Engineering, Srinagar 190006, Jammu and Kashmir, India

© Springer Nature Switzerland AG 2019
A. E. Hassanien et al. (eds.), *Security in Smart Cities: Models, Applications, and Challenges*, Lecture Notes in Intelligent Transportation and Infrastructure,
https://doi.org/10.1007/978-3-030-01560-2_3

1 Introduction

Technology development for Smart cities is presently a hot research area. The chief aims of making a city smart are to provide core infrastructure and give a decent quality of life to the citizens. The focus is on sustainable and inclusive development and the idea is to look at compact areas. The core infrastructure element in a smart city include—Robust I.T connectivity and digitalization. Therefore, Internet of things (IOT) plays an important role in smart cities. IOT is based on a global infrastructure network which connects uniquely identified objects, by exploiting the data captured by the sensors and actuators, and the equipment used for communication and localization. Medical care and health care represent one of the most attractive application areas for the IOT. The IOT has the potential to give rise to many medical applications such as remote health monitoring, fitness programs, chronic diseases, and elderly care. Compliance with treatment and medication at home and by healthcare providers is another important potential application. Therefore, various medical devices, sensors, and diagnostic and imaging devices can be viewed as smart devices or objects constituting a core of the IOT. IOT based healthcare services are expected to reduce costs, increase the quality of life, and enrich the user's experience. In a typical IOT driven healthcare the conventional healthcare is replaced by electronic healthcare. Implementation of an IOT driven healthcare system is accompanied by many challenges—both at administrative and technological levels. From technological point of view security of Electronic Patient Record (EPR) is a core issue that needs attention. Presently a huge medical data such as medical images and EPR, is being transferred globally for the purpose of e-healthcare which includes tele-health and tele-medicine over insecure channels of the internet. The transmission of such sensitive information over insecure channels may lead to easy manipulation which in-turn leads a threat to authentication, and security of the medical data [1, 2]. The various issues pertaining to security and content authentication could be tackled by the use of various technologies like cryptography and data hiding. Cryptography, which is used to secure data during its transit, camouflages the information and thus could lead to arousing suspicion in an eves-dropper thus enhancing chances of an attack. Data hiding has been found to be an alternate and effective solution for securing and authenticating the content transmitted over insecure channels. Data hiding is broadly classified into watermarking and steganography [3–5]. A lot of research work is being carried out on medical image watermarking for the purpose of diagnostic integrity of the medical data [6–11]. Digital watermark used inside a cover medical image could be a hospital logo, doctor's signature, doctor's personal logo etc. In digital watermarking technique the properties like robustness, imperceptibility, payload and security are of the major importance [12–15]. The two main categories of digital image watermarking techniques include spatial domain and Transform domain. In Spatial domain technique, watermark is embedded by directly modifying the pixel value of the host image i.e.., watermark is applied in pixel domain. In this, Simple operations are applied when combining with host image. The methods used in the spatial domain are the least significant bit (LSB), correlation based and spread spectrum techniques. The other technique includes Transform domain. In frequency domain, the host image is converted into frequency domain using any of the transformation methods. The most commonly used transformation methods are DCT, DWT, and DFT. In these techniques, the transformed coefficients are modified instead of directly changing the pixel values. Although, spatial domain embedding is very simple, computationally efficient and has high embedding capacity but the spatial domain-based watermarking techniques are not robust to most of the image processing operations and hence cannot be used for security purposes and especially for EPR which contains the sensitive information of the patient. Transform domain techniques perform better on robustness front. This chapter presents an adaptive technique for EPR hiding in medical images. The scheme has been shown to be capable of providing high quality watermarked images and is robust to various image processing attacks. Given various attributed of our proposed scheme it could be very useful for providing a robust security to EPR shared in an IOT driven healthcare system [16–18]. The main contributions of this work are:

- The scheme is adaptive i.e., modification factor is a function of arithmetic mean of the cover image block under consideration.
- Imperceptibility is good for a fair amount of payload.
- Provides high degree of robustness
- Provides double layer of security to EPR.

2 Related Work

A detailed literature survey reveals that a lot of work is being carried out in the field of EPR hiding in medical images owing to its importance. Ray-Shine et al. [19] have proposed two methods to improve the reliability and robustness. To improve the reliability, for the first method, the principal components of the watermark have been embedded into the host image in Discrete Cosine Transform (DCT); and for the second method, those are embedded into the host image in Discrete Wavelets Transform (DWT). To improve the robustness, the Particle Swarm Optimization (PSO) is used for finding the suitable scaling factors. Lestriandoko et al. [20] have proposed the modified difference expansion watermarking using LSB replacement in the difference of virtual border. The payload has been formed by image hashing using MD5. The proposed technique has been proved and tested for gray scale and color images and as such can be used for important images for military and medical applications. Joining encryption with the watermarking technique, Suganya and Amudha [21] have proposed two algorithms for integrity control of medical images. The authors have used RC4 and AES stream ciphers for the encryption. Osamah and Bee [22] have proposed a robust watermarking technique in which the patient's data has been embedded in ROI part of the medical image and tamper detection and recovery bits are embedded in RONI. The combination of DE and 3- level DWT has been used for embedding purpose. Baisa and Suresh [23] proposed a region based watermarking system. The algorithm is non-blind and using 3 level DWT 64×64 logo has been embedded in 512×512 medical images. The reported PSNR is 48.53 dB. Tanmay et al. [24] proposed a blind watermarking technique using Stationary Wavelet Transform (SWT). The authors have embedded EPR information in color medical images. The images are divided into three planes and then three watermarks are embedded in them.

Lin and Chen [25] proposed a DCT based watermarking algorithm, where watermark has been embedded into LSB of the DCT coefficients of the image. A (64×64) watermark has been used for testing the algorithm. The system has been reported to be robust to some attacks like cropping and uniform noise addition. The robustness of the system for JPEG compression is compromised as a result of quantization carried on low frequency components. Kundu et al. [26], a proposed reversible watermarking technique where EPR is embed in RONI using some advanced encryption standard. Chao et al. [27], proposed a frequency domain technique based on Discrete Cosine Transform (DCT) where an EPR-related data is embedded in quantized DCT coefficients of watermarked image. Acharya et al. [28] put forward a DCT based scheme where graphical ECG signals and encrypted text data of patient information are interleaved with medical images to reduce storage and transmission overheads. Kliem et al. [29] proposed security and communication architecture for networked medical devices in mobility aware e-health environment. Maity et al. [30], proposed reversible contrast mapping watermarking scheme for robustness improvement. To improve robustness; integer wavelet transform based on spread spectrum technique is applied. Ko et al. [31], proposed the nested QIM-based method for reversible watermarking, which is targeted for the healthcare information management systems. However, important properties of watermarking algorithm such as robustness and imperceptibility are not addressed in this scheme, and thus the Ko's watermarking method is not suitable for medical applications. Wang et al. [32] presented a fragile watermarking approach for image authenticity and integrity. The authors have improved the quality of the watermarked and retrieved images. The original image has been divided into 8×8 blocks and DCT is applied on each block. The obtained coefficients are organized using zigzag order, first ten coefficients are adapted into a binary sequence. The watermark generated is based on block content and variances. The watermark is divided into three data chunks and inserted in three different blocks. The tamper blocks are discovered during the extraction phase based on length of the original watermark. Parah et al. [33] have proposed a DWT based watermarking technique for medical images. The payload as reported in the paper is 1024 bits only. The reported average PSNR is 45 dB. The perceptual quality of the watermarked images is better, but payload of the reported technique is very low. Farhana et al. [34], have proposed two EPR hiding algorithms for medical images utilizing discrete cosine transform. The scheme has been shown robust to various images processing and geometric attacks. The scheme however has a very small payload capability.

3 Proposed Work

The block diagram of the proposed scheme has been depicted in Fig. 1. The cover image has been divided into 8×8 non-overlapping blocks and DCT is applied to each block. The mean of each 8×8 DCT block is computed and it is utilized for determining the embedding strength. Since medical information to be embedded in the image is usually sensitive in nature, and as such demands an additional layer of security. This has been provided by encrypting the EPR using concept of chaotic encryption as in [35].

We have tested two variants of the proposed scheme. One, where we have defined an embedding factor (δ) equal to 30 (irrespective of the image characteristics) and in the second approach, mean value of each 8×8 DCT block has been used as embedding factor (δ) for the second algorithm to take advantage of the local characteristics and make the system adaptive. The embedding algorithm is described as follows.

3.1 EPR Embedding

The EPR embedding is carried out in the following step:

- Read the EPR data.
- Encrypt the EPR using a chaotic sequence.

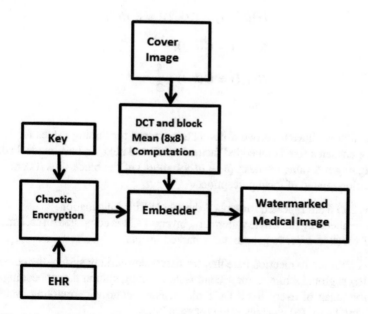

Fig. 1 Block diagram of proposed system

- Read Cover Image.
- Divide cover image in 8×8 blocks.
- Select the two mid-frequency DCT coefficients $C(i, j)$, $C(k, l)$ from an 8×8 DCT block in which EPR bit is to be embedded.
- For embedding bit 1 coefficient $C(i, j)$ is adjusted and made relatively greater then coefficient $C(k, l)$.
- For embedding bit 0 coefficient $C(i, j)$ is adjusted and made relatively smaller than coefficient $C(k, l)$.
- For embedding either bit zero or 1 an addition factor called modification factor is incorporated. This factor can be either taken a fixed value or the mean of corresponding DCT block and is used iteratively as follows during the embedding process to ensure that above defined embedding criterion is met.

For bit 1:

$$if \, |(C(i, j) - C(k, l))| < \delta \; then$$
$$C(i, j) = C(i, j) + \frac{\delta}{2}$$
$$C(k, l) = C(k, l) - \frac{\delta}{2}$$
$$end$$

For bit 0:

$$if \, |C(i, j) - C(k, l)| < \delta \; then$$
$$C(i, j) = C(i, j) - \frac{\delta}{2}$$
$$C(k, l) = C(k, l) + \frac{\delta}{2}$$
$$end$$

Note that we have used two different approaches for finding values for δ, firstly we have chosen a fixed embedded factor $\delta = 30$ and next we have made δ adaptive by assigning it a value of mean ($\delta = block \, mean$) of the block thus it contains the information related of local perception.

- Steps 2 to 6 are repeated for embed the whole EPR information.
- After embedding the whole watermark, inverse DCT is computed on each modified DCT block in order to get the watermarked image.

It is important to mention here that for first watermarking system the embedding factor (α) is global while as for second watermarking system the embedding factor (δ) is the mean of every 8×8 DCT block. For second watermarking system the embedding factor (α) changes value for each block.

3.2 EPR Extraction

The EPR is extracted by the applying the block wise DCT to the watermarked image. The watermarked image is divided into non-over lapping blocks of size 8×8. The block wise comparison is done for the extraction and if the $C1 < C2$ bit 0 is extracted else bit 1 is extracted. The bit sequence then obtained is a randomized form of embedded EPR. The original sequence is then obtained using the same key as that used at the transmitting end.

4 Experimental Results

The proposed work has been carried out using MATLAB R2014a platform for 512×512 medical images. A payload of 4096 bits as EPR has been embedded in the medical images. The objective analysis of different images has been performed by calculating various parameters like, Peak Signal to Noise Ratio (PSNR), Structural Similarity Index (SSIM) and Normalized Cross-Correlation (NCC) between original medical image and watermarked images.

4.1 Imperceptibility Analysis

Imperceptibility refers to ability of an image to successfully hide information without any apparent loss of perceptual quality. Some of the commonly used medical images have been used for analysis of our proposed scheme. Figure 2 shows various medical images used for analysis while as Fig. 3 shows a corresponding subjective quality of watermarked images. The perceptual quality parameters (PSNR, NCC & SSIM) of the watermarked images obtained in both the variants of our system have been represented in Table 1.

From the objective quality analysis it is clear that second algorithm shows better results. The average PSNR obtained by the second algorithm is 4 dB more than that of first algorithm. Furthermore, the average SSIM value also increases drastically from 0.8945 to 0.9861. On the basis of the payload the results have been compared with that of Parah et al. as shown in Table 2. As is evident from Table that despite four times increase in the payload the proposed system is capable of producing good quality watermarked images. The subjective image qualities of various images and their respective watermarked images for first algorithm and second algorithm have been shown in Figs. 2 and 3 respectively.

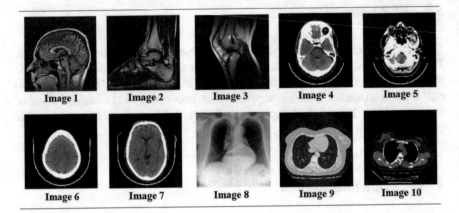

Fig. 2 Original images

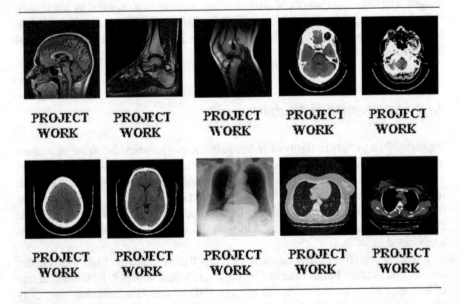

Fig. 3 Watermarked images

4.2 Robustness Analysis

The robustness analysis of the proposed scheme has been carried out by subjecting watermarked images obtained from the system to various attacks like, salt and pepper noise, Gaussian noise, JPEG compression, rotation, different filtering processes

Table 1 Perceptual transparency

Image	($\alpha = 30$)			(α = mean of each block)		
	PSNR (dB)	NCC	SSIM	PSNR (dB)	NCC	SSIM
Image 1	37.7283	0.9964	0.8729	39.3697	0.9961	0.9825
Image2	35.5455	0.9955	0.8758	36.6289	0.9957	0.9806
Image 3	39.0706	0.9966	0.8606	41.5322	0.9965	0.9836
Image 4	37.1718	0.9971	0.8759	38.8601	0.9971	0.9811
Image 5	39.7368	0.9999	0.8744	44.8624	0.9999	0.9898
Image 6	38.4242	0.9999	0.9399	43.2651	0.9999	0.9891
Image 7	38.9879	1.0000	0.9358	44.5974	0.9999	0.9911
Image 8	39.7147	1.0000	0.9211	47.5473	1.0000	0.9910
Image 9	38.2134	1.0000	0.8999	43.047	1.0000	0.9866
Image 10	37.9912	1.0000	0.9190	41.7704	1.0000	0.9909
Average	38.2975	0.9982	0.8945	42.0829	0.9981	0.9861

Table 2 Perceptual transparency comparision

Image	Farah et al.		Proposed scheme		
	Payload (bits)	PSNR (dB)	Payload (bits)	($\delta = 30$) PSNR (dB)	(δ = mean of each block) PSNR (dB)
Image 1	1024	43.9221	4096	37.7283	39.3697
Image 2	1024	42.9560	4096	35.5455	36.6289
Image 3	1024	42.8736	4096	39.0706	41.5322
Image 4	1024	44.1513	4096	37.1718	38.8601
Image 5	1024	46.3413	4096	39.7368	44.8624
Image 6	1024	44.0202	4096	38.4242	43.2651
Image 7	1024	45.2139	4096	38.9879	44.5974
Image 8	1024	45.9632	4096	39.7147	47.5473
Average	1024	44.4302	4096	38.2975	42.0829

(Median filtering, Wiener filtering and Average filtering) and other image processing techniques.

4.2.1 Median Filtering Attack

The test images are subjected to Median filtering of filter size 3×3. The subjective quality analysis of the watermarked images is presented in Fig. 4 when Median filtering is applied on watermarked media. The objective performance parameters corresponding to Median filtering are presented in Table 3.

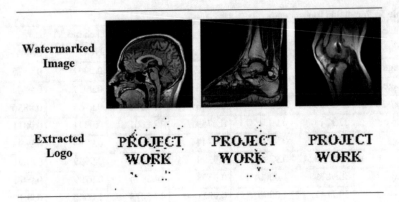

Fig. 4 Watermarked images and their respective logos extracted after median filtering attack

Table 3 Objective performance parameters for watermarked images and extracted logos for median filtering attack

Image name	Embedding factor = 30			With mean		
	PSNR	BER	NCC	PSNR	BER	NCC
Image 1	24.2012	0.0557	0.9451	24.1610	0.0938	0.9057
Image 2	23.9738	0.0659	0.9333	24.1879	0.0908	0.9081
Image 3	23.5281	0.0679	0.9317	23.6785	0.0662	0.9325
Image 4	23.0355	0.0984	0.9011	23.3092	0.1028	0.8960
Image 5	23.1118	0.0923	0.9051	23.1580	0.1052	0.8947
Image 6	23.2651	0.0940	0.9030	23.2493	0.1106	0.8872
Image 7	23.2466	0.0908	0.9070	23.5164	0.0942	0.9025
Image 8	24.5371	0.0620	0.9376	24.4245	0.0613	0.9394
Image 9	23.9977	0.0574	0.9419	23.9021	0.0789	0.9210
Image 10	23.2195	0.0854	0.9145	23.0822	0.1169	0.8840

4.2.2 Salt and Pepper Noise

The Salt and Pepper noise is added to the watermarked images with noise density 0.01. The images distorted by Salt and Pepper noise and the watermarks extracted from them are shown in Fig. 5. The objective performance parameters corresponding to Median filtering have been presented in Table 4.

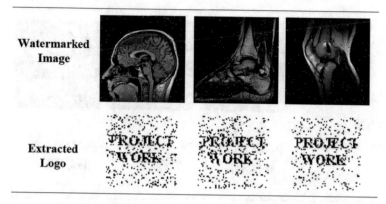

Fig. 5 Watermarked images and their respective logos extracted after salt and pepper attack

Table 4 Objective performance parameters for watermarked images and extracted logos for salt and pepper noise

Image name	Embedding factor = 30			With mean		
	PSNR	BER	NCC	PSNR	BER	NCC
Image 1	38.5128	0	0.9944	39.0439	0.0110	0.9895
Image 2	39.5334	0.0034	0.9965	40.2995	0.0054	0.9944
Image 3	44.3698	0.0002	1.0000	44.3698	0.0002	1.0000
Image 4	34.3866	0.0457	0.9571	34.4635	0.0513	0.9520
Image 5	34.0033	0.0613	0.9435	34.0630	0.0703	0.9354
Image 6	39.2945	0.0256	0.9737	39.5296	0.0308	0.9689
Image 7	34.5124	0.0354	0.9644	34.5988	0.0425	0.9566
Image 8	42.1089	0.0009	0.9989	41.9897	0.0027	0.9976
Image 9	30.2668	0.0542	0.9486	30.2955	0.0798	0.9236
Image 10	33.0718	0.0339	0.9670	33.1724	0.0552	0.9459

4.2.3 Histogram Attack

Watermarked images obtained using the proposed technique were subjected to Histogram Equalization. The subjective and objective results have been shown in Fig. 6 and Table 5. From the results it is clear that the proposed algorithm is completely robust to histogram equalization.

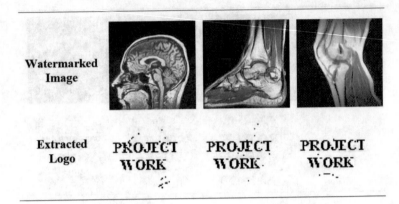

Fig. 6 Watermarked images and their respective logos extracted after Histogram attack

Table 5 Objective performance parameters for watermarked images and extracted logos for histogram attack

Image name	Embedding factor = 30			With mean		
	PSNR	BER	NCC	PSNR	BER	NCC
Image 1	12.4824	0.0042	0.9960	12.4918	0.0059	0.9944
Image 2	9.6420	0.00008	0.9997	9.6912	0.0032	0.9971
Image 3	8.8269	0.0029	0.9968	8.8269	0.0029	0.9968
Image 4	8.5552	0.0020	0.9981	7.6286	0.0054	0.9949
Image 5	7.0610	0.0066	0.9941	6.0882	0.0120	0.9890
Image 6	7.5788	0.0012	0.9987	6.7583	0.0039	0.9960
Image 7	9.2137	0.0012	0.9987	8.3210	0.0039	0.9962
Image 8	13.5954	0.0002	0.9997	13.5789	0.0009	0.9989
Image 9	15.6117	0.0027	0.9976	15.4251	0.0051	0.9949
Image 10	6.4738	0.0032	0.9968	5.8706	0.0056	0.9941

4.2.4 Gaussian Noise Attack

The watermarked images have been distorted with Gaussian noise with 0.0002 variance values. The distorted images and extracted logos from them have been shown in Fig. 7 The objective performance parameters corresponding to Gaussian noise is presented in Table 6.

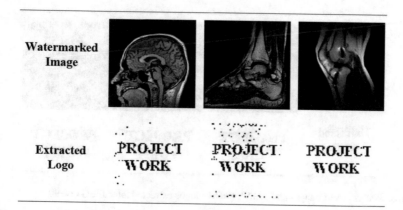

Fig. 7 Watermarked images and their respective logos extracted after addition of Gaussian noise

Table 6 Objective performance parameters for watermarked images and extracted logos for Gaussian noise attack

Image name	Embedding factor = 30			With mean		
	PSNR	BER	NCC	PSNR	BER	NCC
Image 1	34.6711	0	1.0000	35.5135	0.0034	0.9962
Image 2	35.3860	0	1.0000	36.3885	0.0193	0.9788
Image 3	35.5868	0	1.0000	35.5876	0	1.0000
Image 4	34.7365	0.0073	0.9925	35.2890	0.1206	0.8703
Image 5	34.9102	0.0103	0.9901	35.4781	0.1472	0.8430
Image 6	35.8513	0.0056	0.9938	36.6586	0.1362	0.8537
Image 7	34.1874	0.0071	0.9922	34.7203	0.1125	0.8770
Image 8	34.9547	0	1.0000	34.6977	0	1.0000
Image 9	34.0648	0.0004	0.9995	34.4980	0.0364	0.9601
Image 10	35.3502	0.0027	0.9971	36.4603	0.1233	0.8703

4.2.5 JPEG Attack with Different Quality Factor

JPEG is one of the standard compression techniques used to compress the images. The watermarked images have been subjected to JPEG compression with of different quality factors [36, 37]. The compressed images for a quality factor of 90 and the logos extracted from them are shown in Fig. 8 and Table 7 shows the performance parameters for JPEG compression for various images for a quality factor of 90. The objective quality analysis if the scheme for quality factors of 80, 70, 60, 50 and 40 have been respectively shown in Tables 8, 9, 10, 11 and 12.

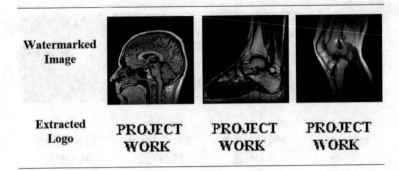

Fig. 8 Watermarked images and their respective logos extracted after JPEG Q = 90

Table 7 Objective performance parameters for watermarked images and extracted logos for JPEG 90

Image name	Embedding factor = 30			With mean		
	PSNR	BER	NCC	PSNR	BER	NCC
Image 1	38.2969	0	1.0000	40.0751	0	1.0000
Image 2	39.1188	0	1.0000	41.4955	0	1.0000
Image 3	40.4008	0	1.0000	40.4008	0	1.0000
Image 4	36.4579	0.0049	0.9949	37.0992	0.0100	0.9901
Image 5	36.5729	0.0081	0.9928	37.2132	0.0149	0.9861
Image 6	38.5812	0.0049	0.9946	39.8014	0.0081	0.9914
Image 7	35.7798	0.0059	0.9936	36.4097	0.0081	0.9912
Image 8	39.2391	0	1.0000	37.9585	0	1.0000
Image 9	34.6960	0	1.0000	35.1991	0.0022	0.9976
Image 10	37.5012	0.0020	0.9979	39.1366	0.0063	0.9936

4.2.6 Low Pass Filtering Attack

The watermarked images have been distorted with low pass filtering attack. The distorted images and extracted logos from them are shown in Fig. 9. The objective performance parameters corresponding to Gaussian noise is presented in Table 13.

Table 8 Objective performance parameters for watermarked images and extracted logos for JPEG Q = 80

Image name	Embedding factor = 30			With mean		
	PSNR	BER	NCC	PSNR	BER	NCC
Image 1	36.1259	0	1.0000	38.6784	0.0002	0.9997
Image 2	37.1680	0	1.0000	39.9795	0.1135	0.8757
Image 3	37.3020	0	1.0000	37.3020	0	1.0000
Image 4	35.6799	0.0088	0.9914	35.9796	0.5854	0.3596
Image 5	35.6577	0.0129	0.9879	35.9132	0.6501	0.2902
Image 6	37.8612	0.0081	0.9914	38.4277	0.6416	0.2969
Image 7	35.0394	0.0095	0.9898	35.4150	0.581	0.3990
Image 8	36.6779	0	1.0000	37.146	0	1.0000
Image 9	32.9958	0.002	0.9997	33.3614	0.1948	0.7864
Image 10	36.5921	0.0029	0.9971	37.5582	0.5798	0.3652

Table 9 Objective performance parameters for watermarked images and extracted logos for JPEG Q = 70

Image name	Embedding factor = 30			With mean		
	PSNR	BER	NCC	PSNR	BER	NCC
Image 1	37.3889	0	1.0000	38.8563	0.3103	0.6610
Image 2	38.2081	0	1.0000	39.8673	0.3406	0.6292
Image 3	42.7980	0	1.0000	42.7980	0	1.0000
Image 4	34.8766	0.0134	0.9863	35.0520	0.5918	0.3529
Image 5	34.7387	0.0203	0.9812	34.8751	0.6567	0.2843
Image 6	36.9914	0.0142	0.9855	37.3101	0.6504	0.2878
Image 7	34.2546	0.0127	0.9866	34.5153	0.5515	0.3958
Image 8	39.1451	0	1.0000	36.0514	0.0076	0.9933
Image 9	32.0516	0.0032	0.9965	32.3294	0.2222	0.7586
Image 10	35.7171	0.0076	0.9922	36.5327	0.6006	0.3441

4.2.7 Sharpening Attack

The watermarked images have been distorted with sharpening attack. The distorted images and extracted logos from them are shown in Fig. 10. The objective performance parameters corresponding to sharpening is presented in Table 14.

Table 10 Objective performance parameters for watermarked images and extracted logos for JPEG Q = 60

Image name	Embedding factor = 30			With mean		
	PSNR	BER	NCC	PSNR	BER	NCC
Image 1	36.9949	0.0004	0.9995	38.3949	0.3613	0.6109
Image 2	37.6366	0.0313	0.9657	39.5806	0.4653	0.4957
Image 3	39.9527	0	1.0000	39.9527	0	1.0000
Image 4	34.2053	0.6462	0.2937	34.3400	0.6040	0.3408
Image 5	33.9800	0.7132	0.2235	34.0542	0.6724	0.2696
Image 6	36.4453	0.6941	0.2404	36.6856	0.6558	0.2827
Image 7	33.7384	0.5830	0.3620	33.9819	0.5657	0.3834
Image 8	38.6872	0.0022	0.9976	38.9488	0.0273	0.9743
Image 9	31.7005	0.1975	0.7838	31.8407	0.2852	0.6935
Image 10	35.2530	0.5867	0.3577	35.9477	0.6199	0.3253

Table 11 Objective performance parameters for watermarked images and extracted logos for Q = 50

Image name	Embedding factor = 30			With mean		
	PSNR	BER	NCC	PSNR	BER	NCC
Image 1	36.1398	0.0027	0.9976	37.7747	0.4109	0.5630
Image 2	37.0981	0.1211	0.8674	39.0863	0.5181	0.4443
Image 3	38.1110	0.0034	0.9962	38.1110	0.0034	0.9962
Image 4	33.3358	0.6648	0.2787	33.7862	0.6746	0.2680
Image 5	32.9621	0.7300	0.2066	33.3391	0.7385	0.1988
Image 6	35.3467	0.7058	0.2296	36.0860	0.7141	0.2211
Image 7	32.9953	0.6023	0.3430	33.4578	0.6130	0.3355
Image 8	37.5351	0.0059	0.9936	37.9812	0.0505	0.9534
Image 9	31.3372	0.2161	0.7647	31.5417	0.3425	0.6305
Image 10	34.4665	0.6033	0.3411	35.4142	0.6475	0.2974

4.2.8 Rotation Attack (10°)

Rotation is one of the important intentional attacks carried out on the watermarked images. To check robustness of our scheme to this attack we have rotated the watermarked images by 10°. The distorted images and extracted logos from them are shown in Fig. 11 The objective performance parameters corresponding to rotation is presented in Table 15.

Table 12 Objective performance parameters for watermarked images and extracted logos Q = 50

Image name	Embedding factor = 30			With mean		
	PSNR	BER	NCC	PSNR	BER	NCC
Image 1	34.8432	0.0078	0.9928	37.4442	0.4998	0.4748
Image 2	36.2475	0.3369	0.6321	38.3419	0.5837	0.3778
Image 3	37.0299	0.1384	0.8489	37.0299	0.1384	0.8489
Image 4	32.8373	0.6846	0.2570	32.9090	0.6934	0.2511
Image 5	32.3938	0.7480	0.1900	32.4156	0.7539	0.1827
Image 6	35.0783	0.7290	0.2101	35.1796	0.7141	0.2211
Image 7	32.5906	0.6316	0.3170	32.7427	0.6521	0.2996
Image 8	35.8425	0.0134	0.9861	36.5638	0.1001	0.9014
Image 9	30.9741	0.2913	.6884	31.1816	0.4397	0.5308
Image 10	34.3293	0.6321	0.3127	34.6436	0.6755	0.2701

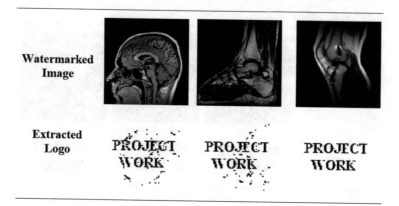

Fig. 9 Watermarked images and their respective extracted logos

4.2.9 Average Filtering Attack

The watermarked images have been distorted with average filtering attack. The distorted images and extracted logos from them are shown in Fig. 12. The objective performance parameters corresponding to average filtering attack is presented in Table 16.

Table 13 Objective performance parameters for watermarked images and extracted logos corresponding to Gaussian Noise

Image name	Embedding factor = 30			With mean		
	PSNR	BER	NCC	PSNR	BER	NCC
Image 1	37.6726	0.0083	0.9914	37.8284	0.0239	0.9775
Image 2	38.7411	0.0063	0.9933	38.9847	0.078	0.9818
Image 3	44.7805	0	1.0000	44.7805	0	1.0000
Image 4	27.3187	0.0583	0.9448	27.3499	0.0679	0.9357
Image 5	26.1129	0.0798	0.9263	26.1410	0.0881	0.9175
Image 6	28.0191	0.0322	0.9681	28.0612	0.0381	0.9626
Image 7	26.1933	0.0442	0.9553	26.2210	0.0515	0.9486
Image 8	32.7259	0.0004	0.9995	32.7096	0.0068	0.9933
Image 9	28.8186	0.0415	0.9601	28.8405	0.0625	0.9416
Image 10	28.7683	0.0430	0.9603	28.8286	0.0696	0.9333

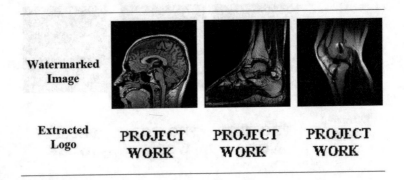

Fig. 10 Watermarked images and their respective extracted logos

4.2.10 Weiner Filtering Attack

The watermarked images have been distorted with weiner filtering attack. The distorted images and extracted logos from them are shown in Fig. 13. The objective performance parameters corresponding to weiner filtering attack is presented in Table 17.

4.3 A Brief Discussion About the Results

This section presents a brief discussion presented in terms of various Figures and Tables as presented above. From the objective quality analysis, it is clear that second

Table 14 Objective performance parameters for watermarked images and extracted logos corresponding to sharpening attack

Image name	Embedding factor = 30			With mean		
	PSNR	BER	NCC	PSNR	BER	NCC
Image 1	25.2080	0	1.0000	26.4786	0	1.0000
Image 2	25.8174	0	1.0000	27.4144	0	1.0000
Image 3	28.2589	0	1.0000	28.2589	0	1.0000
Image 4	23.9090	0.0022	0.9979	24.2602	0.0046	0.9952
Image 5	24.5208	0.0024	0.9979	24.8936	0.0063	0.9941
Image 6	26.1554	0.0004	0.9992	26.8543	0.0032	0.9971
Image 7	23.9461	0.0024	0.9973	24.3995	0.0068	0.9936
Image 8	26.8424	0	1.0000	26.4662	0	1.0000
Image 9	19.7966	0.0004	0.9995	20.0027	0.0022	0.9976
Image 10	23.2024	0.0024	0.9973	23.7883	0.0068	0.9928

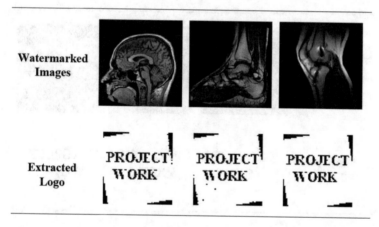

Fig. 11 Watermarked images and their respective extracted logos

algorithm (mean based) shows better results. The average PSNR obtained by the second algorithm is 4 dB more than that of first algorithm. Furthermore, the average SSIM value also increases drastically from 0.8945 to 0.9861. On the basis of the payload the results have been compared to that of Farah et al. as shown in Table 2. As is evident from Table that despite four times increase in the payload the proposed system is capable of producing good quality watermarked images. On the robustness front it has been seen that the system is robust to median filtering as shown by subjective and objective analysis. The robustness performance of median based technique is at par with that where fixed embedding factor of value 30 is used. Similarly, the performance of the proposed scheme to various attacks like salt and pepper attack,

Table 15 Objective performance parameters for watermarked images and extracted logos corresponding to rotation attack

Image name	Embedding factor = 30			With mean		
	PSNR	BER	NCC	PSNR	BER	NCC
Image 1	28.8303	0.0608	0.9333	28.8931	0.065	0.9325
Image 2	27.8615	0.0601	0.9341	27.9141	0.0620	0.9319
Image 3	38.9548	0.0591	0.9352	38.9548	0.0591	0.9352
Image 4	31.0961	0.0830	0.9100	31.1841	0.0925	0.9006
Image 5	29.9587	0.0916	0.9043	30.0312	0.1028	0.8936
Image 6	32.0958	0.0747	0.9188	32.2205	0.0806	0.9126
Image 7	30.1399	0.0847	0.9081	30.2212	0.0898	0.9025
Image 8	13.4669	0.0706	0.9226	13.4663	0.0676	0.9258
Image 9	32.7720	0.0030	0.9314	32.8740	0.0667	0.9282
Image 10	33.0425	0.0759	0.9188	33.2747	0.0894	0.9051

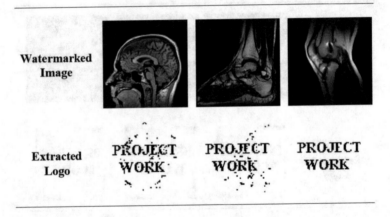

Fig. 12 Watermarked images and their respective extracted logos

histogram equalization, Gaussian noise, and rotation for the two approaches is at par. However, it has been seen that the robustness of the mean-based scheme is slightly less compared to fixed embedding based embedding factor for JPEG attack with various quality factors.

Table 16 Objective performance parameters for watermarked images and extracted logos corresponding to average filtering

Image name	Embedding factor = 30			With mean		
	PSNR	BER	NCC	PSNR	BER	NCC
Image 1	38.0812	0.0083	0.9914	38.2517	0.0234	0.9780
Image 2	38.7708	0.0063	0.9933	39.0194	0.0178	0.9818
Image 3	44.9809	0	1.0000	44.9809	0	1.0000
Image 4	27.3184	0.0583	0.9448	27.3498	0.0679	0.9357
Image 5	26.1127	0.0798	0.9263	26.1409	0.0881	0.9175
Image 6	28.0188	0.0322	0.9681	28.0611	0.0381	0.9620
Image 7	26.1931	0.0442	0.9553	26.2210	0.0515	0.9486
Image 8	43.4364	0.0002	0.9997	43.2427	0.0066	0.9936
Image 9	28.192	0.0415	0.9601	28.8417	0.0625	0.9416
Image 10	28.7680	0.0430	0.9603	28.8286	0.0696	0.9333

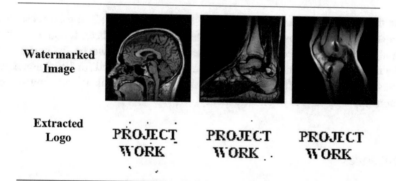

Fig. 13 Watermarked images and their respective extracted logos

5 Conclusion

A secure and robust medical image watermarking scheme capable of securely embedding patient data has been presented in this chapter. The embedding has been carried out based on inter-coefficient comparison between two DCT coefficients in each 8 × 8 DCT coefficient block. Two variants of the embedding scheme have been used for embedding the data. In the first algorithm a global embedding factor has been used for embedding data bit in a block. However, in the second algorithm the local characteristics of every block have been used instead of using a global embedding factor. This has been done by computing mean of every DCT block and utilizing that as embedding factor. This approach thus makes the proposed scheme as adaptive in nature. A detailed perceptual analysis of the two schemes has been carried out which

Table 17 Objective performance parameters for watermarked images and extracted logos corresponding to Weiner filtering

Image name	Embedding factor = 30			With mean		
	PSNR	BER	NCC	PSNR	BER	NCC
Image 1	39.9551	0	1.0000	40.4443	0.0037	0.9965
Image 2	41.0709	0	1.0000	41.8553	0.0009	0.9992
Image 3	45.4625	0	1.0000	45.4625	0	1.0000
Image 4	36.3452	0.0327	0.9686	36.5816	0.0415	0.9601
Image 5	36.4328	0.0452	0.9595	36.7014	0.0574	0.9469
Image 6	39.3213	0.0212	0.9788	39.8286	0.0259	0.9745
Image 7	35.7855	0.0237	0.9759	35.9948	0.0308	0.9695
Image 8	43.6625	0.00024	0.9997	43.4528	0.0046	0.9952
Image 9	32.8757	0.0168	0.9847	32.9550	0.0352	0.9657
Image 10	37.0773	0.0161	0.9855	37.5218	0.0369	0.9638

show that mean based scheme has better perceptual quality. The robustness of the proposed system for most of the attacks has been however been found to be at par with algorithm first (fixed embedding-based approach). However, the performance of the mean-based technique has been found a bit deteriorated for JPEG attack with various quality factors. The security of the scheme has been taken care of by utilizing the non-linear dynamics of chaotic theory.

References

1. Wakatani A (2002) Digital watermarking for ROI medical images by using compressed signature image. In: Proceedings of the 35th annual Hawaii international conference on system sciences (HICSS), pp 7–10
2. Parah SA, Sheikh JA, Bhat GM (2009) On the realization and design of chaotic spread spectrum modulation technique for secure data transmission. In: Proceedings of first IEEE sponsored international conference IMPACT-2009 at AMU Aligarh India, pp 241–244
3. Parah SA, Sheikh JA, Bhat GM (2013) On the realization of a spatial domain data hiding technique based on intermediate significant bit plane embedding (ISBPE) and post embedding pixel adjustment (PEPA). In: Proceedings of IEEE international conference on multimedia signal processing and communication technologies IMPACT, 23–25 Nov 2013, pp 51–55
4. Dey N, Samanta S, Chakraborty S, Das A, Chaudhuri SS, Suri JS (2014) Firefly algorithm for optimization of scaling factors during embedding of manifold medical information: an application in ophthalmology imaging. J Med Imaging Health Inform 4(3):384–394
5. Bhat GM, Parah SA, Sheikh JA (2014) A secure and efficient spatial domain data hiding technique based on pixel adjustment. Am J Eng Technol Res, US Library Congress, (USA) 14(2):38–44

6. Farhana A, Bhat GM, Sheikh JA, Parah SA (2017) Hiding clinical information in medical images: a new high capacity and reversible data hiding technique. J Biomed Inform Elsevier. https://doi.org/10.1016/j.jbi.2017.01.006
7. Loan NA, Sheikh JA, Bhat GM, Parah SA (2017) Utilizing neighborhood coefficient correlation: a new image watermarking technique robust to singular and hybrid attacks. Multidim Syst Sign Process. https://doi.org/10.1007/s11045-017-0490-z
8. Sheikh JA, Parah SA, Farhana A, Loan NA, Bhat GM (2016) Reversible and high capacity data hiding technique for E-healthcare applications. Multimedia Tools Appl. https://doi.org/10.1007/s11042-016-4196-2
9. Akhoon JA, Parah SA, Sheikh JA, Bhat GM (2016) Information hiding in edges: a high capacity information hiding technique using hybrid edge detection. Multimedia Tools Appl. https://doi.org/10.1007/s11042-016-4253-x
10. Assad UI, Parah SA, Sheikh JA, Bhat GM (2016) Realization and robustness evaluation of a blind spatial domain watermarking technique. Int J Electron. https://doi.org/10.1080/00207217.2016.1242162
11. Alhaqbani B, Fidge C (2008) Privacy-preserving electronic health record linkage using pseudonym identifiers. In: Proceedings of 10th international conference e-health networking, applications and services, pp 108–117
12. Parah SA, Sheikh JA, Loan NA, Bhat GM (2016) Robust and blind watermarking technique in DCT domain using inter- block coefficient differencing. Digit Sig Process Elsevier. https://doi.org/10.1016/j.dsp.2016.02.005
13. Peleg M, Beimel D, Dori D, Denekamp Y (2008) Situation-based access control: privacy management via modeling of patient data access scenarios. J Biomed Inform Elsevier 41:1028–1040
14. Sudeb D, Malay KK (2013) Effective management of medical information through ROI-lossless fragile image watermarking technique. Comput Methods Programs Biomed, pp 662–675
15. Parah SA, Sheikh JA, Farhana A, Loan NA, Bhat GM (2015) Information hiding in medical images: a robust medical image watermarking system for E-healthcare. Multimed Tools Appl. https://doi.org/10.1007/s11042-015-3127-y
16. Darwish A Hassanien AE, Elhoseny M, Sangaiah AK, Muhammad K (2017) The impact of the hybrid platform of internet of things and cloud computing on healthcare systems: opportunities, challenges, and open problems. J Ambient Intell Humanized Comput First Online: 29 Dec 2017. https://doi.org/10.1007/s12652-017-0659-1
17. Elhoseny M, Abdelaziz A, Salama A, Riad AM, Sangaiah AK, Muhammad K (2018) A hybrid model of internet of things and cloud computing to manage big data in health services applications. Future Gener Comput Syst, Available online 15 Mar 2018. In Press https://doi.org/10.1016/j.future.2018.03.005
18. Abdelaziza A, Elhoseny M, Salama AS, Riad AM (2018) A machine learning model for improving healthcare services on cloud computing environment. Measurement 119:117–128. https://doi.org/10.1016/j.measurement.2018.01.022
19. Ray SR, Shi JH, Jui LL, Tzong WK, Rong JC (2013) An improved SVD-based watermarking technique for copyright protection. Expert Syst Appl
20. Lestriandoko NH, Wirahman T (2010) Reversible watermarking using difference of virtual border for digital image protection. In: International conference on distributed framework and applications (DFmA), 2–3 Aug 2010
21. Suganya G, Amudha K (2014) Medical image integrity control using joint encryption and watermarking techniques
22. Osamah MA, Bee EK (2011) Authentication and data hiding using a hybrid ROI-based watermarking scheme for DICOM images. J Digit Imaging 24:114–125
23. Baisa LG, Suresh NM (2012) ROI based embedded watermarking of medical images for secured communication in telemedicine. World Acad Sci Eng Technol 6
24. Tanmay B, Sirshendu H, Chaudhuri SRB (2014) A semi-fragile blind digital watermarking technique for medical image file authentication using stationary wavelet transformation. Int J Comput Appl 104:0975–8887

25. Lin SD, Chen CF (2000) A robust DCT-based watermarking for copyright protection. IEEE Trans Consum Electron 46:415–421
26. Kundu MK, Das S (2010) Lossless ROI medical image watermarking technique with enhanced security and high payload embedding. In: Proceedings of the 20th international conference of the pattern recognition Istanbul, Turkey, pp 1457–1460
27. Chao HM, Hsu CM, Miaou SG (2002) A data hiding technique with authentication, integration and confidentiality for electronic patient records. IEEET Trans Inf Technol Biomed 6:46–52
28. Acharya UR, Niranjan UC, Iyenger SS, Kannathal N, Min LC (2004) Simultaneous storage of patient information with medical images in the frequency domain. Comput Methods Programs Biomed 76:13–19
29. Kliem A, Hovestadt M, Kao O (2012) Security and communication architecture for networked medical devices in mobility-aware e-health environments. In: 2012 IEEE first international conference mobile services, pp 112–114
30. Maity HK, Maity SP (2012) Joint robust and reversible watermarking for medical images. In: Procedia technology 6, pp 275–282
31. KO LT, Chen JE, Shieh YS, Sin HC, Sung TY (2012) Nested quantization index modulation for reversible watermarking & its application to healthcare information management systems. Comput Math Methods Med
32. Wang M, Yu J, Jiang G, Peng Z, Shao F, Luo T (2015) New fragile watermarking method for stereo image authentication with localization and recovery. AEU-Int J Electron Commun 69(1):361–370
33. Parah SA, Farhana A, Sheikh JA, Bhat GM (2015) On the realization of robust watermarking system for medical images. In: 12th IEEE India international conference (INDICON) on electronics, energy, environment, communication, computers, control (E3-C3), Jamia Millia Islamia, New Delhi, 17–20 Dec 2015
34. Frahana A, Javaid AS, Bhat GM (2017) Information hiding in medical images: a robust medical image watermarking system for E-healthcare. Multimed Tools Appl. https://doi.org/10.1007/s11042-015-3127-y
35. Dey N, Parah SA, Sheikh JA, Bhat GM (2017) Realization of a new robust and secure watermarking technique using DC coefficient modification in pixel domain and chaotic encryption. J Global Inf Manag 26(4)
36. Elhoseny M, Ramírez-González G, Abu-Elnasr OM, Shawkat SA, Arunkumar N, Farouk A (2018) Secure medical data transmission model for IoT-based healthcare systems. IEEE Access 6:20596–20608. https://doi.org/10.1109/access.2018.2817615
37. Shehab A, Elhoseny M, Muhammad K, Sangaiah AK, Yang P, Huang H, Hou G (2018) Secure and robust fragile watermarking scheme for medical images. IEEE Access 6:10269–10278. https://doi.org/10.1109/access.2018.2799240

A Swarm Intelligence Model for Enhancing Health Care Services in Smart Cities Applications

Ahmed Abdelaziz, Ahmed S. Salama and A. M. Riad

Abstract Cloud computing plays a significant role in healthcare services (HCS) within smart cities due to its the ability to retrieve patients' data, collect big data of patients by sensors, diagnosis of diseases and other medicinal fields in less time and less of cost. However, the task scheduling problem to process the medical requests represents a big challenge in smart cities. The task scheduling performs a significant role for the enhancement of the performance through reducing the execution time of requests (tasks) from stakeholders and utilization of medical resources to help stakeholders for saving time and cost in smart cities. In addition, it helps the stakeholders to reduce their waiting time, turnaround time of medical requests on cloud environment, minimize waste of CPU utilization of VMs, and maximize utilization of resources. For that, this paper proposes an intelligent model for HCS in a cloud environment using two different intelligent optimization algorithms, which are Particle Swarm Optimization (PSO), and Parallel Particle Swarm Optimization (PPSO). In addition, a set of experiments are conducted to provide a competitive study between those two algorithms regarding the execution time, the data processing speed, and the system efficiency. The results showed that PPSO algorithm outperforms on PSO algorithm. In addition, this paper proposes a new PPSO dependent algorithm using CloudSim package to solve task scheduling problem to support healthcare providers in smart cities to reduce execution time of medical requests and maximize utilization of medical resources.

A. Abdelaziz (✉)
Department of Information Systems, Higher Technological Institute, Cairo, Egypt
e-mail: ahmed.aziz.1157@gmail.com

A. M. Riad
Faculty of Computers and Information, Mansoura University, Mansoura, Egypt

A. S. Salama
Information Systems Department, Faculty of Computing and Information Technology,
University of Jeddah, Jeddah, Kingdom of Saudi Arabia

A. S. Salama
Computers and Information Systems Department,
Sadat Academy for Management Sciences, Cairo, Egypt

© Springer Nature Switzerland AG 2019
A. E. Hassanien et al. (eds.), *Security in Smart Cities: Models, Applications,
and Challenges*, Lecture Notes in Intelligent Transportation and Infrastructure,
https://doi.org/10.1007/978-3-030-01560-2_4

Keywords Mobile cloud computing · Internet of things · Smart cities
Intelligent applications

1 Introduction

In recent years, cloud computing has become useful to stakeholders (patients, doctors
and etc.) in several domains for transferring medical services over the Internet. Cloud
computing transmits application, infrastructure services to big numbers of stakeholders with assorted and dynamically changing requirements [1]. Cloud is com-posed of
datacenters, hosts, VMs, resources and etc. Datacenters are containing a big number
of resources and list of different applications. Hosts are composed of a large number
of VMs to store and regain several medical resources to stakeholders.

Cloud computing uses the virtualization technique which permits to share a single
physical instance of a resource or an application among various stakeholders and
enterprises [2]. It does by allocating a logical name to a physical storage and providing
a pointer to that physical resource when requested.

Virtualization consists of hardware virtualization, operating system (OS) virtualization, server virtualization and storage virtualization in cloud computing environment. Hardware virtualization is mainly done for the host platforms, because controlling VMs is much easier than controlling a physical host [3]. Operating System
Virtualization is mainly used for experiment the different applications on different
platforms of OS [4]. Server virtualization can be divided into several physical hosts
the request basis. Storage virtualization is created to recovery purposes. Virtualization plays a very significant role in the cloud computing, stakeholders' share the
medical data in the clouds like medical applications etc. [5].

Cloud computing in smart cities can perform an important role in containing
healthcare costs, optimizing resources and ushering in a new period of innovations.
Current orientation target towards accessing data anytime, anywhere, which can be
accomplished when moving healthcare data to the cloud computing. The benefits of
applying cloud computing on healthcare services are (patient care, lowering costs,
optimizing resources, high performance in retrieving patients' data and etc.). However, there are some risks of cloud computing for healthcare services in smart cities
include (Regulatory, loss of data, Intellectual Property Right, Liability and etc.).

Smart cities are facing a big problem to support healthcare services by managing
big data of patients, intelligent diagnosis of diseases and other medical fields by
cloud computing environment to reduce time and cost. The stakeholders in HCS
suffer from time delay of medical requests on cloud computing environment. Many
reasons that lead to time delay of medical requests such as waiting time, turnaround
time of medical requests, the waste of CPU utilization and the waste of resources
utilization. This paper proposes intelligent algorithms such as PSO and PPSO to find
a suitable solution to solve task scheduling problem that helps the stakeholders to
reduce their waiting time, turnaround time of medical requests on cloud computing

environment, minimize waste of CPU utilization of VMs, and maximize utilization of medical resources.

In order to evaluate the performance of those two algorithms (PSO and PPSO), this paper provides a comparative study to their work in a cloud environment. The comparison aims to see which algorithm can be used to improve the performance of HCS. This paper also selected the optimal algorithm through execution time, speed and efficiency. Depending on that comparison, this paper proposes a new algorithm to implement it on CloudSim simulation package.

This paper is arranged as follows: Sect. 2 introduces related work, Sect. 3 introduces basics and background, Sect. 4 introduces a proposed intelligent model of cloud computing for HCS, Sect. 5 introduces the proposed PSO and PPSO based algorithms for cloud computing, Sect. 6 introduces experimental results and finally, section presents conclusion and future work.

2 Related Work

Through related work, many studies were done on applying and using intelligent techniques such as PSO, ACO, GA, and cloud computing for HCS and how to apply them in smart cities. The following is a presentation of some research related to the subject:

Boulos et al. [6], introduced a new approach to solve a problem of medical data analysis in geospatial in smart cities by internet of things. This study tries to describe and discuss in great detail an Internet-connected web of citizens (people) and electronic sensors/devices (things) that can serve many functions related to public and environmental health surveillance and crisis management applications. Research opinion, internet of things can help to find the optimal solution to analysis of medical data in smart cities.

Alhussein [7] introduced a novel framework to detect Parkinson's disease in smart cities based on cloud computing by support vector machine. This disease can affect to impair the body's balance, damages motor skills, and leads to disorder in speech production. The proposed framework can achieve 97.2% accuracy in detecting Parkinson's disease. Researcher opinion, the proposed framework can detect Parkinson's disease in smart cities based on cloud computing by support vector machine.

Bhunia et al. [8], presented the new method to collect and process sensitive health data to diagnosis of health problems in smart cities to save the patients from death. This study uses a fuzzy scheme to detect critical physiological stages of a human body easily. Researcher opinion, the proposed method may find the optimal solution to determine critical physiological stages of a human body and use it in smart cities.

Islam et al. [9], introduced a new model to collect and process big healthcare data in smart cities efficiently by mobile cloud environment. This study seeks to solve the virtual machine migration problem and provision virtual machine resources by using ant colony optimization. Researcher opinion, the proposed model can solve the virtual machine migration problem efficiently.

Sajjad et al. [10], introduced a novel framework in smart cities to detect and count white blood cells in blood samples for saving the body against both infectious disease and foreign invaders. This study aims to introduce a framework based on mobile-cloud for segmentation and classification of white blood cells in blood samples. Researcher opinion, the proposed framework can detect count white blood cells in blood samples.

Mishra [11] presented a new method to compute task scheduling efficiently based on BCO technique and trying to distribute workloads between the resources equally to reduce the overall execution time for tasks and then, to improve the effectiveness of the whole cloud computing services. Researcher opinion, clarity of proposed technique those implement in task scheduling to reduce the time on cloud environment.

Bhatt et al. [12] presented an evolutionary concept of implementation of PSO over CloudSim and introduced a new way of checking of tasks in future. When cloud model applied with PSO, enhances the efficiency of the result and sometime it gave 100% results also. This study presented the result of implementation of PSO in CloudSim which is the basic idea that it can execute this technique in any new environment or any new simulator. Researcher opinion, clarity of proposed technique those implement in CloudSim to enhance load balancing operation.

Gomathi et al. [13], presented a Hybrid PSO (HPSO) based task scheduling to balance in the load across the entire system while trying to reduce the time of a given tasks. This HPSO is used to improve the resources utilization while trying to detect optimal schedule for given tasks in the cloud computing environment. Researcher opinion, clarity of the proposed task scheduling technique using HPSO has been effective to reduce the time and maximize the resources utilization in cloud computing environment.

Mohana et al. [14], presented a comparison of most successful optimization techniques by Swarm Intelligence: ACO and PSO. A comparison is finished to examine the performance of ACO and PSO techniques. In cloud computing, the optimal task scheduling is used to enhance the overall execution time of the task. This paper analyzes the two most successful algorithms of optimization techniques by Swarm Intelligence: ACO and PSO. Researcher opinion, PSO algorithm in task scheduling on cloud environment is given better outcomes when compared to ACO.

Ajeena et al. [15], introduced a new approach to facilitate task scheduling operation based on PSO technique. This method helps users to obtain the optimal quality of service (QoS) in the lowest time and the lowest cost. Researcher opinion, PSO technique may be will not be able to detect the optimal solution of task scheduling in cloud computing.

Kaur et al. [16], presented various domains based on Cloud computing such as virtualization, distributed computing, grid computing and etc. Task scheduling is a challenging affair to obtain high profit and to expeditiously increase working of cloud computing. This paper introduced a new way to obtain high benefit from resources; optimized resources utilization is important. Researcher opinion, lack of clarity of PSO algorithm to achieve optimized of resources utilization.

El-Sisi et al. [17], presented a PSO algorithm for cloud task scheduling has been proposed. PSO algorithm is a new method may find the optimal solution in task scheduling on cloud computing environment and selecting suitable resources for task implementation and users meet their application's performance requirements with minimum expenditure. Researcher opinion, the proposed technique is minimizing the time of a given jobs set in cloud computing environment.

Bilgaiyan et al. [18], presented a different solutions for cloudlet scheduling problem. Scheduling is an important activity in multi-jobs systems to efficiently manage resources, reduce idle time and increase quality of systems. To reduce the total make span taken, the scheduling principle should aim to reduce the amount of data transmit with minimum cost and ensure balanced distribution of tasks as per processing capability. Researcher opinion, Efficient PSO, ACO and BCO on scheduling of tasks can significantly enhance the viability and applicability of cloud computing systems.

Tawfeek et al. [19], presented a new approach to find the best solution for task scheduling based on ACO technique on cloud computing compared with different techniques such as First Come First Served (FCFS) and Round-Robin (RR). This paper uses ACO, FCFS and RR techniques to obtain the best resource allocation for cloudlets in the dynamic cloud system to minimize the time of cloudlets on the entire system. Then, the task scheduling applied on the CloudSim package. Researcher opinion, ACO algorithm may obtain the best resource allocation for cloudlets in the cloud computing environment.

Salama [20] introduced a new model to improve the access time for the mobile cloud computing services and improving the mobile commerce transactions based on PPSO technique. This paper uses PSO technique to examine the quality of the proposed PSO and PPSO based techniques for mobile cloud computing. Researcher opinion, PPSO algorithm is sharing in minimizing time of tasks and helping to maximize resources utilization in cloud environment.

Awad et al. [21], presented a mathematical framework using Load Balancing Mutation (balancing) a particle swarm optimization (LBMPSO) based allocation for cloud computing that occupies into account accuracy, implementation time, conveyance time, round trip time, conveyance cost and load balancing between cloudlets and virtual machine. Researcher opinion, LBMPSO enhances the accuracy of cloud computing environment and perfect distribution of tasks onto resources compared to other techniques.

Al-Olimat et al. [22], presented a study to minimize make span and maximize resource utilization, HPSO technique is implemented inside of CloudSim to advance the work of the already implemented simple broker. The new approach increases the resource utilization and reduces the time. This study presents a solution for improving the make span scheduled tasks and the utilization of cloud resources through the use of a hybrid variant of popular population-based metheuristic (PBM) algorithm and PSO technique. Researcher opinion, PSO algorithm can minimize the make span of the workload while excelling at optimizing the simulated scheduling results of CloudSim in a way that will maximize the resources utilization and reduce the costs in cloud computing environment.

Priyadarsini et al. [23], presented a study to use Improved Particle Swarm Optimization (IPSO) algorithm for independent (Meta) task scheduling in cloud computing with a proposed cost function. The parameters considered are time and utilization of resource. This paper illustrates a new method called IPSO to achieve the optimal solution in task scheduling on cloud environment with consideration of the performance. Researcher opinion, the proposed IPSO technique realizes better resources utilization and reduces time when compared with the existing standard techniques.

Vidhya et al. [24], presented a study to propose a system considering that distributed storage system is heterogeneous where each node exhibit different online availability. The optimum data assignment policy minimizes the cyclic redundancy check and their related cost. The optimum data redundancy is the smallest amount of redundancy desired to the data availability for accessing data at any time. The redundancy ratio determines how many times the data blocks are replicated. The value of redundancy ratio should be low to memorize minimum redundancy. Researcher opinion, the PPSO technique allocates data blocks according to their online availabilities.

Alkhashai et al. [25], presented a study to specify the cloudlets to multiple resources of computing. Consequently, the total cost of implementation is to minimum and load to be shared between these resources of computing. This paper introduces a new approach to find the optimal solution for independent cloudlet scheduling and finding suitable resources for task implementation based on Best-Fit-PSO (BFPSO) and PSO-Tabu Search (PSOTS). The two techniques have been implemented over CloudSim package. Measuring the performance of cloudlet scheduling based on implementation time, cost and resources utilization. Researcher opinion, the proposed algorithms (i.e., BFPSO and PSOTS) outperform the default PSO with respect to the time, resources utilization and cost.

Smart cities are facing a big challenge to support healthcare services by managing big data of patients, intelligent diagnosis of diseases and other medical fields by cloud computing environment to reduce time and cost. This paper aims to improve healthcare services in smart cities by introducing intelligent model to enhance task scheduling to reduce total execution time of medical requests from stakeholders and maximize utilization of medical resources. Intelligent model is based on different intelligent algorithms such as PSO and PPSO. Most of the related works are based on managing big data of patients, using PSO, ACO and other to enhance task scheduling by using cloud computing. This paper seeks to find the optimal solution to reduce execution time of medical requests for helping patients in smart cities for improving their health by cloud computing. For that, this paper proposes architecture to enhance healthcare services based on cloud computing in smart cities.

3 Basics and Background

This section introduces an overview of swarm intelligence algorithms. Swarm intelligence defines the collective attitude of systems consist of many individuals interacting locally with each other and with their environment. Swarms inherently use forms of decentralized control and self-organization to obtain their objectives. Swarm intelligence techniques such as PSO, PPSO, ant colony optimization (ACO) and bee

Fig. 1 The main steps of the
PSO algorithm

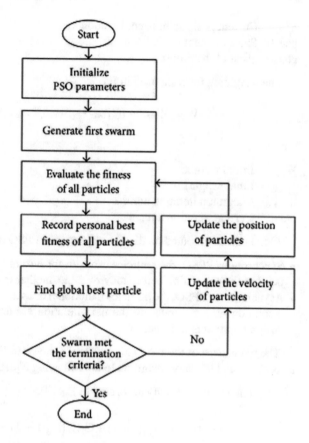

colony optimization (BCO). This study focuses on PSO and PPSO review as swarm
intelligence techniques used in cloud computing for future research interest.

PSO has particles which perform elect solutions of the problem, each particle
seeking for most favorable solution in the search space, each particle or candidate
solution has a position and velocity. A particle updates its velocity and position
based on its inertia, own experience and gained knowledge from other particles in
the swarm, aiming to detect the best solution of the problem. The main steps of the
PSO algorithm are shown in Fig. 1. The particles update its position and velocity
according to the following Eqs. (1, 2) [26, 27]:

$$V_1^{K+1} = WV_1^K + C_1 \, rand_2 \times (pbest_i - S_i^K) + (gbest_i - S_i^K) \tag{1}$$

where:

V_1^{K+1} Velocity of agent i at iteration k,
W   Weighting function,
Rand   Random number between 0 and 1,

S_i^K Current position of agent iteration k,
pbesti   Pbest of agent i,
gbest$_i$ gbest of the group.

The weighting function used in Eq. 2:

$$W = W_{max} - ((W_{max} - W_{min})/iter_{max}) \times iter \qquad (2)$$

where:

W_{max} Initial weight,
W_{min} Final weight,
iter$_{max}$ Maximum iteration number,
iter   Current iteration number.

PPSO can be executed on distributed systems. PPSO composed of two types:

- Synchronous PPSO, the particles can wait for it to finish from analyzing optimal point in the population before moving on to another task.
- Asynchronous PPSO, the particles (solutions) or as aforementioned before implementation time generated in the next iteration are analyzed before the current design iteration is finished.

The main steps of the PPSO algorithm are shown in Fig. 2. The particles update its position and velocity according to the following equation (3, 4) [28, 29]:

(a) Update particle velocity according to Eq. (3).

$$V_{I,j}(t) = W\,V_{I,j}(t-1) + C_1R_1\big(P_{i,j}(t-1) - X_{i,j}(t-1)\big)$$
$$+ C_2R_2\big(P_{s,j}(t-1) - X_{i,j}(t-1)\big)$$
$$+ C_3R_3\big(P_{g,j}(t-1) - X_{i,j}(t-1)\big) \qquad (3)$$

where:

- $X_{i,j}$ = the position of ith particle in jth swarm,
- $V_{I,j}$ = the velocity of ith particle in jth swarm,
- $P_{i,j}$ = the pbest of ith particle in jth swarm,
- $P_{s,j}$ = the swarm best of jth swarm,
- Pg, j = the global best among all the sub swarms,
- W = inertia weight,
- C_1, C_2, C_3 = acceleration parameters,
- R_1, R_2, R_3 = the random variables.

(b) Update particle position according to Eq. (4).

$$X_{i,j}(t) = X_{i,j}(t-1) + V_{I,j}(t) \qquad (4)$$

Fig. 2 The main steps of the PPSO algorithm

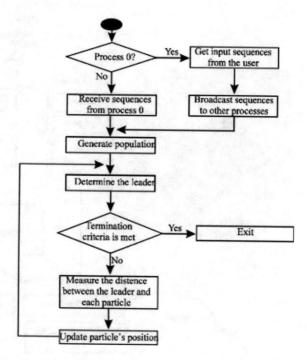

Table 1 Comparison study between PSO and PPSO

No.	Criteria	PSO	PPSO
1	Unit name	Particles	Parallel particles
2	Scientific considerations	Initialize population, inertia weight (W), acceleration constants (C1 and C2)	Initialize population, inertia weight (W), acceleration constants (C1, C2 and C3)
3	Reliability	Based on velocity and position of particles	Based on velocity and position of particles
4	Synchronous and asynchronous	None	Exist
5	Adaptability	Inherently continuous	Inherently concurrent
6	Formulation	Update velocity and position of particles	Update velocity and position of particles

$X_{i,j}(t)$ = the current position of ith particle in jth swarm, $X_{i,j}(t-1)$ = the new position of ith particle in jth swarm, $V_{I,j}(t)$ = the current velocity of ith particle in jth swarm.

Table 1 introduces a comparative study between PSO and PPSO.

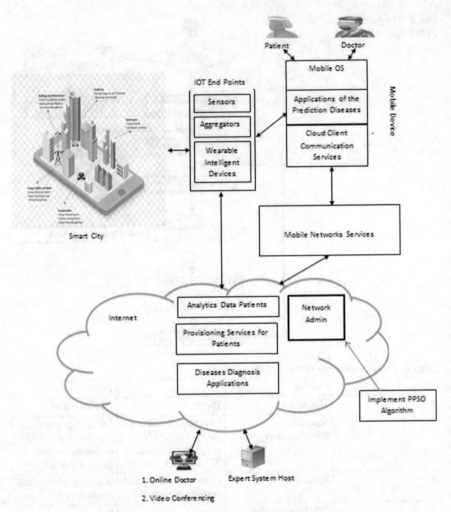

Fig. 3 Intelligent model of cloud computing for HCS

4 The Proposed Intelligent Model of Cloud Computing for HCS

This section describes the architecture of the proposed intelligent cloud computing model for HCS. It consist of five components are stakeholders devices, stakeholders requests (tasks), application layer, smart home network and home gateway as shown in Fig. 3.

Stakeholders' uses a variety of devices (PC, Laptop, Smartphone, Tablet, Digital sensor and etc.) to send a variety of medical requests (tasks) easily through cloud computing to obtain different medical services such as retrieving patient's data, telemedicine, diagnosis of diseases, electronic medical records (EMR) and etc. Each network may have several application hosts = {Host1, Host2… and Hostn} providing the SaaS and can be allocated to execute the cloud stakeholder's requests. Each application host has a set of resources = {R1, R2 … and Rn} that can be allocated for the coming stakeholders requests. Each network has a network administrator that is responsible for the coordination of the communication between the hosts inside the network and between this network and other networks in the clouds. Network administrator is responsible for applying the intelligent technique (PPSO algorithm) that it uses to enhance the task scheduling process which leads to reduce total time of stakeholders' requests (tasks) and maximize utilization of medical resources.

5 The Proposed PSO and PPSO of Cloud Computing for HCS

The proposed architecture is based on PSO and PPSO algorithms to calculate execution time of stakeholders' requests and fitness function to enhance task scheduling through three attributes are CPU utilization, turnaround time and waiting time. This attributes consists of three parameters as follows:

- **Arrival Time (AT)**: time at which the task arrives in the ready queue. (5)
- **Burst Time (BT)**: time required by a task for CPU execution. (6)
- **Completion Time (CT)**: time at which task completes its execution. (7)

With these three parameters, the task scheduling can calculate the three of significant attributes, CPU utilization, turnaround time and waiting time, shown in Table 2.

This paper proposed PSO and PPSO based algorithms of cloud computing for HCS as shown below in Fig. 5. The proposed algorithms try to enhance task scheduling on cloud computing to reduce execution time of requests (tasks) from stakeholders and maximization of resources utilization.

Table 2 Comparison study between PSO and PPSO

S. No.	Criterion	Description	Formula
1	CPU utilization (U)	The percentage of CPU capacity used during a specific period of time	U = 100% (% time spent in the idle task)
2	Turnaround time (TT)	Time difference between completion time and arrival time	TT = CT-AT
3	Waiting time (WT)	Time difference between turnaround time and burst time	WT = TT-BT

Fig. 4 The proposed flow
chart of PSO for cloud
computing

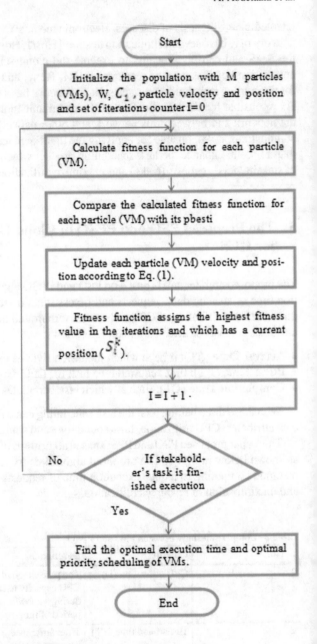

6   The Proposed PSO for Cloud Computing

PSO for Cloud Computing is proposed as follows (A declarative flow chart is shown
in Fig. 4).

Assume that there are M particles (VMs) in all the clouds. Each VM in the cloud(s) is considered a particle which represents a potential solution (VM) that can be allocated for executing the stakeholder's subtasks. PSO on cloud computing can be applied as follows:

(a) Initialize the population with M particles (VMs), W, C1, particle velocity and position and set iteration counter $I = 0$.
(b) While not all stakeholder's subtasks not finished.
(c) Calculating the fitness function (optimal task scheduling) by CPU utilization, turnaround time and waiting time for each particle (VM) to enhance task scheduling and calculating the percentage of utilization resources.
(d) Compare the calculated fitness function of each particle (VM) with its pbesti. If current value is better than pbest, then put the current location as pbest location. Moreover, if current value is better than gbest, then reconstruct gbest to the current index in particle array. Assign the best particle (VM) as gbest.
(e) Update each Particle Velocity and position according to Eqs. (1, 2).
(f) Fitness function assigns the optimal fitness value in the iterations.
 If stakeholder's task is finished execution
 {

   • Find the optimal execution time and optimal task scheduling.

 }

 Else
 {

   • Calculate fitness function (optimal task scheduling) for each particle (VM).

 }
(h) $I = I + 1$
(i) End while
(j) Stop

6.1 The Proposed PPSO for Cloud Computing

In PPSO, parallel processing aims to produce the same results that achievable using multiple processors with the goal of reducing the run time. The same steps described in PSO will be applied, but:

(a) PPSO will define how many group of processors needed for the fitness function (optimal priority scheduling of tasks) to be executed, because it can be designed to be 2n sets.
(b) Barrier synchronization stops the algorithm from move to the next step until the fitness function (optimal priority scheduling of tasks) has been reported, which is required to maintain algorithm coherence. Ensure that all of the particle (VM)

Table 3 PSO parameters

No.	Parameters	Values
1	Number of particles	100–1000
2	C1	2
3	C2	2
4	W_{max}	0.9
5	W_{min}	0.3
6	Number of iterations	10

Table 4 PSO results

Number of trials	Particles	ET
1	100	3.08
2	200	4.91
3	300	6.82
4	400	8.91
5	500	11.76
6	600	14.92
7	700	18.91
8	800	21.07
9	900	24.80
10	1000	26.02

fitness evaluations have been completed and results reported before the velocity and position calculations can be executed.

(c) Update particle velocity and position according to Eqs. (3, 4).

7 Experimental Results

This section discusses the experimental results and the implementation details of our proposed algorithm.

7.1 Implementation of PSO and PPSO

This paper provides the implementation of the PSO, and PPSO and comparison between these algorithms through execution time, speed and efficiency. PSO uses fitness function and some parameters to enhance task scheduling, as shown in Table 3 and refer to Table 2 (fitness function).

Table 4 showed the results of the execution of PSO algorithm and composed of three parts, the first part iterations, particles and execution time (ET).

Fig. 5 Relationship between particles and ET in PSO

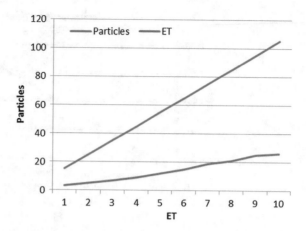

Table 5 PPSO results

Number of trials	Particles	ET
1	100	5.05
2	200	6.01
3	300	6.68
4	400	7.77
5	500	8.98
6	600	10.11
7	700	11.14
8	800	13.99
9	900	14.15
10	1000	15.78

Table 5 PPSO results

Figure 5 showed positive relationship between the number of particles and execution time, whenever the increased the number of particles which reflects the increased in the execution time.

Table 5 showed the results of the execution of PPSO algorithm.

Figure 6 showed positive relationship between the number of particles and execution time, whenever the increased the number of particles which reflects the increased in the execution time.

Figure 7 showed the relationship between PSO and PPSO where execution time in PPSO decreased compared with PSO.

7.2 Implementation of PPSO Algorithm on CloudSim

This section proposes PPSO algorithm on CloudSim to enhance task scheduling to minimize execution time of requests (tasks). The first implementation is default

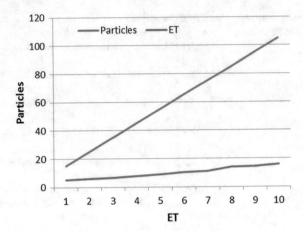

Fig. 6 Relationship between particles and ET in PPSO

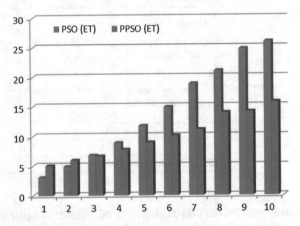

Fig. 7 Relationship between PSO and PPSO

```
========== OUTPUT ==========
Cloudlet ID    STATUS    Data center ID    VM ID    Time    Start Time    Finish Time
     0         SUCCESS         2             0       800       0.1           800.1
     1         SUCCESS         2             1      1200       0.1          1200.1
     3         SUCCESS         2             3      8000       0.1          8000.1
     2         SUCCESS         2             2     16000       0.1         16000.1

BUILD SUCCESSFUL (total time: 3 seconds)
```

Fig. 8 Results of default CloudSim

CloudSim where first task takes the first VM; the second task takes the second VM and etc. Total time to build successful cloudlets is 3 s as shown in Fig. 8.

The second implementation is PPSO algorithm on CloudSim to enhance task scheduling where first task may take the second VM; the second task may take the

```
========== OUTPUT ==========
Cloudlet ID    STATUS    Data center ID    VM ID    Time     Start Time    Finish Time
     0         SUCCESS         2              1      1600        0.1          1600.1
     1         SUCCESS         2              1      2000        0.1          2000.1
     3         SUCCESS         2              3      8000        0.1          8000.1
     2         SUCCESS         2              2     16000        0.1         16000.1

BUILD SUCCESSFUL (total time: 1 second)
```

Fig. 9 Sample of results of PPSO algorithm on CloudSim

Fig. 10 Proposed PPSO leads to minimized TT with increasing in no. of processors

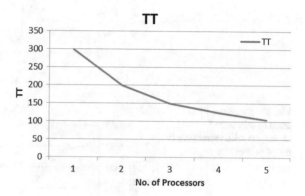

Fig. 11 Proposed PPSO leads to minimized WT with increasing in no. of processors

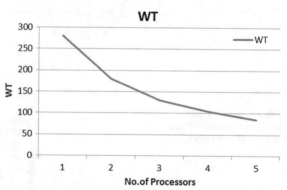

fourth VM, depending on the number of processors, task status and VM status. Total time to build successful cloudlets is 1 s. PPSO algorithm outperform on CloudSim default in order to total time in build successful cloudlets as shown in Fig. 9.

Simultaneously, Figs. 10, 11, 12, and 13 illustrate the inverse relationship between the number of processors and the turnaround time, the waiting time, CPU utilization, and the Makespan (time), respectively.

Figure 14 showed inverse relationship between the number of processors and resource utilization, whenever the increased the number of processors which reflects the decreased in the resource utilization.

Fig. 12 Proposed PPSO
leads to minimized waste of
CPU utilization with
increasing in no. of
processors

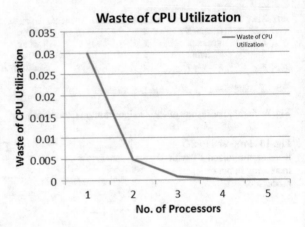

Fig. 13 Proposed PPSO
leads to minimized
makespan with increasing in
no. of processors

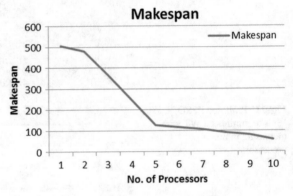

Fig. 14 Relationship
between resources utilization
and no. of processors

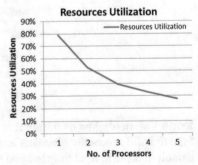

Not to overuse the processors when execution of PPSO algorithm on cloud computing environment, whenever the increased the number of processors which reflects the decreased in the resource utilization.

The proposed model have a high flexibility through change the number of processors, tasks, VMs and etc. easily. The robust of the proposed model is clear concerning the efficiency of PPSO that outperformed the efficiency of PSO and it succeeded to enhance task scheduling on cloud environment.

8 Conclusion and Future Work

This paper presented intelligent model based on PSO and PPSO to enhance task scheduling in smart cities by cloud environment. This paper showed PPSO algorithm outperforms on PSO in order to speed, execution time and efficiency. Therefore, this paper proposes the application of PPSO algorithm to find the optimal task scheduling to help stakeholders to reduce execution time, turnaround time and waiting time of medical requests (tasks), maximize utilization of resources and applied this algorithm on MATLAB tool and CloudSim package. In the future work, we are planning to apply new optimization algorithms to find the optimal task scheduling as well as applying our model in different applications [30–46].

References

1. Singh A, Hemalatha M (2013) Cluster based Bee Algorithm for virtual machine placement in cloud data centre. JATIT 57(3):1–10
2. Chen L, Zhang J, Cai L, Meng T (2015) MTAD: a multitarget heuristic algorithm for virtual machine placement. Int J Distrib Sens Netw 2014:1–14
3. Camati R, Calsavara A, Lima L (2014) Solving the virtual machine placement problem as a multiple multidimensional Knapsack problem. IARIA, IEEE, pp 253–260
4. Suseela B, Jeyakrishnan V (2014) A multi-objective hybrid Aco-Pso optimization algorithm for virtual machine placement in cloud computing. IJRET 3(4):474–476
5. Zhao J, Hu L, Ding Y, Xu G, Hu M (2014) A heuristic placement selection of live virtual machine migration for energy-saving in cloud computing environment. PloS ONE 9(9):1–13
6. Boulos MN, Al-Shorbaji NM (2014) On the internet of things, smart cities and the WHO healthy cities. Int J Health Geograph 13:2–6
7. Alhussein M (2017) Monitoring Parkinson's disease in smart cities", special section on advances of multisensory services and technologies for healthcare in smart cities. IEEE, 5:19835–19841
8. Bhunia SS, Dhar SK, Mukherjee N (2014) iHealth: a fuzzy approach for provisioning intelligent health-care system in smart city. e-Health Pervasive Wirel Appl Serv IEEE 14:187–193
9. Islam M, Razzaque A, Hassan MM, Nagy W, Song B (2017) Mobile cloud-based big healthcare data processing in smart cities. IEEE 0(0):1–12
10. Sajjad M, Khan S, Jan Z, Muhammad K, Moon H, Kwak JT, Rho S, Baik SW, Mehmood I (2016) Leukocytes classification and segmentation in microscopic blood smear: a resource-aware healthcare service in smart cities. IEEE 0:1–15
11. Mishra R, Jaiswal A (2012) Bees life algorithm for job scheduling in cloud computing. ICCIT 3:186–191
12. Bhatt K, Bundele M (2013) CloudSim estimation of a simple particle swarm algorithm. IJARC-SSE 3(8):1279–1287
13. Gomathi B, Krishnasamy K (2013) Task scheduling algorithm based on hybrid particle swarm optimization in cloud computing environment. JATIT 55(1):33–38

14. Mohana SJ, Saroja M, Venkatachalam M (2014) Comparative analysis of swarm intelligence optimization techniques for cloud scheduling. IJISET 1(10):15–19
15. Beegom AS, Rajasree MS (2014) A particle swarm optimization based pareto optimal task scheduling in cloud computing. ICSI 2:79–86
16. Kaur G, Sharma S Er. (2014) Optimized utilization of resources using improved particle swarm optimization based task scheduling algorithms in cloud computing. IJETAE 4(6):110–115
17. El-Sisi AB, Tawfeek MA, Keshk AE, Torkey FA (2014) Intelligent method for cloud task scheduling based on particle swarm optimization algorithm. ACIT, 39–44
18. Bilgaiyan S, Sagnika S, Das M (2014) An analysis of task scheduling in cloud computing using evolutionary and swarm-based algorithms. IJCA 89(2):11–18
19. Tawfeek M, El-Sisi A, Keshk A, Torkey F (2015) Cloud task scheduling based on ant colony optimization. IAJIT 12(2):129–137
20. Salama AS (2015) A swarm intelligence based model for mobile cloud computing. IJITCS 2:28–34
21. Awad AI, El-Hefnawy NA, Abdel_kader HM (2015) Enhanced particle swarm optimization for task scheduling in cloud computing environments. ICCMIT 65:920–929
22. Al-Olimat HS, Alam M, Green R, Lee JK (2015) Cloudlet scheduling with particle swarm optimization. ICCSNT, IEEE 31:991–995
23. Priyadarsini RJ, Arockiam L Dr (2015) An improved particle swarm optimization algorithm for meta task scheduling in cloud environment. IJCST 3(4):108–112
24. Vidhya M, Sadhasivam N (2015) Parallel particle swarm optimization for reducing data redundancy in heterogeneous cloud storage. IJTET 3(1):73–78
25. Alkhashaiand HM, Omara FA (2016) BF-PSO-TS: hybrid heuristic algorithms for optimizing task scheduling on cloud computing environment. IJACSA 7(6):207–212
26. Abdelaziz A, Elhoseny M, Salama AS, Riad AM, Hassanien A (2017) Intelligent algorithms for optimal selection of virtual machine in cloud environment, towards enhance healthcare services. In: Proceedings of the international conference on advanced intelligent systems and informatics, vol 639. Springer, pp 23–37
27. Elhoseny M, Salama AS, Abdelaziz A, Riad A (2017) Intelligent systems based on cloud computing for healthcare services: a survey. Int J Comput Intell Stud Indersci 6(2/3):157–188
28. Abdelaziz A, Elhoseny M, Salama AS, Riad AM (2018) A machine learning model for improving healthcare services on cloud computing environment. Measurement 119:117–128
29. Elhoseny M, Abdelaziz A, Salama AS, Riad AM, Muhammad K, Sangaiah AK (2018) A hybrid model of Internet of Things and cloud computing to manage big data in health services applications. Future generation computer systems
30. Tharwat A, Elhoseny M, Hassanien AE, Gabel T, Arun Kumar N (2018) Intelligent Beziér curve-based path planning model using chaotic particle swarm optimization algorithm. Cluster Comput, 1–22. https://doi.org/10.1007/s10586-018-2360-3
31. Tharwat A, Mahdi H, Elhoseny M, Hassanien AE (2018) Recognizing human activity in mobile crowdsensing environment using optimized k-NN algorithm, expert systems with applications. Available online 12 Apr 2018, https://doi.org/10.1016/j.eswa.2018.04.017
32. Hosseinabadi AAR, Vahidi J, Saemi B, Sangaiah AK, Elhoseny M (2018) Extended genetic algorithm for solving open-shop scheduling problem. Soft Comput. https://doi.org/10.1007/s00500-018-3177-y
33. El Aziz MA, Hemdan AM, Ewees AA, Elhoseny M, Shehab A, Hassanien AE, Xiong S (2017) Prediction of Biochar yield using adaptive neuro-fuzzy inference system with particle swarm optimization. In: 2017 IEEE PES PowerAfrica conference, June 27–30, Accra-Ghana, IEEE, 2017, pp 115–120. https://doi.org/10.1109/powerafrica.2017.7991209
34. Ewees AA, El Aziz MA, Elhoseny M (2017) Social-spider optimization algorithm for improving ANFIS to predict biochar yield. In: 8th international conference on computing, communication and networking technologies (8ICCCNT), 3–5 July, Delhi-India, IEEE
35. Elhoseny M, Tharwat A, Yuan X, Hassanien AE (2018) Optimizing K-coverage of mobile WSNs. Expert Syst Appl 92:142–153. https://doi.org/10.1016/j.eswa.2017.09.008

36. Sarvaghad-Moghaddam M, Orouji AA, Ramezani Z, Elhoseny M, Farouk A, Arun Kumar N (2018) Modelling the spice parameters of SOI MOSFET using a combinational algorithm. Cluster Comput. https://doi.org/10.1007/s10586-018-2289-6
37. Rizk-Allah RM, Hassanien AE, Elhoseny M (2018) A multi-objective transportation model under neutrosophic environment. Comput Electr Eng. https://doi.org/10.1016/j.compeleceng.2018.02.024
38. Batle J, Naseri M, Ghoranneviss M, Farouk A, Alkhambashi M, Elhoseny M (2017) Shareability of correlations in multiqubit states: optimization of nonlocal monogamy inequalities. Phys Rev A 95(3):032123. https://doi.org/10.1103/PhysRevA.95.032123
39. Elhoseny M, Nabil A, Hassanien AE, Oliva D (2018) Hybrid rough neural network model for signature recognition. In: Hassanien A, Oliva D (eds) Advances in soft computing and machine learning in image processing. Studies in computational intelligence, vol 730. Springer, Cham. https://doi.org/10.1007/978-3-319-63754-9_14
40. Elhoseny M, Tharwat A, Farouk A, Hassanien AE (2017) K-coverage model based on genetic algorithm to extend WSN lifetime. IEEE Sens Lett 1(4):1–4. IEEE. https://doi.org/10.1109/lsens.2017.2724846
41. Yuan X, Elhoseny M, El-Minir HK, Riad AM (2017) A genetic algorithm-based, dynamic clustering method towards improved WSN longevity. J Netw Syst Manage 25(1):21–46. https://doi.org/10.1007/s10922-016-9379-7
42. Elhoseny M, Nabil A, Hassanien AE, Oliva D (2018) Hybrid rough neural network model for signature recognition. In: Hassanien A, Oliva D (eds) Advances in soft computing and machine
43. Elhoseny M, Tharwat A, Farouk A, Hassanien AE (2017) K-coverage model based on genetic algorithm to extend WSN lifetime. IEEE Sens Lett 1(4):1–4. https://doi.org/10.1109/lsens.2017.2724846
44. Elhoseny M, Shehab A, Yuan X (2017) Optimizing robot path in dynamic environments using genetic algorithm and Bezier Curve. J Intell Fuzzy Syst 33(4):2305–2316. IOS-Press https://doi.org/10.3233/jifs-17348
45. Elhoseny M, Tharwat A, Hassanien AE (2017) Bezier Curve based path planning in a dynamic field using modified genetic algorithm. J Comput Sci. https://doi.org/10.1016/j.jocs.2017.08.004
46. Metawaa N, Kabir Hassana M, Elhoseny M (2017) Genetic algorithm based model for optimizing bank lending decisions. Expert Syst Appl 80:75–82. https://doi.org/10.1016/j.eswa.2017.03.021

A Machine Learning Model for Predicting of Chronic Kidney Disease Based Internet of Things and Cloud Computing in Smart Cities

Ahmed Abdelaziz, Ahmed S. Salama, A. M. Riad and Alia N. Mahmoud

Abstract Cloud computing and internet of things (IOT) plays an important role in health care services especially in the prediction of diseases in smart cities. IOT devices (digital sensors and etc.) can be used to send big data onto chronic kidney diseases (CKD) to store it in the cloud computing. Therefore, these big data are used to increase the accuracy of prediction of CKD on cloud environment. The prediction of dangerous diseases such as CKD based cloud-IOT is considered a big problem that facing the stakeholders of health cares in smart cities. This paper focuses on predicting of CKD as an example of health care services on cloud computing environment. Cloud computing is supported patients to predict of CKD anywhere and anytime in smart cities. For that, this paper proposes a hybrid intelligent model for predicting CKD based cloud-IOT by using two intelligent techniques, which are linear regression (LR) and neural network (NN). LR is used to determine critical factors that influence on CKD. NN is used to predict of CKD. The results show that, the accuracy of hybrid intelligent model in predicting of CKD is 97.8%. In addition, a hybrid intelligent model is applied on windows azure as an example of a cloud computing environment to predict of CKD to support patients in smart cities. The proposed model is superior to most of the models referred to in the related works by 64%.

A. Abdelaziz (✉) · A. N. Mahmoud
Department of Information Systems, Higher Technological Institute, Cairo, Egypt
e-mail: ahmed.aziz.1157@gmail.com

A. M. Riad
Faculty of Computers and Information Sciences, Mansoura University, Mansoura, Egypt

A. S. Salama
Information Systems Department, Faculty of Computing and Information Technology, University of Jeddah, KSA, Jeddah, Saudi Arabia

A. S. Salama
Computers and Information Systems Department, Sadat Academy for Management Sciences, Cairo, Egypt

© Springer Nature Switzerland AG 2019
A. E. Hassanien et al. (eds.), *Security in Smart Cities: Models, Applications, and Challenges*, Lecture Notes in Intelligent Transportation and Infrastructure, https://doi.org/10.1007/978-3-030-01560-2_5

Keywords Cloud computing · Internet of things · Smart cities
Chronic kidney disease · Linear regression · Neural network

1 Introduction

Cloud computing has become critical in healthcare services for transmitting medical services over the Internet. Cloud computing is sending applications, infrastructure services to huge numbers of patients with assorted and dynamically changing requirements. Cloud is consisting of datacenters, servers, VMs, resources and etc [1]. Datacenters are containing a big number of resources and list of different applications. Hosts are consisting of a large number of virtual machines (VMs) to store and regain several medical resources to stakeholders (doctors, patients and etc.).

Cloud computing introduces many services and more benefits for millions of users across the world over the internet. It aims at using IOT devices to send patients' data (structured medical data, semi structured medical data and unstructured medical data) on cloud environment to configure big data of patients. For that, cloud computing in healthcare services helps stakeholders in retrieving big data of patients, managing of patients' medical record, telemedicine, applications of diseases prediction and etc. To get the best performance for applications of prediction of diseases, it needs the intelligent devices (IOT devices) to collect big data for patients, store it in the cloud environment and analyze it.

IoT is a system of connected computing devices and objects that are provided with unique identifiers and the ability to transmit big data of patients over a network without requesting people-to-people or people-to-computer interaction. IOT devices are used to send a huge of patients' data and stored it in the cloud environment. In this way its availability is increased, providing a raise of force to analytical and predictive tools, providing an increase in accuracy for predicting diseases.

IOT architecture is composed of four stages. Stage 1 of IoT architecture consists of networked things, typically wireless sensors and actuators. Stage 2 includes sensor data aggregation systems and analog-to-digital data conversion. In Stage 3, edge IT systems perform preprocessing of the data before it moves on to the data center or cloud. Finally, in Stage 4, the data is analyzed, managed, and stored on traditional back-end data center systems.

Many applications are used in diagnosis or predictions of diseases do not help the stakeholders to use them at anytime and anywhere [3]. Applications are built to predict of diseases on cloud computing to help patients to use them at anytime, anywhere and less cost, as shown below:

- A new method presented to predict cervical cancer by using decision tree [4].
- A new model introduced for predicting hepatitis disease by using support vector machine [5].
- Heart disease prediction has been executed by using Naïve Bayes on Google App Engine [6].
- A new framework presented to detect breast cancer by using fuzzy logic [7].

- A new model presented to diagnosis lung cancer by using fuzzy logic [8].
- A new method introduced to diagnosis diabetes disease by using support vector machine [9].

Smart cities are facing a big problem to support healthcare services by managing big data of patients, intelligent diagnosis of diseases and other medical fields by cloud-IOT to reduce time and cost. This paper aims to improve healthcare services in smart cities by introducing a hybrid intelligent model to predict of CKD of cloud-IOT to help stakeholders to improve their health.

This paper focuses on prediction of CKD by using intelligent techniques on cloud environment. CKD is one of the most serious diseases facing the world where the latest statistics are recorded 2.5–11.2% across Europe, Asia, Australia and North America are suffering from CKD. The United States of America has 27 million people and 50 thousand people in Egypt suffering from CKD. It proposes hybrid intelligent model on cloud environment to predict of CKD. Hybrid intelligent model is composed of LR and NN. LR is used to determine critical factors of CKD. NN is used to predict of CKD. A hybrid intelligent model is executed on windows azure.

This paper is arranged as follows: Sect. 2 introduces related work, Sect. 3 introduces basics and background, Sect. 4 introduces the proposed architecture of cloud-IOT for CKD, Sect. 5 introduces the proposed LR and NN based algorithms for cloud computing, Sect. 6 introduces experimental results and finally, section presents conclusion and future work.

2 Related Work

This section introduces a set of studies based diagnosis of diseases on cloud computing environment, as shown below:

Jena et al. [10], introduced new method to predict kidney failure disease based classification techniques of data mining. Multilayer perception outperforms on Naïve Bayes, support vector machine, conjunctive rule and decision table. Accuracy of multilayer perception is 99.75%. Classification techniques were applied on WEKA mining tool. Researcher opinion, multilayer perception can predict kidney failure disease.

Batra et al. [11] presented a survey to diagnosis chronic kidney failure based on predictive analytics method by exploiting the potential of hadoop and map/reduce tool. This study introduced many researches based on diagnosis or predicting chronic kidney failure and factors that influence in it. Researcher opinion, the results showed that the critical factors influence on chronic kidney failure.

Boukenze et al. [12], introduced a comparative study between support vector machine, decision tree and Bayesian network to predict chronic kidney failure. Decision tree outperforms on support vector machine and Bayesian network in predicting of chronic kidney failure. Accuracy of decision tree is 97%. The intelligent techniques are applied on hadoop and map/reduce tool. Researcher opinion, decision tree may predict of chronic kidney failure appropriately.

Padmanaban et al. [13], introduced a comparative study between Naïve Bayes and decision tree for predicting chronic kidney failure. Decision tree outperforms on Naïve Bayes in predicting of chronic kidney failure. Accuracy of decision tree is 91%. The intelligent techniques are applied on WEKA mining tool. Researcher opinion, decision tree can predict of chronic kidney failure appropriately.

Salekin et al. [14], presented a comparative study between K-Nearest Neighbors and Random Forest to predict of chronic kidney failure. K-Nearest Neighbors outperforms on Random Forest in predicting of chronic kidney failure. Accuracy of K-Nearest Neighbors is 99.3%. The intelligent techniques are applied on WEKA mining tool. Researcher opinion, K-Nearest Neighbors may predict of chronic kidney failure.

Maithili et al. [15], introduced a new framework to predict cancer disease by using NN on cloud computing environment. The aim at the study is to improve efficiency and accuracy of diagnosis of diseases on cloud computing. Researcher opinion, NN technique may effect in diagnosis of cancer on cloud environment.

Aruna et al. [16], introduced a new model to predict breast cancer disease by using digital mammograms on cloud computing environment. The aim at the study is to improve efficiency and accuracy of breast cancer disease on cloud computing. Researcher opinion, digital mammograms can influence in diagnosis of breast cancer on cloud environment.

Abdulbaki et al. [17], introduced a new method of diagnosis skin disease by using back-propagation NN on cloud computing environment. This study seeks to improve efficiency and accuracy of skin disease on cloud computing. Researcher opinion, back-propagation NN can diagnosis of skin disease on cloud environment appropriately.

Al-Ghamdi et al. [18], introduced a new system for predicting diabetes disease by using artificial NN on cloud computing platforms. This study seeks to improve efficiency and accuracy of diabetes disease on cloud computing. The artificial NN technique is applied on Google App Engine. Researcher opinion, artificial NN may effect in predicting of diabetes disease on cloud environment.

Sandhu et al. [19], introduced a new model for predicting breast cancer by using artificial NN on cloud computing. This study seeks to improve efficiency and accuracy of breast cancer on cloud computing. The artificial NN technique is applied on Google App Engine. Researcher opinion, artificial NN may affect in predicting of breast cancer on cloud environment.

Kumar et al. [20], introduced a survey for diagnosis and analysis diabetes disease on cloud computing platform. This study seeks to improve efficiency and accuracy of patient's blood glucose control at remote areas on cloud computing. Researcher opinion, a survey helps to cover most researches about patient's blood glucose control at remote areas on cloud environment.

Aswin et al. [21], introduced a new method of predicting diabetes disease by using fuzzy logic and uncertainty factors on cloud computing. This study seeks to improve efficiency and accuracy of diabetes disease on cloud computing. The fuzzy logic and uncertainty factors are applied on Google App Engine. Researcher opinion, fuzzy logic and uncertainty factors can influence in predicting of diabetes disease on cloud environment.

Boulos et al. [22], introduced a new approach to solve a problem of medical data analysis in geospatial in smart cities by internet of things. This study tries to describe and discuss in great detail an Internet-connected web of citizens (people) and electronic sensors/devices (things) that can serve many functions related to public and environmental health surveillance and crisis management applications. Research opinion, internet of things can help to find the optimal solution to analysis of medical data in smart cities.

Alhussein [23] introduced a novel framework to detect Parkinson's disease in smart cities based on cloud computing by support vector machine. This disease can affect to impair the body's balance, damages motor skills, and leads to disorder in speech production. The proposed framework can achieve 97.2% accuracy in detecting Parkinson's disease. Researcher opinion, the proposed framework can detect Parkinson's disease in smart cities based on cloud computing by support vector machine.

Bhunia et al. [24], presented the new method to collect and process sensitive health data to diagnosis of health problems in smart cities to save the patients from death. This study uses a fuzzy scheme to detect critical physiological stages of a human body easily. Researcher opinion, the proposed method may find the optimal solution to determine critical physiological stages of a human body and use it in smart cities.

Islam et al. [25], introduced a new model to collect and process big healthcare data in smart cities efficiently by mobile cloud environment. This study seeks to solve the virtual machine migration problem and provision virtual machine resources by using ant colony optimization. Researcher opinion, the proposed model can solve the virtual machine migration problem efficiently.

Sajjad et al. [26], introduced a novel framework in smart cities to detect and count white blood cells in blood samples for saving the body against both infectious disease and foreign invaders. This study aims to introduce a framework based on mobile-cloud for segmentation and classification of white blood cells in blood samples. Researcher opinion, the proposed framework can detect count white blood cells in blood samples.

Smart cities are facing a big challenge to support healthcare services by managing big data of patients, intelligent diagnosis of diseases and other medical fields by cloud-IOT to reduce time and cost. This paper aims to improve healthcare services in smart cities by introducing hybrid intelligent model to predict of CKD by cloud-IOT to help stakeholders to improve their health. Intelligent medical applications are based on different intelligent algorithms such as neural network, fuzzy logic, Naive Bayes, clustering, support vector machine and etc. Most of the related works are based on managing big data of patients, diagnosis of diseases and other medical fields by using cloud computing. This paper seeks to find the optimal solution to predict of CKD for helping patients in smart cities for improving their health by cloud-IOT. For that, this paper proposes intelligent architecture to predict of CKD based on cloud-IOT in smart cities.

Table 1 Preliminary factors that influence on CKD

No	Factor	Information factor	Description
1	Age	Numerical	Years
2	Blood pressure	Numerical	Mm/Hg
3	Specific gravity	Nominal	1.005,1.010,1.015,1.020,1.025
4	Albumin	Nominal	0.1.2.3.4.5
5	Sugar	Nominal	0.1.2.3.4.5
6	Red blood cells	Nominal	Normal, abnormal
7	Pus cell	Nominal	Normal, abnormal
8	Pus cell clumps	Nominal	Present, notpresent
9	Bacteria	Nominal	Present, notpresent
10	Blood glucose random	Numerical	Mgs/dl
11	Blood urea	Numerical	Mgs/dl
12	Serum creatinine	Numerical	Mgs/dl
13	Sodium	Numerical	mEq/L
14	Potassium	Numerical	mEq/L
15	Hemoglobin	Numerical	Gms
16	Packed cell volume	Numerical	Gms
17	White blood cell count	Numerical	Cells/cumm
18	Red blood cell count	Numerical	Millions/cmm
19	Hypertension	Nominal	Yes, no
20	Diabetes mellitus	Nominal	Yes, no
21	Coronary artery disease	Nominal	Yes, no
22	Appetite	Nominal	Good, poor
23	Pedal edema	Nominal	Yes, no
24	Anemia	Nominal	Yes, no
25	Class	Nominal	CKD, NOCKD

3　Basics and Background

This section introduces an overview of CKD, LR and NN as follows:

A.　*Chronic Kidney Disease*

CKD is one of the most serious diseases of the world. There are many factors that effect of CKD, shown in Table 1.

B.　*Linear Regression Analysis*

LR is used to determine critical factors of CKD. Simple regression is composed of one dependent variable and one independent variable. Multiple regression analysis is composed of one dependent variable and more than independent variable. It is formulated as follows [1, 27, 28]:

$$y = \beta_0 + \beta_1 x_1 + \beta_2 x_2 + \cdots + \beta_n x_n + \varepsilon \qquad (1)$$

where:

y	represents the dependent variable
$x_1, x_2, \ldots, x_n$	are the independent variables
β_i	is the regression coefficient
ε	is the random error component
β_0	is the y intercept

Y represents the dependent variable (degree of influence on CKD process) and $x_1, x_2 \ldots x_n$ are the independent variables (factors that influencing on CKD).

LR introduced regression statistics such as mean absolute error (MAE), root mean squared error (RMSE), relative absolute error (RAE), relative squared error (RSE) and coefficient of determination (CD). MAE is a quantity used to measure how close predictions are to the eventual outcomes. RMSE can be compared between models whose errors are measured in the same units. RAE can be compared between models whose errors are measured in the different units. RSE can be compared between models whose errors are measured in the different units. CD summarizes the explanatory power of the regression model. MAE, RMSE, RAE, RSE and CD are calculated as follows [6]:

$$\text{MAE} = \frac{1}{n} \sum_{i=1}^{n} |f_i - y_i| \qquad (2)$$

where f_i, is the prediction and y_i the true value.

$$\text{RMSE} = \sqrt{\frac{\sum_{i=1}^{n}(f_i - y_i)^2}{n}} \qquad (3)$$

$$\text{RAE} = \frac{\sum_{i=1}^{n}|f_i - y_i|}{\sum_{i=1}^{n}|\bar{y} - y_i|} \qquad (4)$$

where $\bar{y}$ is the mean of y_i.

$$\text{RSE} = \frac{\sum_{i=1}^{n}(f_i - y_i)^2}{\sum_{i=1}^{n}(y - y_i)^2} \qquad (5)$$

Coefficient of Determination,

$$\left(\text{R}^2\right) = \frac{SSR}{SST} = 1 - \frac{SSE}{SST} \qquad (6)$$

Sum of Squares Regression,

$$\text{SSR} = \sum_{i}(f_i - \bar{y}) \qquad (7)$$

Fig. 1 The anatomy of a NN

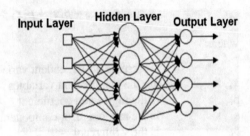

Sum of Squares Total,

$$SST = \sum_i (y_i - \bar{y}_i)^2 \tag{8}$$

Sum of Squares Error,

$$SSE = \sum_i (y_i - f_i)^2 \tag{9}$$

C. *Neural Network Overview*

NN is a novel process of programming computers. It is represented of the human brain's information processing mechanism. NN is applied on more applications such as pattern recognition, diagnosis of diseases and data classification, through a learning process. NN has many types of networks such as feed-forward network and back-propagation network, as shown in Fig. 1. NN is composed of input layers (factors that influence on CKD), hidden layers and output layers (decision, (CKD or NOCKD)), as follows:

- Input Layer—The input units represent the raw data that is fed into the network.
- Hidden Layer—The hidden unit is specified by the input units and the weights on the connections between the input and the hidden units.
- Output Layer—The attitude of the output units depends on the activity of the hidden units and the weights between the hidden and output units.

4   The Proposed Architecture of Cloud-IOT for CKD

This section describes the architecture of cloud-IOT for CKD. It consists of four components are stakeholders' devices, IOT end points, cloud broker and public cloud, as shown in Fig. 2.

Stakeholders use a variety of devices (PC, Laptop, Smartphone, Tablet, Digital sensors, etc.) to send a variety of medical readings and requests (tasks) easily through cloud computing to obtain different medical services such as diagnosis of diseases (CKD) as an example of healthcare services. Cloud broker is responsible to send and receive requests (tasks) from the cloud service. IOT end points are delivered big

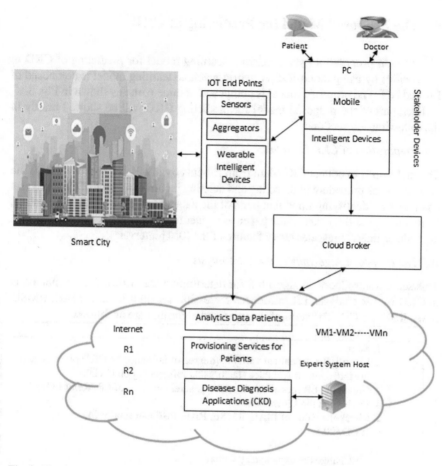

Fig. 2 The proposed architecture of cloud–IOT for CKD

medical data to clouds through the internet to configure a huge medical database to improve prediction of CKD. Stakeholders' are used different devices to deliver different medicinal requests (tasks) easily through cloud environment to get a set of medicinal services such as diagnosis of diseases (CKD) as an example of healthcare services. Network communication services are responsible to deliver and receive stakeholders' requests from the cloud services. Each network contains many application hosts = {T1, T2 ... and TN} that can be assigned to implement the cloud stakeholder's tasks. Each application host (expert system application, online doctor and etc.) has a variety of resources = {E1, E2 ... and EN} that can be assigned for the coming stakeholders tasks. Each network has a network director that is responsible for the assortment of the connection between the servers inside the network and other networks in the clouds. This architecture introduced to stakeholders' expert system application (hybrid machine learning model based cloud-IOT) for predicting CKD as an example of diagnosis of diseases, towards improves healthcare services.

5   The Proposed Model for Predicting of CKD

This section proposed a hybrid machine learning model for predicting of CKD in smart cities by using cloud-IOT. A hybrid machine learning model is composed of LR and NN. The proposed model is composed of three parts: as shown in Fig. 3.

Each part of the proposed model for predicting CKD will be showed briefly in the following sub sections:

A. *Preparation of CKD Data Set*

The first step is to collect CKD data from patients by intelligent sensors. CKD data cleaning is the procedure of detecting and deleting incorrect records of CKD data set and refers to identifying imperfect parts of the data and then changing, modulating, or removing the dirty data. This paper uses data normalization based on min-max normalizer linearly rescales every feature of the [0, 1] interval.

B. *The Proposed Algorithm of the LR Analysis*

This section introduces an algorithm for determining the critical factors that affect in CKD by LR analysis. LR introduces regression statistics such as MAE, RMSE, RAE, RSE and CD. The steps of the proposed algorithm are as follows:

LR Algorithm
1. Start
2. Identify the dependent variable (degree of influence on CKD process) and the independent variables (factors that influencing on CKD).
3. Create the LR model based on the general equation of LR. Refer to Eq.1
4. Evaluation the LR model.
5. Check the value of MAE, RMSE, RAE, RSE and finally CD.
6. If CD < 0.5
{
-    Change the explanatory factors
-    Evaluation the LR model
}
Else
{
-    Approve the model
}
7. Check the feature weights (FW) for each variable to determine the critical factors list of CKD
8. If FW < 0.05
{
-    Determination the critical factors list of CKD
}
Else
{
-    refuse the other factors
}
9. Stop

Fig. 3 The proposed model
for predicting CKD

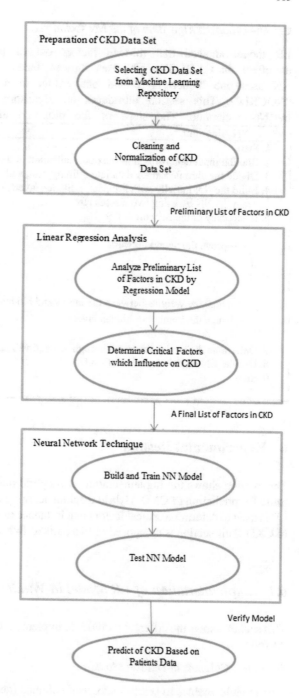

C. *The Proposed Algorithm of the NN Technique*

LR model showed that thirteen factors out of twenty-four of them had an effect on CKD. The NN uses thirteen factors as inputs in a network. NN uses one hidden layer and one output in a network (class, "CKD", "NOCKD"). This section introduces an algorithm for predicting of CKD by NN technique. The steps of the proposed algorithm are as follows:

NN Algorithm
1. Start
2. Run the input data using min-max normalization in the range [0.0-1.0].
3. Divide the chronic kidney data into training, test, and validate.
4. Build the NN initially with P inputs, L hidden layer(s) and T outputs.
5. Train the NN by using two classes NN.
6. If accuracy of the network > 0.9
{
-Approve the neural network
}
Else
{
-Change the weights between the inputs and hidden layers
-Change the number of hidden layers
}
7. Determine the model accuracy through test and validate the network.
8. Use the NN model to predict of CKD.
9. Stop

6 Experimental Results

This section shows the implementation of a hybrid machine learning on windows azure for predicting of CKD. Hybrid machine learning is composed of LR and NN. LR is used to determine critical factors that influence on CKD. NN is used to predict of CKD. This section is composed of two parts as follows:

6.1 *Implementation of LR Model in Windows Azure*

This section shows the different modules to implement LR model in windows azure. As follows:

A. *Select Columns in Dataset Module*

This module enables to select a subset of columns (dependent variables and independent variables of preliminary list of CKD data) to use in LR operation. Refer to Table 1.

B. *Missing Values Scrubber Module*

This module enables to provide some basic procedures for processing missing values. This module is composed of three options. The first option is replace any missing value to zero. The second option is keep of columns that contain values missing. The third option doesn't generate missing value indicator columns.

C. *Split Data Module*

This module enables to divide a dataset into two different sets. This module enables to split data into training and testing sets. The dataset is divided into 70% training data and 30% testing data.

D. *Linear Regression Module*

This module is composed of two important options. The first option is ordinary least squares that assume strong linear relationship between the inputs and the dependent variable. The second option is L2 Regularization to enhance performance of the model. It is set to 0.001 by default.

E. *Train Model Module*

This module enables to determine critical factors list of CKD. Table 3 gives the feature weights that indicate the contribution of each independent variable to the LR model. If FW < 0.05, the factor is statistically important. For example, the FW for blood urea (FW = 0.001) is less than 0.05, therefore, this factor should be accepted, as shown in Table 2.

After the screening of FW-values, the most critical factors which influence on CKD as presented in Table 3.

F. *Evaluate Model Module*

This module shows MSE = 0.40, RMSE = 0.60, RAE = 0.46, RSE = 0.41 and finally CD = 0.58. The CD indicates that 0.58 of the variance in the dependent variable (degree of influence on CKD process) can be explained by the independent variables (selected factors), while the rest (0.42) is explained by other causes, shown in Fig. 4.

6.2 *Implementation of NN Model in Windows Azure*

This section shows the different modules to implement NN model in windows azure. As follows:

A. *Select Columns in Dataset Module*

This module enables to select critical factors list of CKD data to use as inputs in NN model. Refer to Table 3.

Table 2 Sample of factors that influencing on CKD

No	Factor name	FW
1	Age	−0.001
2	Blood pressure	0.0006
3	Specific gravity	−0.001
4	Sugar	0.01
5	Blood glucose random	−0.0005
6	Diabetes mellitus	0.9008
7	Coronary artery disease	0.6012
8	Appetite	0.0801
9	Pedal edema	0.0616
10	Blood urea	0.001
11	Serum creatinine	−0.04
12	Sodium	0.002
13	Potassium	0.004
14	Hemoglobin	0.03
15	Packed cell volume	−0.001
16	White blood cell count	−0.00004
17	Anemia	0.01

Table 3 Summary of critical factors that influencing on CKD

No	Factor name	FW
1	Age	−0.001
2	Blood pressure	0.0006
3	Specific gravity	−0.001
4	Sugar	0.01
5	Blood glucose random	−0.0005
6	Blood urea	0.001
7	Serum creatinine	−0.04
8	Sodium	0.002
9	Potassium	0.004
10	Hemoglobin	0.03
11	Packed cell volume	−0.001
12	White blood cell count	−0.00004
13	Anemia	0.01

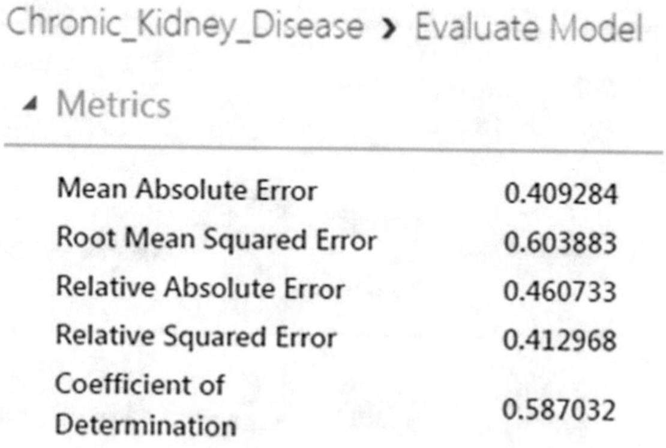

Fig. 4 Regression statistics

Chronic_Kidney_Disease ❯ Evaluate Model

◢ Metrics

Mean Absolute Error	0.409284
Root Mean Squared Error	0.603883
Relative Absolute Error	0.460733
Relative Squared Error	0.412968
Coefficient of Determination	0.587032

Table 4 MSE for the network

Number of hidden layers	MSE values
2	0.3563
1	**0.2675**
3	0.3071
4	0.6501

B. *Split Data Module*

This module enables to divide a dataset into two different sets. This module enables to split data into training and testing sets. The dataset is divided into 70% training data and 30% testing data.

C. *Two Class Neural Network Module*

This unit enables predicting of CKD. This module can be used to predict a target that contains only two values. The NN model contains thirteen inputs, one hidden layer and one output. In this sample, the mean squared error (MSE) is used to evaluate the performance of the model. MSE is calculated using Eq. (10). Whenever, MSE is small that indicate better Performance of the NN model. The NN model uses one hidden layer with MSE = 0.2675. The accuracy of the NN model with different hidden layers is presented in Table 4.

$$\text{MSE} = \sum_{j=0}^{P} \sum_{i=0}^{N} \left(t_{ij} - y_{ij} \right)^2 \div NP \tag{10}$$

where:

P   is the number of output possessing elements.
N   is the number of observations.
t_{ij} is the target outputs.
y_{ij} is the actual outputs.

Chronic_Kidney_Disease - Copy ❯ Score Model ❯ Scored dataset

rows   columns
200   16

hemoglobin	packed cell volume	white blood cell count	anemia	class	Scored Labels	Scored Probabilities
11.4	0	0	no	ckd	ckd	0.000022
9.8	28	8000	no	ckd	ckd	0
14.1	45	9400	no	notckd	no	0.959307
0	0	0	no	ckd	ckd	0.008337
11.8	37	10200	no	ckd	ckd	0.417251
13.7	47	9200	no	notckd	no	0.937067
12.4	44	6900	no	ckd	ckd	0

Fig. 5 Sample of scored dataset

D. *Train Model Module*

This module shows some settings such as learning rate $= 0.1$, number of learning iteration $= 100$, the initial learning weight $= 0.1$ and type of normalizer (min-max). The min-max normalizer linearly rescales every feature to the [0, 1] interval. Min-max normalizer is calculated using Eq. (11).

$$Z = \frac{x - \min(x)}{[\max(x) - \min(x)]} \tag{11}$$

E. *Score Model Module*

This module enables to predict a value for the class, as well as the probability of the predicted value. This module shows scored labels and scored probabilities. The Scored Probabilities indicates that the closer we get to zero, the greater the risk of CKD. On the contrary, the closer we get to one, the lower the probability of CKD, shown in Fig. 5.

F. *Evaluate Model Module*

This module shows true positive (TP), false positive (FP), true negative (TN), false negative (FN) accuracy, precision, recall and F1 score of the NN model. TP is cor-

True Positive	False Negative	Accuracy	Precision	Threshold		AUC
51	0	0.978	0.962	0.5		0.997

False Positive	True Negative	Recall	F1 Score
2	39	1.000	0.981

Fig. 6 Performance attributes of NN model

rectly data identified. FP is incorrectly data identified. TN is correctly data rejected. FN is incorrectly data rejected. Accuracy is 0.978 of the total number of correct predictions. The performance of the NN model is based on accuracy. Precision is 0.962 of positive cases that were correctly identified. Recall is 1.000 of actual positive cases which are correctly identified. F1 score is 0.981 the harmonic mean of precision and Recall, shown below in Fig. 6. Accuracy, precision, recall and F1 score are calculated using Eqs. 12, 13, 14 and 15.

$$\text{Accuracy} = \frac{(\text{TP} + \text{TN})}{(\text{TP} + \text{TN} + \text{FP} + \text{FN})} \qquad (12)$$

$$\text{Precision} = \frac{\text{TP}}{(\text{TP} + \text{FP})} \qquad (13)$$

$$\text{Recall} = \frac{\text{TP}}{(\text{TP} + \text{FN})} \qquad (14)$$

$$F1 \text{ score} = \frac{2\text{TP}}{(2\text{TP} + \text{FP} + \text{FN})} \qquad (15)$$

G. *Set up Web Services Module*

This module is composed of web service input and web services output. Web service input enables stakeholders to enter data in the NN model. Web service output enables stakeholders to return data (predicting of CKD), when the web service is accessed.

6.3 Case Study

Three cases of patients were selected for predicting of CKD in the NN model on windows azure. Stakeholders' uses web services on windows azure to predict of CKD, as shown in Fig. 7. Table 5 shows thirteen critical factors to predict of CKD and actual data of three cases of patients with CKD.

Patient one has CKD with probability (0.00004). The closer the probability is to zero, the greater the risk of CKD. Patient two hasn't CKD (NOCKD) with probability (0.99992). The closer the probability is to one, the lower the risk of CKD. Patient three hasn't CKD (NOCKD) with probability (0.99998), shown in Fig. 8.

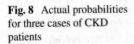

Fig. 7 Sample of patient data to predict of CKD

Fig. 8 Actual probabilities
for three cases of CKD
patients

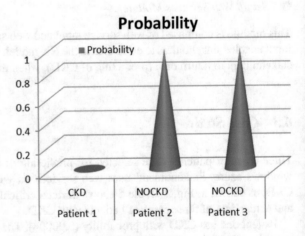

Figure 9 shows a comparison between the proposed model and the state-of-the art methods regarding to the model accuracy. The results show that the proposed model greatly improves the accuracy of prediction by 64%.

Table 5 Actual data of three cases of CKD patients

No	Factor name	Patient 1	Patient 2	Patient 3
1	Age	48	58	28
2	Blood pressure	70	80	100
3	Specific gravity	1.005	1.025	1.1
4	Sugar	0	0	1
5	Blood glucose random	117	131	120
6	Blood urea	56	18	20
7	Serum creatinine	3.8	1.1	1.2
8	Sodium	111	141	130
9	Potassium	2.5	3.5	2.1
10	Hemoglobin	11.2	15.8	10
11	Packed cell volume	32	53	50
12	White blood cell count	6700	6800	5000
13	Anemia	Yes	No	Yes
14	Class	CKD	NOCKD	NOCKD
15	Probability	0.00004	0.99992	0.99998

Fig. 9 The accuracy of the proposed model compared to the state-of-the art models

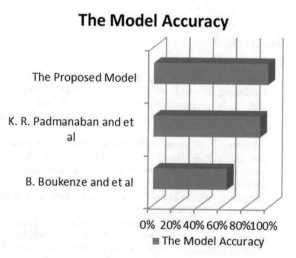

The proposed hybrid machine learning model is applied on azure machine learning services.

7 Conclusion and Future Work

The stakeholders are facing a big challenge to during their interaction with HCS applications due to the limited resource and the time consumption. This paper proposes a novel model based on cloud IOT to predict of CKD. This paper also introduces a hybrid intelligent model based on LR and NN to predict of CKD on cloud computing. The experimental results show the thirteen critical factors out of twenty-four factors that influence on CKD and the accuracy of hybrid intelligent model in predicting CKD is 97.8%. Three cases of patients were selected for predicting CKD in the hybrid intelligent model on windows azure. Patient one has CKD with probability (0.00004). Patient two hasn't CKD with probability (0.99992). Patient three hasn't CKD with probability (0.99998). The results also show that the proposed model greatly improves the accuracy of prediction by 64%.

Future work will focus on applying the new intelligent algorithms such as deep neural network to improve accuracy of diseases diagnosis, analyze big data of patients' records efficiently and maximize utilization of medicinal resources. Moreover, additional Applications and techniques [29–45] will be demonstrated to measure the performance of the proposed method.

References

1. Elhoseny M, Salama AS, Abdelaziz A, Riad A (2017) Intelligent systems based on cloud computing for healthcare services: a survey. Int J Comput Intell Stud Indersci 6(2/3):157–188
2. Sun G, Chang V, Ramachandran M, Sun Z, Li G, Yu H, Liao D (2017) Efficient location privacy algorithm for internet of things (IoT) services and applications. J Netw Comput Appl 89:3–13 (Elsevier)
3. Abdelaziz A, Elhoseny M, Salama AS, Riad AM, Hassanien A (2017) Intelligent algorithms for optimal selection of virtual machine in cloud environment, towards enhance healthcare services. In: Proceedings of the international conference on advanced intelligent systems and informatics, vol 639. Springer, Berlin, pp 23–37
4. Sharma S (2016) Cervical cancer stage prediction using decision tree approach of machine learning. IJARCCE 5(4):345–348
5. Kumar CB, Kumar MV, Gayathri T, Kumar SR (2014) Data analysis and prediction of hepatitis using support vector machine (SVM). IJCSIT 5(2):2235–2237
6. Prerana T, Shivaprakash N, Swetha N (2015) Prediction of heart disease using machine learning algorithms-naïve Bayes, introduction to PAC algorithm. Comp Algorithms HDPS IJSE 3(2):90–99
7. Tintu P, Paulin R (2013) Detect breast cancer using fuzzy c means techniques in Wisconsin prognostic breast cancer (WPBC) data sets. IJCAT 2(5):614–617
8. Hamad AM (2016) Lung cancer diagnosis by using fuzzy logic. IJCSMC 5(3):32–41
9. Arjun C, Anto S (2015) Diagnosis of diabetes using support vector machine and ensemble learning approach. IJEAS 2(11):68–72
10. Jena L, Kamila N (2015) Distributed data mining classification algorithms for prediction of chronic-kidney-disease. IJERMT 4(11):110–118
11. Batra A, Singh V (2016) A review to predictive methodology to diagnosis chronic kidney disease. In: International conference on computing for sustainable global development, vol 4. IEEE, New York, pp 2760–2763

12. Boukenze B, Mousannif H, Haqiq A (2016) Performance of data mining techniques to predict in healthcare case study: chronic kidney failure disease. IJDMS 8(3):1–9
13. Padmanaban KR, Parthiban G (2016) Applying machine learning techniques for predicting the risk of chronic kidney disease. IJST 9(29):1–5
14. Salekin A, Stankovic J (2016) Detection of chronic kidney disease and selecting important predictive attributes. ICHI 8:1–9
15. Maithili A, Kumari RV, Rajamanickam S (2012) Neural networks cum cloud computing approach in diagnosis of cancer. IJERA 2(2):428–435
16. Aruna S, Nandakishore LV, Rajagopalan SP (2012) Cloud based decision support system for diagnosis of breast cancer using digital mammograms. IJCA 1(1):1–3
17. Abdulbaki AS, Khadim SA, Najim SM (2016) Eczema disease detection and recognition in cloud computing. Comput Technol Appl 7:57–64
18. Al-Ghamdi AA, Wazzan MA, Mujallid FM, Bakhsh NK (2011) An expert system of determining diabetes treatment based on cloud computing platforms. IJCSIT 2(5):1982–1987
19. Sandhu IK, Nair M, Shukla H, Sandhu SS (2015) Artificial neural network: as emerging diagnostic tool for breast cancer. IJPBS 5(3):29–41
20. Kumar PS, Chaithra AS (2015) A survey on cloud computing based health care for diabetes: analysis and diagnosis. IOSR-JCE 17(4):109–117
21. Aswin V, Deepak S (2012) Medical diagnostics using cloud computing with fuzzy logic and uncertainty factors. In: International symposium on cloud and services computing, vol 17. IEEE, New York, pp 107–112
22. Boulos MN, Al-Shorbaji NM (2014) On the internet of things, smart cities and the WHO healthy cities. Int J Health Geogr 13:2–6
23. Alhussein M (2017) Monitoring Parkinson's disease in smart cities, special section on advances of multisensory services and technologies for healthcare in smart cities, vol 5. IEEE, New York, pp 19835–19841
24. Bhunia SS, Dhar SK, Mukherjee N (2014) iHealth: a fuzzy approach for provisioning intelligent health-care system in smart city. In: e-Health pervasive wireless applications and services, vol 14. IEEE, New York, pp 187–193
25. Islam M, Razzaque A, Hassan MM, Nagy W, Song B (2017) Mobile cloud-based big healthcare data processing in smart cities. IEEE, New York, pp 1–12
26. Sajjad M, Khan S, Jan Z, Muhammad K, Moon H, Kwak JT, Rho S, Baik SW, Mehmood I (2016) Leukocytes classification and segmentation in microscopic blood smear: a resource-aware healthcare service in smart cities. IEEE, New York, pp 1–15
27. Abdelaziz A, Elhoseny M, Salama AS, Riad AM (2018) A machine learning model for improving healthcare services on cloud computing environment. Measurment 119:117–128
28. Elhoseny M, Abdelaziz A, Salama AS, Riad AM, Muhammad K, Sangaiah AK (2018) A hybrid model of Internet of Things and cloud computing to manage big data in health services applications. Future Gener Comput Syst (2018)
29. Tharwat A, Elhoseny M, Hassanien AE, Gabel T, Arun kumar N (2018) Intelligent Beziér curve-based path planning model using chaotic particle swarm optimization algorithm, cluster computing. Springer, Berlin March 2018, 1–22 https://doi.org/10.1007/s10586-018-2360-3
30. Tharwat A, Mahdi H, Elhoseny M, Hassanien AE (2018) Recognizing human activity in mobile crowdsensing environment using optimized k-NN algorithm. Expert Systems With Applications, Available online 12 April 2018 https://doi.org/10.1016/j.eswa.2018.04.017
31. Hosseinabadi AAR, Vahidi J, Saemi B, Sangaiah AK, Elhoseny M (2018) Extended genetic algorithm for solving open-shop scheduling problem. Soft Comput April 2018 https://doi.org/10.1007/s00500-018-3177-y
32. El Aziz MA, Hemdan AM, Ewees AA, Elhoseny M, Shehab A, Hassanien AE, Xiong S (2017) Prediction of biochar yield using adaptive neuro-fuzzy inference system with particle swarm optimization. In: 2017 IEEE PES PowerAfrica conference, June 27–30, Accra-Ghana. IEEE, New York, pp 115–120 https://doi.org/10.1109/powerafrica.2017.7991209
33. Ewees AA, El Aziz MA, Elhoseny M (2017) Social-spider optimization algorithm for improving ANFIS to predict Biochar yield. In: 8th international conference on computing, communication and networking technologies (8ICCCNT), July 3–5, Delhi-India. IEEE, New York

34. Elhoseny M, Tharwat A, Yuan X, Hassanien A (2018) Optimizing K-coverage of mobile WSNs. Expert Syst Appl 92:142–153. https://doi.org/10.1016/j.eswa.2017.09.008
35. Sarvaghad-Moghaddam M, Orouji AA, Ramezani Z, Elhoseny M, Farouk A, Arun kumar N (2018) Modelling the spice parameters of SOI MOSFET using a combinational algorithm. Cluster Computing, Springer, Berlin, March 2018 In Press https://doi.org/10.1007/s10586-018-2289-6
36. Rizk-Allah RM, Hassanien AE, Elhoseny M (2018) A multi-objective transportation model under neutrosophic environment. Comput Electr Eng (Elsevier) In Press https://doi.org/10.1016/j.compeleceng.2018.02.024
37. Batle J, Naseri M, Ghoranneviss M, Farouk A, Alkhambashi M, Elhoseny M (2017) Shareability of correlations in multiqubit states: optimization of nonlocal monogamy inequalities. Phys Rev A 95(3):032123. https://doi.org/10.1103/PhysRevA.95.032123
38. Elhoseny M, Nabil A, Hassanien AE, Oliva D (2018) Hybrid rough neural network model for signature recognition. In: Hassanien A, Oliva D (eds) advances in soft computing and machine learning in image processing. studies in computational intelligence, vol 730. Springer, Cham https://doi.org/10.1007/978-3-319-63754-9_14
39. Elhoseny M, Tharwat A, Farouk A, Hassanien AE (2017) K-coverage model based on genetic algorithm to extend WSN lifetime. IEEE Sens Lett 1(4):1–4. https://doi.org/10.1109/lsens.2017.2724846 (IEEE)
40. Yuan X, Elhoseny M, El-Minir HK, Riad AM (2017) A genetic algorithm-based, dynamic clustering method towards improved WSN longevity. J Netw Syst Manage 25(1):21–46. https://doi.org/10.1007/s10922-016-9379-7 (Springer, US)
41. Elhoseny M, Nabil A, Hassanien AE, Oliva D (2018) Hybrid rough neural network model for signature recognition. In: Hassanien A, Oliva D (eds) Advances in soft computing and machine
42. Elhoseny M, Shehab A, Yuan X (2017) Optimizing robot path in dynamic environments using genetic algorithm and Bezier curve. J Intell Fuzzy Syst 33(4):2305–2316. https://doi.org/10.3233/jifs-17348 (IOS-Press)
43. Elhoseny M, Tharwat A, Hassanien AE (2017) Bezier curve based path planning in a dynamic field using modified genetic algorithm. J Comput Sci. https://doi.org/10.1016/j.jocs.2017.08.004) (Elsevier)
44. Metawaa N, Hassana K, Elhoseny M (2017) Genetic algorithm based model for optimizing bank lending decisions. Expert Syst Appl 80:75–82. https://doi.org/10.1016/j.eswa.2017.03.021 (Elsevier)
45. Metawa N, Elhoseny M, Kabir Hassan M, Hassanien AE (2016) Loan portfolio optimization using genetic algorithm: a case of credit constraints. In: Proceedings of 12th international computer engineering conference (ICENCO). IEEE, New York, pp 59–64 https://doi.org/10.1109/icenco.2016.7856446

Part II
Secure Models for Smart Cities Applications

Data Security and Challenges in Smart Cities

I. S. Farahat, A. S. Tolba, Mohamed Elhoseny and Waleed Eladrosy

Abstract Day by day cities become more intelligence because governments move slowly to convert each thing to become smarter. These cities are built with the goal of increasing liveability, safety, revivification, and sustainability by building smart services like smart education, smart government, smart mobility, smart homes and e-health but it is important to build these services along with the method for securing and maintaining the privacy of citizen's data. Citizens can build their own services that meet their requirements and needs. This book chapter discusses the internet of things and its applications in smart cities then discusses smart cities and challenge that faces smart cities and describes how to protect citizen data by securing the WiFi based data transmission system that encrypts and encodes data before transfer from source to destination where the data is finally decrypted and decoded. The proposed system is embedded with authentication method to help the authorized people to access the data. The proposed system first compresses data with run-length encoding technique then encrypt it using the AES method but with a rotated key then the source transfers the encoded and encrypted data to the destination where the data is decrypted then decoded to restore the original data then the original data is upload to the destination's website.

Keywords Smart cities · Wireless security · Smart cites · Threats and attacks Real time data security · Smart sensors data encoding · Decoding · Encryption Decryption and authentication

I. S. Farahat
Department of Computer Science, Faculty of Computers and Information,
South Valley University, Luxor Branch, Qena, Egypt

A. S. Tolba · M. Elhoseny (✉) · W. Eladrosy
Department of Computer Science, Faculty of Computers and Information,
Mansoura University, Mansoura, Egypt
e-mail: mohamed_elhoseny@mans.edu.eg

© Springer Nature Switzerland AG 2019
A. E. Hassanien et al. (eds.), *Security in Smart Cities: Models, Applications,*
and Challenges, Lecture Notes in Intelligent Transportation and Infrastructure,
https://doi.org/10.1007/978-3-030-01560-2_6

1 Introduction

The Internet of Things is considered as the third surge in the growth of the Internet and it is famous for IoT. IoT is a system of interconnected digital machines, computing devices, physical objects, animals and people by providing a unique identifier for each of them [1]. IoT is a thing to thing communication that arises around cloud computing. The main role of the sensors in IoT is gathering, measuring, and evaluating data. The power of the IoT stems from the measured data and from using it to meet the requirements of the industry [2].

It can be characterized as associating objects/system, for example smart TVs, smart mobiles and smart sensors to the internet to make any imaginative way to communicate this object with each other and with people [3]. Smart cities are the most important application of internet of things [3]. Smart cities intend to build intelligent services by using wireless sensors, present communication network and smart management system [4]. The goal of building smart cities is ameliorating the livability of cities [5] but still now 46% of the population at the world live in rustic areas instead of urban areas [6] and there is the country that individual lives with small cites rather than lives in large cities. Smart cities goal is to increase livability, safety, revivification, and sustainability of a town and help the individual to live in smart large urban cities [7]. Attackers who want to steal or modify data between source and destination are spread around the world. So the help net security website [8], shows the service that smart cities introduce and the risk percentage that each service is facing.

Researchers thought that there is 10-Step Cyber security Checklist for Smart Cities to protect the smart city services. The 10 steps of the check list include [9]. Perform quality inspection and penetration testing, Prioritize security in Smart city adopters should draft service level agreements (SLAs) for all vendors and service providers, Establish a municipal computer emergency response team (CERT) or computer security incident response team (CSIRT), Ensure the consistency and security of software updates, Plan around the life cycle of smart infrastructures, Process data with privacy in mind, Encrypt, authenticate, and regulate public communication channels, Always have a manual override ready, Design a fault-tolerant system and Ensure the continuity of basic services. This chapter is concerned with the built prototype for a smart city WiFi security system that consists of the source, destination, and LM35 temperature sensor and then build an authentication encryption encoded method with low cost hardware and open source software.

This chapter is organized as follows. Section 2 introduces the background to the internet of things, applications of internet of things, smart cities and challenges facing smart cities. Section 3, discusses some of the previous work then, Sect. 4 describes the proposed prototype system. Section 5 presents results and their discussion. Finally, Sect. 6 provides conclusions and hints for future research.

2 Background

Defining the term IoT can be somewhat difficult because it has many definitions depending on who is describing the term [10]. The essential concept of IoT is to connect things together, consequently enabling these "things" to interact with each other and allowing the people to communicate with them [11]. What these things are, varies depending on which context the term is used and the aim of using the thing. The interconnection of embedded devices is anticipated to lead in automation in about all fields, while also enabling high-level applications like a smart grid, and extending to areas like smart cities [3]. The Internet of Things (IoT) is a framework of interrelated processing devices, mechanical and computerized machines, objects, creatures or individuals that are furnished with special identifiers and the capable of exchange information over a system without needing human-to-human or human-to PC communication. It can be characterized as associating objects/gadgets, for example, Internet TVs, smartphones, and sensors to the Internet in order to link the devices together in an imaginative way to allow new communication forms between devices, system components, and people [12].

Goldman Sachs in his new report [13] shows that the IoT may soon fuel an upheaval that may save $305 billion for the industry. The basic element of IoT is the devices called sensors. These devices range from consumer-based device to track fitness to external wearable devices like insulin pumps, cultivated devices like defibrillators, pacemakers etc., and steady devices like fatal monitors, home-monitoring devices etc. So the most important question that comes to mind is: what qualifies a device to be an IoT device? Verizon says that a sensor-equipped "thing" must have three qualities. It must be aware; it must be able to sense and collect data [14]. And we added that it must be autonomous and actionable. For example, if the blood pressure or blood sugar levels of an individual are at a dangerous level, the device must automatically give an alert to initiate clinical action.

A. IoT Architecture

The architecture of IoT consists of various layers, starting from the edge technology layer at the rear to the application layer at the top, as shown in Fig. 1 [15, 16]. The two lower layers contribute to data capturing, while the two higher layers are responsible for data utilization in applications.

B. Application of IoT

Internet of things (IoT) exists in anything in peoples' life from intelligence building to smart wearable. The appearance of IoT help in developing the simple application and links it to the internet. Nowadays IoT moves towards to improve the well-being of the individual. IoT achieves the concept of connecting all objects over the world and this object can talk with each other without intervention of the human so it is important to know what the applications of the IoT are. Table 1 discusses the main important application that IoT field covered.

Smart cities are built in civilized areas by connecting some distributed wireless sensors to collect data that help individuals to manage and benefit from these cities

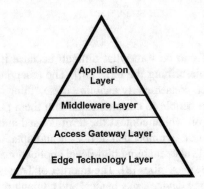

Fig. 1 Layered architecture of IoT [15]

Table 1 IoT application

IoT application	Descriptions
Smart home	Is the application of IoT that concern with building the home that has the services that help individuals to protect and control their home [17]
Wearables	IoT is consisting of some sensor that takes with each other. Wearables are built sensors in the human body or in the things that human wear like glasses, watches, wrists and clothes [18]
Smart cities	IoT is concerned with building cities with the services that avoid traffic congestion, noise, pollution and crime which raises the level of the life of the individuals [19]
Connected car	IoT tries to use the benefits of connecting object with each other in building the smart car that can move without the need of human or connect cars with each other or help smart cars to know the best route [20]
IoT in agriculture	World population extremely increases so the need for the food increases. IoT introduces the method to help farmers to increase food production by using smart agriculture [21]
Healthcare	Healthcare field becomes the most important field in the IoT because the health of the person is an important aspect that people need to live. All massive organizations endure to use healthcare application and smart medical sensors to raise the health of individuals. Research proves that internet of things will be enormous in next year's especially the field of Internet of medical things [22]

[23]. Collected data is collected from all things in these cities like schools, hospital, roads, agriculture, industrial and citizen to manage and monitor all thing around these cities [24, 25]. In the smart city, the service which is delivered to the citizen must be efficient and citizen must know what happens and this can happen by integrating the physical sensors with information and communication technology (ICT) [26, 27]. The smart city allows entities to communicate with each other without intervention of the human.

The objective of moving towards smart cities provides the services that improve the livability of the individuals, increase the value of the life by providing smart

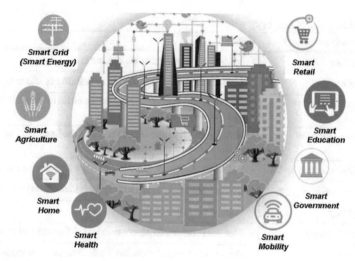

Fig. 2 Elements of a smart city

application such as e-health, smart government and smart education and provide the clean environment that helps individuals to participate their activities each day. Smart cities push Mohanty [28] to say that a smart city is a place which its services is more flexible and efficient because smart cities can help citizen to build their services and modify existing services so smart cities are friendlier, secure, grow faster and greener. To build Smart cities, the main concern is the smart grid (smart energy), smart agriculture, smart retail, smart home, smart education, smart government, smart health and smart mobility which are called the elements of the smart city. Figure 2 shows the elements of smart cities.

Each element in the smart city is an internet of thing application so any element has a model that makes it. For example, we illustrate the IT healthcare model. Health care system must contain patient system and physician system and network between them. The following figure shows the elements of a remote health care system (Fig. 3).

First, we need to describe body area network (BAN) which consists of a network of sensors that are built in body and connect this sensors to Arduino kit, mobile phone or computer (Fig. 4).

Any element consists of some of the sensors and should include an E-health shield that collects all sensors in one kit. Table 2 describe some smart city sensor categories [29].

Table 3 describes the sensor needed in our example of health care system prototype.

In smart cities, sensors are facing some security challenge for example sensor that used to detect traffic congestion, collect data and send data to a centralized server. This data can be stolen or modified in the way from source to destination [30]. Table 4 discusses some security challenges that face the smart city.

Table 2 Categories of smart city sensors

Component (sensor)	Uses
Smart parking	– Sensing about available park space in the city
Structural health	– Measure the material condition of building as and bridges, etc.
Noise urban maps	– Record sound in centric zones of the cities to detect if there is noise or not
Smartphone detection	– Monitor all smartphones and detect if the individual use WIFI or Bluetooth
Electromagnetic field levels	– Detect the energy of radio frequency which device radiate
Traffic congestion	– Detect traffic congestion to help individuals to take best routes
Smart lighting	– Control the lights of the street
Waste management	– Monitor rubbish containers to detect if they are full or not
Smart roads	– Detect accident and monitors roads to detect best roads

Table 3 Healthcare sensors and shield

Component (sensor)	Uses
Patient position sensor (accelerometer)	– Keeps an eye on five various patient positions using a threefold axis accelerometer
Blood pressure sensor	– Gauge human blood pressure – Gauge systolic, diastolic and mean arterial pressure using the oscillometric technique
Pulse and oxygen in blood sensor (spo2)	– Measure pulse rate and the quantity of Oxygen melted in blood
Body temperature sensor	– Monitors body temperature
Breathing (airflow) sensor	– Measures airflow rate – Monitors the patient's breathing
Electrocardiogram sensor (ECG)	– Gauges the electrical and muscular functions of the heart
Electromyography sensor (EMG)	– Measures the electrical activity of muscles
Galvanic skin response sensor (GSR sweating)	– Gauge the electrical response of the skin
E-health shield	– Collects all IoMST sensors in one kit

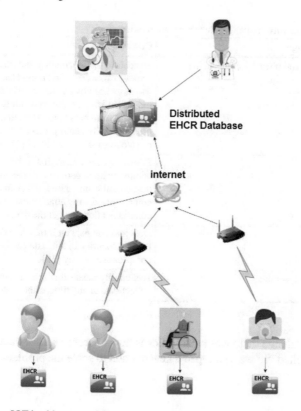

Fig. 3 Remote IOT healthcare model

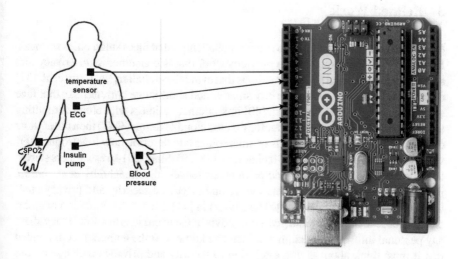

Fig. 4 BAN: body area network

Table 4 Some security challenges that faces smart cities

Security challenge	Description
Confidentiality and integrity compromise	To achieve Confidentiality and integrity it must achieve two things, first thing is authentication which means no one can access data except the individual who has the credentials like username and password. The other thing is privacy which means protect data which is private to specific individual [31–34]
Eavesdropping	Eavesdropping means that if the communication between source and destination is unsecured, data can be compromised. Any attacker can intercept the communication and get the data [31, 33, 35]
Data loss	Data loss can happen if this data doesn't have enough security so this data can be compromised easily [36]
Availability compromise	Availability means that sensors continue to collect data at any time without stopping [30]

The big challenge that faces smart cities is how to secure sensor data from being stolen or modified. So we developed a new method to solve the problems of security and privacy.

3 Related Work

Researchers must find solutions to security challenges that faces smart cities services. In this section, we discuss some previous work that is concerned with privacy and security challenges and develop methods that avoid these challenges. Batty et al. [37] discuss the factor that makes the city smarter and show some current risks that face smart cities. In 2017, Petrolo et al. [40] link internet of things with cloud computing technology to develop a new technology called cloud of things (CoT) then the authors discuss the challenge and benefits of smart cities and describe how to introduce service of smart cities by using CoT technology [38, 39]. Since 2014, researchers started to be concerned with security and privacy challenges so, Elmaghraby et al., built a model with major services of smart cities and discussed security and privacy challenge that faced smart cities [40]. The authors in [41] discuss the limitation of smart cities in the side of privacy and give some advisement to citizens to know if they share any personal information, another citizen can know it so the authors recommended that it must think about giving a solution to security and privacy challenge before building the smart city. Until now researchers discussed some security problems and gave some simple solution to it [42, 43] but some researchers moved towards securing data transfer through channels [29]. At the end of 2017, Usman et al./mix between Feistel method and uniform substitution-permutation method to provide a new mechanism that is used to solve the problem of security and privacy issues and called it secure IoT (SIT) [44].

4 Mathematical Model of AES and RLE

The proposed system consists of two methods: encryption and encoding. This section presents the mathematical model of the two methods.

A. *Mathematical model of AES encryption*

The block length and key length of AES can be determined by 128 bits, 192 bits, or 256 bits, and the corresponding round time is 10, 12 or 14. Each round consists of 4 transformations: the S-box substitution (ByteSub), ShiftRow, MixColumn, and AddRoundKey. With AES starting from the AddRoundKey, with 9 rounds of iteration, in the final round, all operations happen except MixColumn step. In this chapter, the data length is specified by 256 and the key length is 256 bits [9]. Before implementation of AES, the key is rotated.

a. *Rotate key*

The key is rotated in every encryption as follow:

$$k_i = k_{i-n} \tag{1}$$

where, i is index of data and n is number of rotations.

By repeating the pervious equation by the length of array, the key is rotated

b. *SubBytes*

In a sub byte process, AES uses nonlinear substation that is invertible which is called a S-Box

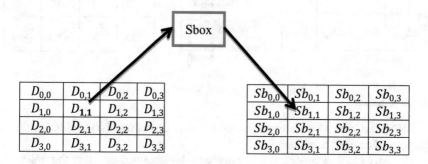

For Each Byte of Input, {XY}

1. Let {AB}:= the multiplicative inverse of {XY} in GF (2^8) as follow:

$$GF(2^8) = GF(2)[b]/(x8 + x4 + x3 + x + 1),\qquad(2)$$

2. Let $\{XY\}' := $ An affine transform of $\{AB\}$ as follow:
   $b' = La \times b \times$ "63"

$$
\begin{bmatrix} b'_0 \\ b'_1 \\ b'_2 \\ b'_3 \\ b'_4 \\ b'_5 \\ b'_6 \\ b'_7 \end{bmatrix}
=
\begin{bmatrix}
1 & 00 & 01 & 11 & 1 \\
1 & 10 & 00 & 11 & 1 \\
1 & 11 & 00 & 01 & 1 \\
1 & 11 & 10 & 00 & 1 \\
1 & 11 & 11 & 00 & 0 \\
0 & 11 & 11 & 10 & 0 \\
0 & 01 & 11 & 11 & 0 \\
0 & 00 & 11 & 11 & 1
\end{bmatrix}
\begin{bmatrix} b_0 \\ b_1 \\ b_2 \\ b_3 \\ b_4 \\ b_5 \\ b_6 \\ b_7 \end{bmatrix}
+
\begin{bmatrix} 1 \\ 1 \\ 0 \\ 0 \\ 0 \\ 1 \\ 1 \\ 0 \end{bmatrix}
\qquad(3)
$$

The choice of constant "63" is to ensure the S-box is not a stable point $(a) = b$; and an oppositely fixed point $(a) = b$.

c. *Shift rows*

SR (Shift Row): shift i bytes to the left for the ith row of the matrix as follows:

$Sb_{0,0}$	$Sb_{0,1}$	$Sb_{0,2}$	$Sb_{0,3}$
$Sb_{1,0}$	$Sb_{1,1}$	$Sb_{1,2}$	$Sb_{1,3}$
$Sb_{2,0}$	$Sb_{2,1}$	$Sb_{2,2}$	$Sb_{2,3}$
$Sb_{3,0}$	$Sb_{3,1}$	$Sb_{3,2}$	$Sb_{3,3}$

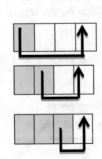

$Sb_{0,0}$	$Sb_{0,1}$	$Sb_{0,2}$	$Sb_{0,3}$
$Sb_{1,1}$	$Sb_{1,2}$	$Sb_{1,3}$	$Sb_{1,0}$
$Sb_{2,2}$	$Sb_{2,3}$	$Sb_{2,0}$	$Sb_{2,1}$
$Sb_{3,3}$	$Sb_{3,0}$	$Sb_{3,1}$	$Sb_{3,2}$

d. *Mix column*

MC (MixColumn) transforms the independent operation of each column for the purpose of causing confusion.

In a Mix Column process a new polynomial is obtained by using the following polynomial equation:

$$a(x) = \{03\}x^3 + \{01\}x^2 + \{01\}x^1 + \{02\}\qquad(4)$$

The general form of equation:

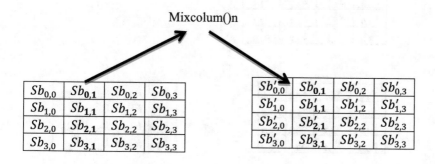

Mixcolum()n

$$\begin{bmatrix} s'_{0,1} \\ s'_{1,1} \\ s'_{2,1} \\ s'_{3,1} \end{bmatrix} = D \cdot \begin{bmatrix} s_{0,1} \\ s_{1,1} \\ s_{2,1} \\ s_{3,1} \end{bmatrix} \text{ for } 0 \le c < NB \quad (5)$$

where,

$$D = \begin{bmatrix} 02 & 03 & 01 & 01 \\ 01 & 02 & 03 & 01 \\ 01 & 01 & 02 & 03 \\ 03 & 01 & 01 & 02 \end{bmatrix}$$

e. *Add round key*

Add Round Key is used to obtain a new form of data by doing bitwise EXOR operation between a round key and state.

Round key can be generated by shift key and key schedule. The Round Key must have the same length as the block length Nb.

It can be denoted as

$$s' = s \oplus K, \quad (6)$$

where K is the round key.

$s'_{0,0}$	$s'_{0,1}$	$s'_{0,2}$	$s'_{0,3}$
$s'_{1,0}$	$s'_{1,1}$	$s'_{1,2}$	$s'_{1,3}$
$s'_{2,0}$	$s'_{2,1}$	$s'_{2,2}$	$s'_{2,3}$
$s'_{3,0}$	$s'_{3,1}$	$s'_{3,2}$	$s'_{3,3}$

$=$

$s_{0,0}$	$s_{0,1}$	$s_{0,2}$	$s_{0,3}$
$s_{1,0}$	$s_{1,1}$	$s_{1,2}$	$s_{1,3}$
$s_{2,0}$	$s_{2,1}$	$s_{2,2}$	$s_{2,3}$
$s_{3,0}$	$s_{3,1}$	$s_{3,2}$	$s_{3,3}$

$\otimes$

$k_{0,0}$	$k_{0,1}$	$k_{0,2}$	$k_{0,3}$
$k_{1,0}$	$k_{1,1}$	$k_{1,2}$	$k_{1,3}$
$k_{2,0}$	$k_{2,1}$	$k_{2,2}$	$k_{2,3}$
$k_{3,0}$	$k_{3,1}$	$k_{3,2}$	$k_{3,3}$

f. *Key schedule algorithm*

Key schedule is composed of two modules: key expansion and round key selection. The data length N_b and key length N_k and the unit is a 4-byte word. That is, $N_b = \frac{block\,length}{32}$ and $N_k = \frac{key\,length}{32}$ the number of iterations (round) is denoted by.

Where, AES-256, $N_b = 8$, $N_k = 8$, and $= 14$. The key expansion of AES-256 is to extend eight 4-byte keywords into 90 4-bytewords [·], where [0], ... , W [7] is the cipher key.

g. *Key expansion*

The Expanded Key is a linear array of 4-byte words and is denoted by

$$k_0 = w_i : k_{i-1} : w_i \oplus SD(k_i : w_{i-1}) \tag{7}$$

To compute the first element in the word and,

$$k_i : w_0 = k_{i-1} : w_0 \oplus SD(k_{i-1} : w_3 \gg 8 \oplus recon_i) \tag{8}$$

To compute any element in the word.
The mathematical model for decryption is the reverse of all operation.

B. *Mathematical model of run length code*

A run length code change each sequential appearance of symbols to individual symbol and attach with this symbol the count of sequential appearance of this symbol and by repeating this method on the all data, will result in data compression.

First we count the consecutive repeating symbols by:

$$N_i = N_{i-1} + 1 \tag{9}$$

When the next symbol equal to current symbol this is denoted by:

$$AD_i = AD_{i-1}, \tag{10}$$

where AD is original data but converted to ASCII code and i is the index of the position of current symbol.

The output of run length code is the symbol plus number of repetitions and can be denoted by:

$$E_i = N_i + AD_i \tag{11}$$

The mathematical model for decoding can first compute the number of repetitions for symbol, the element that have odd index is number of repetition so the number of repetition can be stored by the following equation:

$$n = AD_i, \tag{12}$$

where i is odd index

And the even index is the symbol and can be stored by this equation:

$$S_i = AD_{i+1} \tag{13}$$

The original data can be denoted by:

$$D = S \tag{14}$$

But repeat S by the number of n.

Table 5 System hardware specification

Component type	Name
Microcontroller kit	Esp2866wifi module
Temperature sensor	Lm35 sensor
Breadboard	Breadboard
Electric cable	Hook up wires

Fig. 5 Server side of the proposed system

Fig. 6 Client side of the
proposed system

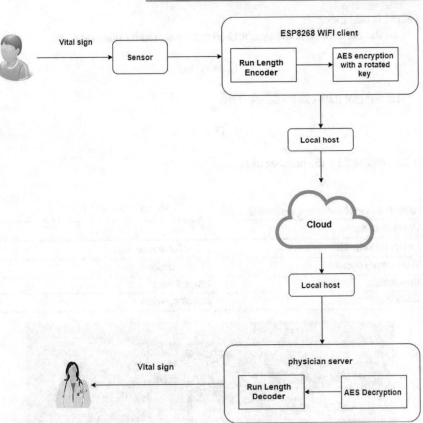

Fig. 7 Security system architecture

5 WIFI Security System

Table 5, shows the components of the proposed system.

Figure 5, shows the destination side of our system.

Figure 6, shows the source side of the proposed system.

Figure 7 shows the block diagram of the proposed security system.

The patient record security is implemented using two algorithms one for encryption and encoding in the device that is owned by the patient and other for decryption, decoding and authentication and it is owned by the physician.

A. Sensor Data authenticated encoding and Encryption Algorithm

The physician D_i needs to monitor patient's P_i status so at the client side C_i we need to encode and encrypt data with AES method with the rotated key R_i before being sent to the physician server S_i.

The following table describes each notation that used in algorithm (Table 6) [45].

The following steps present *the sensor data authenticated encoding encryption algorithm:*

1. *Acquire sensor data such as e.g. temperature.*
2. *The ESP8266 client side C_i encodes data OD_i with run length coding (RLE).*
3. *Client C_i encrypts data by a synchronizer key k_i using AES method but with a rotated key k_i.*
4. *Client side C_i sends encrypted data EED_i to local Wi-Fi through ESP2866 Wi-Fi module*
5. *Local Wi-Fi forwards encrypted data EED_i to the remote physician server S_i hosted in the second ESP6288*

Table 6 Notations

Notations	Meaning
D_i	Physician i
P_i	Patient i
C_i	Client side
S_i	Physician server
R_i	Rotated key i
V_i	Patient vital signs of P_i
T_i	Temperature T_i
SE_i	Sensor
OD_i	Original data
AD_i	ASCII of OD_i
B_i	Binary data of AD_i
N_i	Number of successive equal binary numbers

(continued)

Table 6 (continued)

Notations	Meaning
E_i	Encoded data
AE_i	ASCII of Encoded data E_i
k_i	Key of encryption
N	Shift magnitude of k_i
$ST_{i,j}$	State of data (data in two dimension array form)
$R_{i,j}$	Round key
$SB_{i,j}$	Equivalent Data of R_i in lockup table (s-box)
SD	SubBytes process
$SR_{i,j}$	Data after shift row process
L	Shift magnitude of shift row process
R	Number if rows
$CM_{i,j}$	Is a fixed matrix $\begin{bmatrix} 2 & 3 & 1 & 1 \\ 1 & 2 & 3 & 1 \\ 1 & 1 & 2 & 3 \\ 3 & 1 & 1 & 2 \end{bmatrix}$ used to perform the mix column process
$EE_{i,j}$	Encoded encryption data in a form of two dimensional array
EED_i	Encoded encryption data
k_{n-i}	Last key use in encryption
$CM_{i,j}^{-1}$	Is a inversed matrix of $CM_{i,j}$ $\begin{bmatrix} E & B & D & 9 \\ 9 & E & B & D \\ D & 9 & E & B \\ B & D & 9 & E \end{bmatrix}$
$MC_{i,j}$	Data after mix column process
V_i	Value of encoded data E_i
RE_i	Length of encoded data
AED_i	ASCII of E_i
DD_i	Data after decoding
ID_i	Username of D_i stored in code
PW_i	Password of D_i stored in code
H	Validity check
U_i	Username that D_i entered
$PASS_i$	Password that D_i entered

Authenticated encoding and Encryption Algorithm consists of 3 phases: sensor data acquisition phase, encoding phase and encryption phase.

Acquisition phase

In this phase sensor SE_i acquire data and send it to client side C_i.

(A1)	Sensors SE_i detect patient vital signs V_i such as e.g. temperature T_i
(A2)	Sensors SE_i send this data OD_i to client side C_i
(A3)	C_i received OD_i

Encoding phase

In this phase data od_i is be encoded with RLE encoding method.

(E1)	C_i Convert OD_i to the equivalent ASCII code AD_i
(E2)	C_i computes $N_i = N_{i-1} + 1$ if $AD_i = AD_{i-1}$ then C_i repeat $N_i = N_{i-1} + 1$ until $AD_i \neq AD_{i-1}$
(E3)	C_i computes $E_i += $ "AD_i" + "N_i"
(E4)	Repeat step from E3 to E4 until length $(AD_i) < 0$
(E5)	C_i Convert E_i to the equivalent ASCII code AE_i

Encryption phase

In this phase encoded data E_i is encrypted with AES method but with a rotated key k_i.

(EE1)	C_i compute new key $k_i = k_{i-n}$ then C_i repeat this step if $i > $ length (k_i)
(EE2)	C_i compute $ST_{i,j} = AE_i$
(EE3)	C_i computes $R_{i,j} = ST_{i,j} \oplus k_i$ then C_i compute $SB_{i,j} = SD(R_{i,j})$
(EE4)	C_i computes $SR_{i,j} = SB_{i,j-L}$ while $L=0$ at first then repeat with increase L by this equation $L=L+1$ if $l<R$
(EE5)	C_i computes $EE_{i,j} = SR_{i,j} \otimes CM_{i,j}$ then C_i compute $ST_{i,j} = EE_{i,j}$
(EE6)	C_i computes a key expansion $k_i : w_0 = k_{i-1} : w_0 \oplus SD(k_{i-1} : w_3 \gg 8 \oplus recon_i)$ then compute $k_0 = w_i : k_{i-1} : w_i \oplus SD(k_i : w_{i-1})$ then C_i repeat steps from (EE3) to (EE6) until a number of round determined but if I is a last round escape step (EE5)
(EE7)	C_i computes $EED_i = EE_{i,j}$ then C_i send EED_i to local Wi-Fi

After encryption phase local wifi send data to server side S_i where data can decrypted and decoded.

B. *Sensor Data authenticated decoding and Decryption Algorithm*

The physician server S_i receives encoded and encrypted data EED_i so we need to decrypt this data with the same key K_i that rotates every round and then decoded it to become easy for the physician D_i to access the patients P_i record. But there is another problem that anyone can access this data so we need to implement authenticated access. The second decryption and authentication is implemented as follows:

1. Physician server S_i decrypts data EED_i sent from patient side C_i with the rotated key K_i
2. Physician server S_i decodes the received data E_i to recover the original message OD_i with run length decoding method.
3. Decoded data is upload to an authenticated website built in the Physician server S_i
4. The doctor D_i accesses the website of the patients' P_i record through his credentials
5. After authentication occurs, the doctor D_i can monitor his patients P_i

Authenticated decoding and Decryption Algorithm is comprised into three phases' decryption phase decoding phase, and authentication phase.

Decryption phase

In this phase server side S_i decrypts, encryption encrypted data EED_i WITH AES method with a rotated key k_i.

(D1)	Local wifi send EED_i to S_i
(D2)	S_i computes new key $k_i = k_{i-n}$. then S_i repeat this step if i>length (k_i)
(D3)	S_i computes $ST_{i,j} = EED_i$
(D4)	S_i computes $R_{i,j} = ST_{i,j} \oplus k_n$ then S_i compute $SR_{i,j} = R_{i,j+L}$ while L=0 at first then repeat with increase L by this equation L=L + 1 if l<R
(D5)	S_i compute $SB_{i,j} = SD(SR_{i,j})$
(D6)	S_i computes $R_{i,j} = SB_{i,j} \oplus k_{n-i}$ then S_i computes $MC_{i,j} = R_{i,j} \otimes CM_{i,j}^{-1}$
(D7)	S_i compute $SR_{i,j} = MC_{i,j+L}$ while L=0 at first then repeat with increase L by this equation L=L + 1 if l<R
(D8)	S_i compute $EE_{i,j} = SD(SR_{i,j})$ then S_i compute $ST_{i,j} = EE_{i,j}$
(D9)	C_i computes a key expansion $k_{n-i} : w_0 = k_{n-i+1} : w_n \oplus$ $SD(k_{n-i+1} : W_{n-3} \ll 8 \oplus recon_i$ then compute $k_n = w_{n-i} : k_{n-i+1} : w_{n-i} \oplus SD$ $(k_{n-i} : w_{n-i+1})$ then s_i repeat steps from (E5) to (D7) until a number of n−i<0
(D10)	S_i computes $E_i = EE_{i,j}$

Decoding phase

In this phase server side S_i Decode encoded data E_i With RLE method.

(DD1)	S_i Convert E_i to the equivalent ASCII code AED_i
(DD2)	S_i compute $v_i = AED_i$ then S_i compute $RE_i = AED_{i+1}$
(DD3)	S_i compute $DD_j + =$ "v_j" then repeat step (DD3) and every step increase j by 1 until $j < RE_i$
(DD4)	Repeat step from (DD2) to (DD3) until $I >$ length (AED_i)
(DD5)	S_i Convert DD_j to the equivalent ASCII code OD_i
(DD6)	S_i Upload OD_j to website that is created in S_i

Authentication phase

In this phase doctor d_i require authentication to monitor patient through a website that located in a server side.

(AU1)	D_i Enter name of website or ip of website e.g. 192.168.1.80
(AU2)	Website need authentication by ID_i and PW_i
(AU3)	D_i enter U_i and $PASS_i$ in the website authentication form then S_i compute $H = U_i \oplus ID_i \&\& PASS_i \oplus PW_i$
(AU4)	IF $H = 0$ then S_i permit D_i to monitor p_i if $H \neq 0$ repeat from step (AU2)

6 Experimental Results and Discussion

We have two components in our system esp2866 kit with LM35 sensor for the client and esp2866 kit for the server. The encoding and encryption results of the client side and decoding and decryption results of the server side will be presented in this section.

Results at client side

At the client-side for example, a temperature sensor measures patients temperature then sends the measured data to the esp2866 kit which decodes the data with run length encoding method and then convert the code to ASCII code and encrypts the data with the AES method but we rotated key every time we encrypted data to the server to prevent any attack especially man in the middle attack. Table 7 shows the

Table 7 Some results from client side

Step	Result	
Original message	26.75	23.85
RLE encoded message	15377165211121	15376229211137
ASCII of RLE encoded message	ʕʕmʕ	ʕLʕH̦
Message send to the server	ʕʕ~6?=rfʕo_	ʕHŜJx¢<ghʕʕʕ

Table 8 Some results from client side

Step	Result	
Original message	24.81222222222	25.14547230000
RLE encoded message	1537737211113714020 0306503000	1537710121125164214 105551537623019782
ASCII of RLE encoded message	ʕM%H̦	ʕMeʕ
Message send to the server	\DʕʕAʕ"ʕʕ gʕʕʕ	4ʕ8ʕʕZykwᵔ^}ʕʕ

Table 9 Some results from server side

Step	Result	
Message received from the patient's client	ʕʕ~6?=rf☐o_	ʕHŜJx¢<ghʕʕʕ
Decrypted Message	ʕʕmʕ	ʕLʕH̦
ASCII Conversion and RLE decoding	153777165211121	15376229211137
Original message	26.75	23.85

temperature that the sensor measures and the shape of data after encoding, conversion to ASCII and encryption.

Table 8 shows encryption of encoded data.

Figure 8 is the result form the client side.

Result in server side

When the server receives encrypted and encoded data, the server needs first to decrypt data with the same key that was rotated. Patient data is encrypted and sent to the physician and then converted from ASCII code and then decoded to restore the original message as shown in Table 9.

Table 10 shows decryption decoding data for large data size.

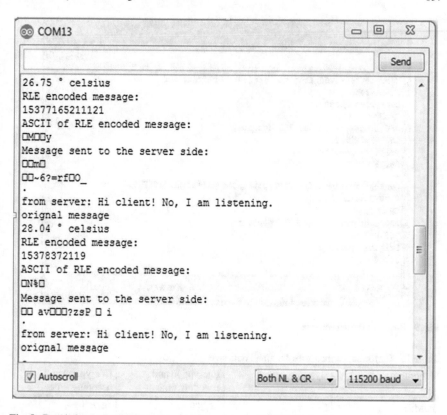

Fig. 8 Result from client side

Table 10 Some results from server side for large data size

Step	Result	
Message received from the patient's client	\DՐՐAՐ"ՐՐ gՐՐՐ	4Ր8ՐՐZykwᴅ^}ՐՐ
Decrypted Message	ՐM%Ḥ	ՐMeՐ
ASCII Conversion and RLE decoding	15377372111371402003065 03000	15377101211251642141055 51537623019782
Original message	24.81222222222	25.14547230000

If a physician wants to access the patient' record, he must enter his user name and password before accessing the website. After the physician enters user name and password, he can access the decrypted data. Figure 9 is the result at the server side.

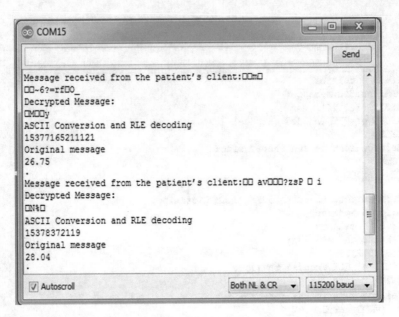

Fig. 9 Result at the server side

Table 11 Performance measures for small data size

Algorithm	Bytes processed	Encryption and encoding time	Decryption and decoding time
IOMST	5	14,289	14,756
AES	5	14,243	11,808
DES	5	17,848	17,844
3DES	5	53,544	53,536

7 Performance Evaluation

The two main properties that distinguish and differentiate one encryption algorithm from another is its capacity to secure the protected data versus attacks and its speed and adequacy in doing so. The first advantage of the proposed system is that the encryption key in both the client and the server is embedded in ESP8266 hardware modules. Additionally, the website is hosted on these modules to protect the system from any hacker. Hacking the hardware is impossible because hacker need human to open his malicious software and then become victim but our system doesn't let any person control it. The second advantage of our system is that we combine both encryption and encoding data so it's difficult to hack. Table 11 compares the proposed encryption method with other methods.

Table 12 show the performance evaluation for large data size (in microsecond).

Table 12 Performance measures of large data size

Algorithm	Bytes processed	Encryption and encoding time	Decryption and decoding time
IOMST	5120	26,583 to 29,881	27,205 to 29,581
AES	5120	30,386	29,941
DES	5120	40,405	39,325
3DES	5120	134,947	134,814

Fig. 10 Comparison between the different encryption techniques

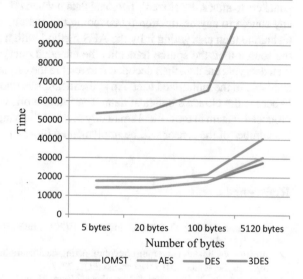

In the previous table, the result of AES with rotated key and RLE encoding is shown in two cases when the data contain repeated data and the second represents random data. The table shows that the proposed method takes less time than the other methods except for AES method as we use a compression method that means less time requirements than the AES method. Figure 10 shows the difference between the different encryption methods in time that each method takes to encrypt data for different data sizes.

In the chapter of SIT method [26] authors show that AES method takes time less than their method so with large data size the proposed method is faster than the SIT method.

The computational complexity of Encoding and Encryption algorithm is $O(n + mc + k)$, where k is key length, n is original data length, m is number of rounds and c is a scaling factor depending on whether 192 or 256 bit keys are used. Since m and c is constant so time complexity at client side is $O(n + k)$.

Computational complexity of decoding and decryption algorithm is $O(n * E + mc + k)$ where E is length of encoded data. Since m and c is constant so time complexity at client side is $O(n * E + k)$.

8 Conclusion

This book chapter introduces the internet of things and discusses its applications and layers then focuses on discussing the mobile security of smart city applications. A low cost WiFI security prototype is presented in detail. We introduce an authentication encoded encryption technique which solves some of the security challenges that face IoT especially in smart cities and its' services. The proposed system avoids any attackers to attack the citizen's personal data it when it is wirelessly transmitted from any source to any destination by encoding this data using the run length encoding technique then encrypting it by the AES method with a rotated key before it leaves the source then the source transmits the encoded encryption data. The destination then decrypts the data then decoded it to restore the original sensor data and makes it available to the authorized user. An authentication method is used to prevent anyone to access the citizen's personal data. The future work will focus on integrating the proposed system in the critical smart city services. Solving The problem of increasing the number of the sensor will be implemented [46–50].

References

1. Osseiran A, Elloumi O, Song J, Monserrat JF (2017) Internet of things. IEEE Commun Stand Mag 1(2):84
2. https://www.linkedin.com/pulse/revolutionizing-healthcare-internet-things-madhuri-vegaraju. September 2015 (last access 2 Oct 2017)
3. Zanella A, Bui N, Castellani A, Vangelista L, Zorzi M (2014) Internet of things for smart cities. IEEE Internet Things J 1(1):22–32
4. Clarke RY (2013) Smart cities and the internet of everything: the foundation for delivering next-generation citizen services. Alexandria, VA, Tech. Rep
5. Mitchell S, Villa N, Stewart-Weeks M, Lange A (2013) The internet of everything for cities: connecting people, process, data and things to improve the livability of cities and communities. San Jose Cisco
6. http://www.pewresearch.org/fact-tank/2017/07/10/rural-and-urban-gun-owners-have-different-experiences-views-on-gun-policy/. July 2017 (last access Jan 2018)
7. Sun Y, Song H, Jara AJ, Bie R (2016) Internet of things and big data analytics for smart and connected communities. IEEE Access 4:766–773
8. https://www.helpnetsecurity.com/2016/10/20/smart-cities-cyber-threats/. October 2016 (last access 23 Dec 2017)
9. https://www.trendmicro.com/us/iot-security/special/221. June 2017 (last access 18 Nov 2017)
10. http://postscapes.com/internet-of-things-definition. 2015 (last accessed 18 Mar 2017)
11. Vermesan O, Friess P (2014) Internet of things-from research and innovation to market deployment, vol 29. River Publishers Aalborg
12. Kortuem G, Kawsar F, Sundramoorthy V, Fitton D (2010) Smart objects as building blocks for the internet of things. IEEE Internet Comput 14(1):44–51
13. http://www.engagemobile.com/goldman-sachs-report-how-the-internet-of-things-can-save-the-american-healthcare-system-305-billion-annually. June 2016 (last access 2 Oct 2017)
14. http://www.hhnmag.com/articles/3438-how-the-internet-of-things-will-affect-health-care,June. 2015 (last access 15 Sept 2017)
15. Santucci G et al (2009) From internet of data to internet of things. In: International conference on future trends of the internet, vol 28

16. Atzori L, Iera A, Morabito G (2010) The internet of things: a survey. Comput networks 54(15):2787–2805
17. Stojkoska BLR, Trivodaliev KV (2017) A review of internet of things for smart home: challenges and solutions. J Clean Prod 140:1454–1464
18. Sun W, Liu J, Zhang H (2017) When smart wearables meet intelligent vehicles: challenges and future directions. IEEE Wirel Commun 24(3):58–65
19. Crooks A, Schechtner K, Dey AK, Hudson-Smith A (2017) creating smart buildings and cities. IEEE Pervasive Comput 16(2):23–25
20. Van Raemdonck W et al (2017) Building connected car applications on top of the world-wide streams platform. In: Proceedings of the 11th ACM international conference on distributed and event-based systems, pp 315–318
21. Stoces M, Vanek J, Masner J, Pavlák J (2016) Internet of things (IoT) in agriculture-selected aspects. AGRIS on-line Pap Econ Inf 8(1):83
22. https://www.hhnmag.com/articles/3438-how-the-internet-of-things-will-affect-health-care. June 2015 (last accessed 4 Feb 2017)
23. https://www.computerworld.com/article/2986403/internet-of-things/just-what-is-a-smart-city.html. October 2015 (last access 27 Oct 2017)
24. McLaren D, Agyeman J (2015) Sharing cities: a case for truly smart and sustainable cities. MIT Press, Cambridge, MA
25. http://www.academia.edu/21181336/Smart_City_Roadmap. January 2016 (last access 28 Dec)
26. https://www.fastcompany.com/3047795/the-3-generations-of-smart-cities. October 2015 (last access Jan 2018)
27. Peris-Ortiz M, Bennett D, Yábar DP-B (2016) Sustainable smart cities: creating spaces for technological, social and business development. Springer, Berlin
28. Mohanty SP, Choppali U, Kougianos E (2016) Everything you wanted to know about smart cities: the internet of things is the backbone. IEEE Consum Electron Mag 5(3):60–70
29. Baig ZA et al (2017) Future challenges for smart cities: cyber-security and digital forensics. Digit Investig 22:3–13
30. Khan MA, Iqbal MM, Ubaid F, Amin R, Ismail A (2016) Scalable and secure network storage in cloud computing. Int J Comput Sci Inf Secur 14(4):545
31. Mukundan R, Madria S, Linderman M (2014) Efficient integrity verification of replicated data in cloud using homomorphic encryption. Distrib Parallel Databases 32(4):507–534
32. Ziegeldorf JH, Morchon OG, Wehrle K (2014) Privacy in the internet of things: threats and challenges. Secur Commun Networks 7(12):2728–2742
33. Pardeshi PM, Borade DR (2015) Improving data integrity for data storage security in cloud computing. Int J Comput Sci Netw Secur 15(6):75
34. Mantelero A, Vaciago G (2015) Data protection in a big data society. Ideas for a future regulation. Digit Investig 15:104–109
35. Baig ZA (2014) Securing the internet of things infrastructure–standards and techniques
36. Mishra N, Lin C-C, Chang H-T (2015) A cognitive adopted framework for IoT big-data management and knowledge discovery prospective. Int J Distrib Sens Networks 11(10):718390
37. Batty M et al (2012) Smart cities of the future. Eur Phys J Spec Top 214(1):481–518
38. Petrolo R, Loscri V, Mitton N (2017) Towards a smart city based on cloud of things, a survey on the smart city vision and paradigms. Trans Emerg Telecommun Technol 28(1)
39. Tawalbeh L, Haddad Y, Khamis O, Aldosari F, Benkhelifa E (2015) Efficient software-based mobile cloud computing framework. In: 2015 IEEE international conference on cloud engineering (IC2E), pp 317–322
40. Elmaghraby AS, Losavio MM (2014) Cyber security challenges in smart cities: safety, security and privacy. J Adv Res 5(4):491–497
41. Ben Ahmed K, Bouhorma M, Ben Ahmed M (2014) Age of big data and smart cities: privacy trade-off. arXivPrepr. arXiv1411.0087
42. AlDairi A (2017) Cyber security attacks on smart cities and associated mobile technologies. Procedia Comput Sci 109:1086–1091

43. Zhang K, Ni J, Yang K, Liang X, Ren J, Shen XS (2017) Security and privacy in smart city applications: challenges and solutions. IEEE Commun Mag 55(1):122–129

44. Usman M, Ahmed I, Aslam MI, Khan S, Shah UA (2017) SIT: a lightweight encryption algorithm for secure internet of things. arXivPrepr. arXiv1704.08688

45. Park Y, Park K, Lee K, Song H, Park Y (2017) Security analysis and enhancements of an improved multi-factor biometric authentication scheme. Int J Distrib Sens Networks 13(8):1550147717724308

46. Sajjad M, Nasir M, Muhammad K, Khan S, Jan Z, Sangaiah AK, Elhoseny M, Baik SW (2018) Raspberry Pi assisted face recognition framework for enhanced law-enforcement services in smart cities. Future Gener Comput Syst. https://doi.org/10.1016/j.future.2017.11.013 (Elsevier)

47. Abdelaziza A, Elhoseny M, Salama AS, Riad AM (2018) A machine learning model for improving healthcare services on cloud computing environment. Measurement 119:117–128. https://doi.org/10.1016/j.measurement.2018.01.022

48. Elhoseny M, Abdelaziz A, Salama A, Riad AM, Sangaiah AK, Muhammad K (2018) A hybrid model of internet of things and cloud computing to manage big data in health services applications. Future Gener Comput Syst. Accepted March 2018, In Press (Elsevier)

49. Elhoseny H, Elhoseny M, Riad AM, Hassanien AE (2018) A framework for big data analysis in smart cities. In: Hassanien A, Tolba M, Elhoseny M, Mostafa M (eds) The international conference on advanced machine learning technologies and applications (AMLTA2018) AMLTA 2018. Advances in intelligent systems and computing, vol 723. Springer, Cham. https://doi.org/10.1007/978-3-319-74690-6_40

50. Elhoseny M, Ramírez-González G, Abu-Elnasr OM, Shawkat SA, Arunkumar N, Farouk A (2018) Secure medical data transmission model for IoT-based healthcare systems. IEEE Access, vol 6, pp 20596–20608. https://doi.org/10.1109/access.2018.2817615

Engineering Large Complex Critical Infrastructures of Future Smart Cities as Self-adaptive Systems

Hosny Abbas, Samir Shaheen and Mohammed Amin

Abstract Smart cities are the expected result of the intensive use of Information and Communication Technologies (ICT) in our daily life. They are the culmination of the great progress that has been made in ICT in the last few years. Smart cities aim to efficiently, sustainably, securely, and reliably manage critical infrastructures such as power systems, transportation, water systems, etc., for the sake of mankind prosperity, comfort and security. However, the critical infrastructures of future smart cities are expected to be large complex critical infrastructures (LCCIs) that need to be autonomous or self-adaptive to be able to survive in critical unpredictable situations. This chapter presents a new approach based on self-organizing multi-agent systems (So-MAS) for modeling and engineering future self-adaptive LCCIs. So-MAS are MAS that have the capability to dynamically and autonomously reorganize themselves to adapt the dynamic changes of work environment. Further, the proposed approach adopts a novel MAS organizational model (called as NOSHAPE) provided in [1, 2] designed to enable large-scale MAS to adaptively self-organize. Based on a simulation environment, a performance evaluation has been conducted that shows how the proposed approach is able to deal with the dynamic unanticipated work environment behaviors. Up to 47% of performance improvement has been achieved with the proposed approach compared to conventional organizational techniques in MAS (i.e. federation). Further future research is still required to address important issues of future smart cities and their associated LCCIs such as security, and resiliency.

Keywords Smart cities · Critical infrastructures
Large-scale complex critical infrastructures
Self-organizing multi-agent systems · NOSHAPE MAS

H. Abbas (✉) · M. Amin
Assiut University, Assiut, Egypt
e-mail: hosnyabbas@aun.edu.eg

M. Amin
e-mail: mhamin@aun.edu.eg

S. Shaheen
Cairo University, Giza, Assiut, Egypt
e-mail: sshaheen@eng.cu.edu.eg

© Springer Nature Switzerland AG 2019
A. E. Hassanien et al. (eds.), *Security in Smart Cities: Models, Applications, and Challenges*, Lecture Notes in Intelligent Transportation and Infrastructure,
https://doi.org/10.1007/978-3-030-01560-2_7

143

1 Introduction

The LCCIs of future smart cities are considered as open systems, highly distributed, heterogeneous, large-scale, and unpredictable uncertain environments [3–6]. Generally, LCCIs are complex industrial networks used in advanced countries to control and monitor critical infrastructures such as smart grids, water distribution, transportation, etc. They are currently a blueprint of modern civilization and future smart cities. It is required to engineer this type of systems in a way that enables them to be self-adaptive against disruptions and perturbations. Two new emerged paradigms are definitely adopted by LCCIs. The first one is the internet-of-things (IoT) paradigm [7] which is getting great attention and it is expected that in the near future most of nations' critical infrastructure utilities and industrial activities will be connected globally using the Internet as the underlying communication network. The second emerging paradigm is SoS (Systems-of-Systems) [8], which are large-scale concurrent and distributed systems, their components are complex systems themselves. SoS are not only complex and large-scale but also they are characterized to be decentralized, distributed, networked and contain compositions of heterogeneous and autonomous components.

LCCIs are considered as SoS because their increase in scale and complexity have reached a point that imposes qualitatively new demands on their engineering approaches. The SoS paradigm deals with increased scale and complexity by decomposing the system into a number of systems interact with each other to from the whole system. One of the important objectives of LCCIs is providing an effective data dissemination for remote supervisory and management of critical infrastructures through global control centers. Control centers are the places where operators exist to supervise and control industrial processes. Each control center is assigned a number of industrial processes to remotely manage and control. The physical industrial processes are composed into groups and each group is assigned to a local control center. Similarly, local control centers can be composed together into groups and each group is assigned to a global control center and so on. This system is considered as a large-scale distributed control center comprises a number of medium-scale control centers. From the power systems domain a control center is the central nerve system of the power system. It senses the pulse of the power system, adjusts its condition, coordinates its movement, and provides defense against exogenous events [9]. In this application, each control center controls, supervises, and monitors a large number of control processes that geographically distributed within the surrounding local environment of each control center. It is required that not only the operators in one control center can monitor and supervise their local control processes but also they are able to supervise, and monitor any control process even if it is located in the local environment of another control center. This system is characterized to be open, highly distributed, and complex as the number of its subsystems can dynamically evolve horizontally (geographical distribution) and vertically (information layers). To build self-adaptive systems, Multi-Agent Systems (MAS) have proven to be the most adequate artificial intelligence system for handling complexity and distribution of modern application domains [10]. What makes MAS distinguishable from other engineering paradigms is their unique capability to support simultaneously many requirements of modern real-life applications which are becoming highly distributed, highly dynamic and uncertain. Compared to conventional SCADA, which adopt the client-server approach solutions, a multi-agent system offers many advantages. For instance, each function or task of the system (i.e., managing a single system component) can be assigned to a separate agent; of course that will increase the system modularity.

A MAS organizational model called NOSHAPE was selected to provide a way to dynamically reorganize the provided large-scale MAS, which might contain a huge number of agents, to coordinate agents' interactions and to enable the MAS to adapt the environment dynamic changes. The main objectives of NOSHAPE are targeted towards adaptivity, modularity, agent localization, agents' interactions coordination. The NOSHAPE model concerns modular MAS, where individual agents live in higher order entities called agent organizations. Any two agents can only interact if they are situated in the same agent organization. Agent organizations are static and explicit entities that have dynamic structural behaviors. An agent organization has a home space and can overlap with other agent organizations that have their own home spaces to share dynamic roles (agents) which are abstract descriptions of the individual agents' activities. The structural behaviors of an agent organization is *doHome, doOverlap, doRetire* (the offline state), etc., they are logically implemented as interaction protocols. They are used to coordinate the interactions of individual agents situated within many separated agent organizations. For detailed information about NOSHAPE MAS, interested reader are invited to read [1, 2, 11]. The next section identifies the NOSHAPE decisions taken to develop this complex large-scale self-adaptive SCADA system.

The remaining of this chapter is organized as follows. Section 2 demonstrates the taken NOSHAPE decisions according the alternative design methods provided in [2]. Section 3 presents the adopted development process of the proposed application, it includes 3 phases: analysis, design, and implementation phases. Further, Sect. 4 evaluates the performance of the proposed approach based on a simulation environment. Finally, Sect. 5 concludes the chapter and highlights future research intentions.

2 NOSHAPE Decisions

The physical architecture of the system-to-be is shown in Fig. 1. According to the operational view of the NOSHAPE model presented [2], there are some important decisions, which should be taken by the designer before starting the development lifecycle of the system-to-be. Similar architecture can be adopted in modern large-scale real-life applications such as healthcare [12–14], industrial forensic investigation [15], and so forth.

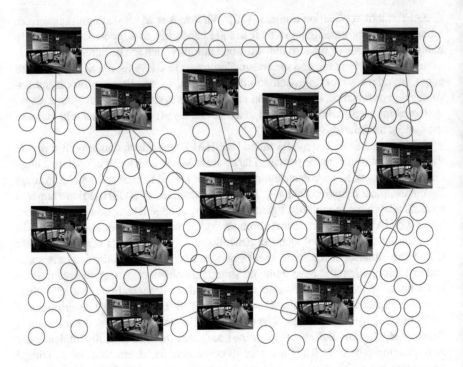

Fig. 1 A large-scale distributed control center

For this case study the taken decisions are presented in Table 1. As demonstrated in the table, the decisions are classified into two types: the NOSHAPE decisions and the design decisions. The system-to-be is complex and highly distributed; therefore it is adequate to model it using MAS and based on NOSHAPE. The system-to-be will be modeled as an OOM NOSHAPE application (OOM stands for One-universe One-world Multiple-organization). The universe and world levels are ignored and only the organization level is considered. The simplified version of the NOSHAPE model will be used to model the organizational behaviors. Further, JADE agents development platform [16, 17] will be used to implements the system-to-be agents, the agents are behavior-based or reactive agents. Also, the acquaintance model proposed in [2] will be used to enable NOSHAPE agent organizations to find each other and dynamically update their address books. Furthermore, one LS agent will be created inside each agent organization to monitor and locally supervise functional agents. Security and agents' mobility will not be supported in the proposed case study. Moreover, the organizational ontology proposed in [2] will be used to model the organizational constructs and their behaviors. Another functional ontology for modeling the application domain knowledge needs to be created too. Other NOSHAPE design decisions have been taken as demonstrated in Table 1.

Table 1 NOSHAPE and design decisions

Type of decision		Decision
NOSHAPE decisions	Adoption of NOSHAPE	Adequate
	System scale	OOM
	Structural behaviors	Home/Overlap/Retire
	Acquaintance model	The proposed model in [2]
	Local supervisors	One LS agent: Monitors, Recovers, and receives triggers from dynamic agents
	Agents mobility	Not used
	Ontology	The NOSHAPE organizational ontology proposed in [2], and the application domain Ontology proposed in this chapter
	Security	Not supported
	Agent platform	JADE
	Agent architecture	JADE-style
Design decisions	Organizational changes implementation	Finite State Machine (FSM)
	Triggers	Ontological concept: Type: {TRJA, TFJA, etc.} Service: String
	Overlap consequences	Register to a remote DF
	Structural behaviors implemented as	Interaction protocols
	NOSHAPE usage	Organizational, matchmaking, coordination, decomposition, aggregation

3 The Development Methodology

The development methodology adopted for developing this case study comprises three main phases: Analysis Phase, Design Phase, and Implementation Phase. These phases are provided in this section in details. The Performance Evaluation Phase will be provided in the next chapter.

3.1 Analyses Phase

The functionality of the system-to-be can be captured using the use cases artifact. A use case describes a required functional scenario in the system. The system higher

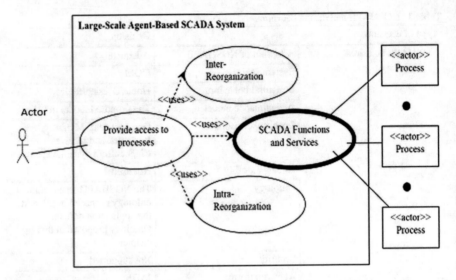

Fig. 2 The higher level use case of the proposed ICN

level use-cases are shown in Fig. 2. There are only two real actors in the system-to-be, the operators and the control processes or industrial plants (through the OPC process communication protocol [18]). The higher level use cases concerns providing the remote operator with an access to the required control process. If the required control process belongs to the same organization with the remote operator agent, then no inter-organization dynamic reorganization is required and local interaction will be enough. The supervisory and control activities use-case (shown in bold border in Fig. 2) is a composite one comprises many other use-cases as presented in Fig. 3. The next step is to identify the types of the system functional agents and the number of each agent instances (cardinality) in each agent organization (control center). Figure 4 presents the final *agent-type diagram*, which presents the required agents types within the system-to-be. As shown in the figure, the system-to-be comprises only two agent types, the operator agent and the control agent, in addition to system agents such as DF, which provides the yellow page service to other agents. The system contains more than one control agent, each of them is associated with a control system (i.e., PLC). Table 2 presents the agents' types and cardinality inside each organization according the NOSHAPE model specification and suggests the responsibilities of each agent type. The responsibilities of each agent type are divided into two types: interaction protocols for interaction with other agents and internal activities for executing the agent assigned functional tasks.

Table 2 Responsibilities of each agent type inside each organization

Agent	Responsibilities	Cardinality per organization
Global Supervisor GS	1. Read configuration files 2. Create local supervisory agents (LS) 3. Configure local supervisory agents 4. Listening to local supervisory agents 5. Monitoring local supervisory agents 6. Manage organization structure 7. Search for required remote services 8. Organizationally Interact with other organizations 9. Interact with administrator 10. Propagate updated address book	One
Local Supervisor LS	1. Receive configuration from global supervisory 2. Creating dynamic agents 3. Configuring dynamic agents 4. Register services to local directory facilitator (DF) 5. Listening to dynamic agents 6. Monitoring dynamic agents 7. Recreating and reconfiguring dead dynamic agents 8. Executing load balance algorithms on dynamic agents 9. Receiving remote agents requests	One
Control Agent CA	1. Receiving configuration from local supervisory 2. Register services to local directory facilitator (DF) 3. Connecting to its assigned control processes 4. Respond to remote operator agents 5. Provide higher level control algorithms 6. Send notifications to local supervisor 7. Respond to local supervisor	Many
Remote Operator Agent RA	1. Searching and subscribing to local DF for dynamic agents 2. Receiving notifications from local DF 3. Interacting with dynamic agents 4. Notifying LS if the required services are not available 5. Present process data to operators (textual, tabular, graphical, etc.) 6. Send operators set points to dynamic agents	Many

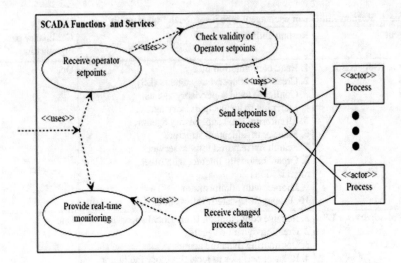

Fig. 3 Use case modeling of supervisory and control activities

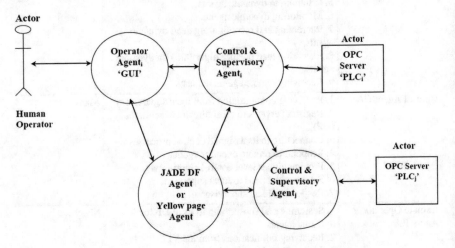

Fig. 4 The agent types diagram with acquaintance relations

3.2 Design Phase

The design phase concerns the transfer from the problem space (analysis phase) to the solution space. It aims to specify the software solution to the problem. Figure 5 presents the NOSHAPE architecture of the system-to-be, as shown each control center is modeled as a NOSHAPE organization assigned a number of control processes or plants controlled by PLCs. The PLCs can be considered as the environment resources of the control center. The initial acquaintance network for each organization is also included in the figure (line connections). For the sake of simplicity and because our

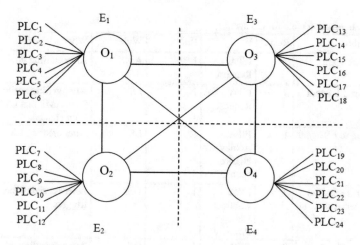

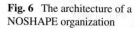

Fig. 5 NOSHAPE modeling of the proposed ICN into 4 initial organizations

Fig. 6 The architecture of a
NOSHAPE organization

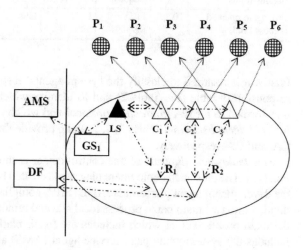

intention is to provide a prototype MAS based on the new proposed organizational model as a proof of concept, the initial number of system organizations is supposed to be four and the initial number of control processes assigned to each organization is supposed to be six, but with real applications these two numbers can be large. The internal structure of each agent organization (control center) is shown in Fig. 6 where P_i represents the control processes (PLCs), C_i and R_i represent the dynamic agents (control agents and operator agents respectively), LS is the local supervisor, GS is the organization global supervisor, and AMS and DF are the agent management (white pages service) and directory facilitator (yellow pages service) respectively.

As illustrated in Table 1, it is decided to implement the proposed agent-based complex ICN using a FIPA-Compliant platform such as the JADE platform. Therefore, the design phase target is to map the analysis phase artifacts to FIPA constructs.

Table 3 Interaction table for GS agent

Interaction	Resp.	IP	Role	With	When
Create LS agent	2	FIPA Request	I	AMS	After starting up
Configure LS agent	3	FIPA Request	I	LS	After getting notification from AMS that LS has been created.
Listening to LS agent	4	FIPA Achiev-eRERe-sponder	R	LS	After creating LS
Search for required remote services	7	FIPA Contract Net Protocol	I	GSs	After receiving local triggers
Overlap with other organizations	8	FIPA Request	I	GS	After accepting other GS proposal
Propagate updated address book	10	FIPA Request	I	GSs	After receiving new address book

Firstly, it is required to classify the system agents' responsibilities provided in the responsibility table (see Table 2) and to identify which of them is suitable to be transformed to a FIPA *interaction protocol* and which of them is considered as an *internal agent behavior*. Tables 3, 4, 5, and 6 provide this classification for GS, LS, CA, and RA, respectively.

The *deployment diagram* of one control center is shown in Fig. 7 in UML notation [19]. The presented deployment diagram is created based on the assumption that the development agent platform is JADE (FIPA-compliant platform). The system is divided into two main parts, process local site and remote site. The former includes the main process server which includes the JADE platform main container which includes the system white page service agent (AMS) and the yellow page service agent (DF). Furthermore, the main process server includes JADE peripheral containers for hosting control agents. Also, the OPC server may be hosted on a separate PC in the local process site as shown in the figure. On the other hand, the remote site is supposed to include the operator stations, it can be another LAN connected to the local process LAN or it can be a web-based remote connection. The next step in the design phase is to create the application domain *ontology*. The ontology is a set of concepts, predicates and agent actions referring to a given domain. Figure 8 shows the designed ontology for the proposed agent-based ICN. A concept is a complex structure defined by a template specified in terms of a name and a set of slots whose values must be of a given type. A predicate is a relation between domain concepts and its value can be true or false. An agent action is a function the agent is required to perform. The ontology components require a content language to be manipulated and exchanged among agents. A content language is the tool that a message receiver used to decode or parse the message to extract specific information; therefore, the

Table 4 Interaction table for LS agent

Interaction	Resp.	IP	Role	With	When
Receive configuration from GS	1	FIPA Inform	R	GS	After starting up
Creating dynamic agents	2	FIPA Request	I	AMS	After receiving process configuration
Configuring dynamic agents	3	FIPA Request	I	DA	After creating DAs
Register services to local directory facilitator (DF)	4	FIPA Request	I	DF	After creating and configuring DAs
Listening to dynamic agents	5	FIPA Achiev-eRERe-sponder	R	DAs	After registering to local DF
Monitoring dynamic agents	6	FIPA AMSSub-scriber	I	AMS	After creating and configuring DAs
Receiving remote agents requests	7	FIPA Request	R	RAs	When an RA sent a trigger to LS

Table 5 Interaction table for an operator agent

Interaction	Resp.	IP	Role	With	When
Search for process service	1	FIPA Request	I	DF	After starting up
Subscribe to a control agent	2	FIPA Request	I	A control agent	After discovering services
Receive process actual values	3	FIPA Inform	R	A control agent	Always
Receives process setpoints	3	FIPA Inform	R	A control agent	Always
Receives notifications and alarms	7	FIPA Inform	R	A control agent	Always
Send a setpoint to a variable	6	FIPA Request	I	A control Agent	When the operator submit a setpoint through his GUI

system agents need to agree on a certain content language to understand each other. For the sake of openness and interoperability, the FIPA-SL content language is used in the proposed ICN application. As an example for illustrating the JADE support of ontology and content languages consider these examples:

1. A remote agent (R1) sends a request message to a control agent (C1) contains a request to write a process variable to a control process as a setpoint:

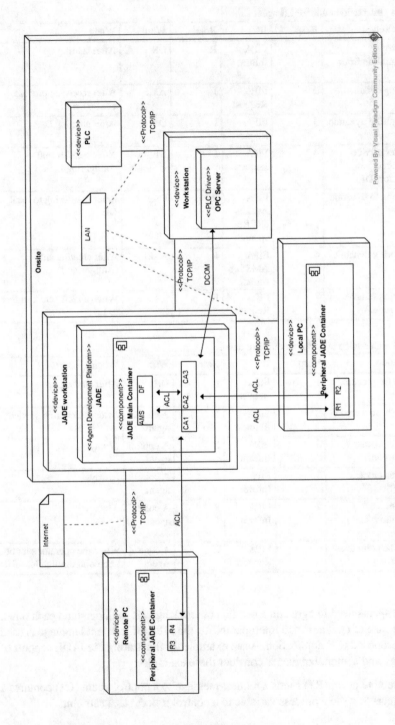

Fig. 7 The proposed ICN deployment diagram in UML

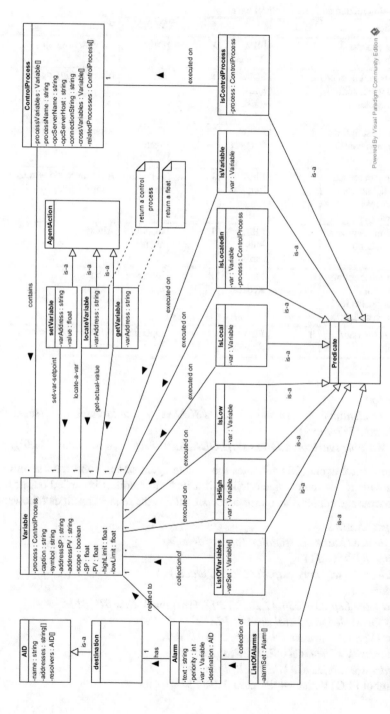

Fig. 8 The application domain ontology in UML

Table 6 Interaction table for a control agent

Interaction	Resp.	IP	Role	With	When
Register process services	1	FIPA Request	I	DF	After starting up
Subscribe to DF to be notified when control agents register their services.	2	FIPA Sub-scribe	I	DF	After Starting up
Handle subscriptions from related control agents	3	FIPA Request	R	Control agents	After initializing
Subscribe to related control agents for cross process variables	4	FIPA Request	I	A control agent	After discovering related control agents by DF
Receive cross process data from control agents	5	FIPA Sub-scribe	I	Control agent	Always
Handle subscription requests from operator agents for local process data	6	FIPA Request	R	Operator agent	Always
Receives operator setpoints	8	FIPA Request	R	Operator agent	When operator send a setpoint

> *((action*
> *(agent-identifier   :name   c1@SCADA   :addresses   (sequence http://scada:7778/acc))*
> *(SetVariable :variableAddress s7:[LOCALSERVER]db1,w26 :value 334.0)))*

2. The control agent (C1) validates the remote agent setpoint and send an inform message to the remote agent telling it if its action request is carried our or not, the message contains an alarm concept contains the request result as follows:

> *((ListOfAlarms*
> *(sequence (Alarm :destination (agent-identifier*
> *:name R1@SCADA*
> *:addresses (sequence http://scada:7778/acc))*
> *:priority 2*
> *:text "Tue Sep 23 08:34:11 2014 \'PLC1Variable4' New SP (334.0)*
> *was forwarded to control process PLC1"*
> *:var (Variable :lowLimit 0.0 highLimit 1000.0*
> *:addressPV s7:[LOCALSERVER]db1,w6*
> *:addressSP s7:[LOCALSERVER]db1,w26*
> *:symbol PLC1Variable4 :PV 360.0 :SP 334.0)))))*

Table 7 Mapping NOSHAPE concepts to JADE constructs and services

NOSHAPE concept	JADE construct
NOSHAPE organization	JADE platform
White page service	JADE AMS
Yellow page service	JADE DF
Global supervisor	JADE agent
Local supervisor	JADE agent
Functional AGENTS	JADE agents
Finite State Machine (FSM)	JADE FSMBehaviour
Dynamic structural behaviors	JADE interaction protocols
Overlap meaning	Registering to a remote DF
Functional activities	JADE agent behaviors
NOSHAPE species and their relations	JADE Ontology classes

The above two examples are given based on the proposed application domain ontology presented in Fig. 8 and they use the FIPA-SL content language. It is not necessary to write messages in text form as presented the above two examples because it is possible to use the JADE agent *ContentManager* construct for creating these messages and let the developer creates and manipulates content expressions as Java objects. Ontology is essentially a collection of schemas that typically doesn't evolve during an agent lifetime. The JADE agent development platform provides the developer with the required tools and classes to create his application ontology, but this way is being cumbersome with large Ontologies. Fortunately, it is possible to define the ontology using the Protégé tool [20], and then, let the Bean-Generator add-on [21] to automatically create the ontology definition class plus the predicates, agent actions and concepts classes. The proposed organizational and functional ontologies have been designed and created by Protégé.

3.3 Implementation Phase

The implementation phase is concerned with moving the system-to-be from the development status to the production status. It includes the deployment of the new system to its target work environment. Table 7 provides a possible mapping from the NOSHAPE concepts to JADE constructs. The proposed ICN can be implemented as a set of independent JADE platforms interact together through a LAN and/or a WAN (Internet). JADE as a middle-ware framework provides all the required low level services to enable flexible agents' interactions. Each JADE platform can be distributed across many machines.

It is intuitively obvious that each NOSHAPE organization in the system-to-be can be modeled as an independent JADE platform and each organization white page and yellow page services can be matched to the JADE platform AMS and DF services respectively. The NOSHAPE organization ontology presented in [2] will be used for the first NOSHAPE application prototype. The ontology concerns the structural behaviors of the higher order NOSHAPE entities; this ontology is created using the protégée tool. This ontology is supposed to be used within any large-scale NOSHAPE MAS. The following subsections highlight the important implemented features for each agent type in the proposed adaptive ICN.

3.3.1   Global Supervisor (GS)

The GS agent is considered as the controller of a NOSHAPE organization. It is concerned with dynamically managing the organization structural behaviors relative to other NOSHAPE organizations. The GS, just after starting execution, will read the XML configuration files (control process configuration and address book) which are provided by the system administrator. Then it will ask the platform AMS to create the local supervisor which in turn will create the dynamic agents and will assign them their services. The GS receives service triggers from the organization dynamic agents and changes the organization structure accordingly by establishing the overlapping relationships with other organizations in the application world. Figure 9 presents the GS engineering and monitoring GUI and Fig. 10 provides the control loop executed by an organization GS. The GS instantiates a JADE *FSMBehaviour* behavior for each organization it knows (included in its address book); The JADE *FSMBehaviour* class implements a composite behavior that schedules its children according to a FSM whose states correspond to the *FSMBehaviour* children. The *FSMBehaviour* class provides methods to register sub-behaviors as FSM states and to register the transitions between states. To share information such as service triggers among the *FSMBehaviour* states (which are implemented as JADE behaviors), the concept of blackboard is used, which is implemented using the *DataStore* class included in the *jade.core.behaviours* package. All the FSM states share the blackboard except the Retire state which represents the offline state of a NOSHAPE organization (the Offline state may be triggered by the system administrator for maintenance).

When the organization GS receives a trigger of type TRJA from dynamic agents it processes the trigger to extract the required service name, then it uses the Contract Net interaction protocol to interact with its acquaintances asking who has the required service. One of the organization acquaintances will respond and send a proposal that it has the required service. Then the GS agent will push the trigger into the blackboard assigned to the FSM assigned to this acquaintance organization. According to the current state of the FSM and the trigger just pushed in the blackboard, the proper transition will be activated and a state change will take place. Each state is assigned a JADE *OneShotBehaviour* encapsulates the required activities and interaction protocols for inter-organization interactions.

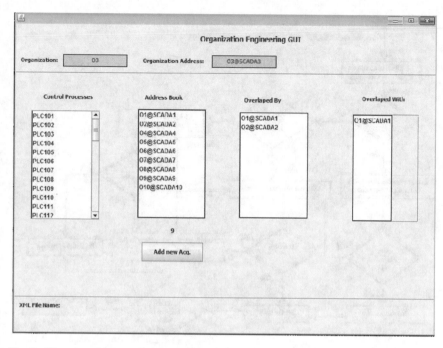

Fig. 9 GS engineering and monitoring GUI

3.3.2 Local Supervisor (LS)

The LS agent is created and configured by the organization GS agent. After initializing, it waits for the configuration message from the GS and after receiving the message it extracts the configuration information. The configuration information includes the control processes or industrial plants and their possible interfaces. Then the LS will create the dynamic control agents which considered as the system interfaces to the automation world. Then the LS agent provides each control agent with its assigned control processes. The LS has many other responsibilities, for example it continuously monitors the control agents' health and applies load balancing algorithms on them if there is a need. The system developer can add many other responsibilities for this agent according to the application domain. But the important job of the LS is receiving dynamic agents' triggers and bypassing them to the organization GS agent. All these responsibilities are implemented as JADE behaviors and interaction protocols.

3.3.3 Control Agent (CA)

Similar to worker bees in a beehive, CAs are the worker agents inside a NOSHAPE organization. They are assigned functional activities according to the application

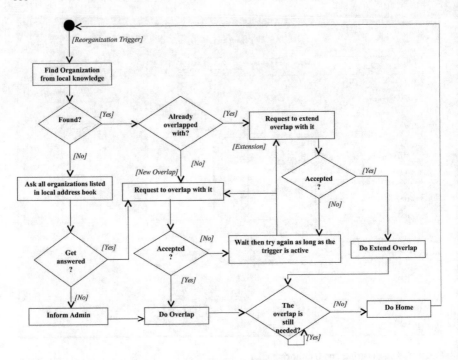

Fig. 10 The control loop of each NOSHAPE organization

domain. They are dynamic which means that they can be dynamically created, removed, moved, shared, and cloned. Further, they are designed by the developer with the suitable interfaces to the environment resources i.e. databases and/or physical non-agent resources such as PLCs. In the proposed ICN, they are prepared to be connected to control processes (PLCs) through the OPC protocol. The JADE agent platform will be used which is a Java-based agent development platform, therefore to establish a connection between a Java JADE agent and an OPC server, a Java-COM bridge or adapter is required. The *JEasyOPC Client* [22] is selected for this mission. In this case study, the CA agents are designed to provide an implementation of the functionalities presented in the use-case diagram shown in Fig. 3. Their main mission is to respond to the remote operator agents requests such as subscriptions for changed process values, changing process variables setpoints and related issues. They are the system service providers. Further, an agent not only interacts with other agents but also it carries out a set of *Internal Behaviors* according to its interaction results with other agents or according to the changes that take place in its environment. In the proposed ICN, the agents' internal behaviors can be extracted from the agent *responsibility table* and the *use case* diagram. The complete list of internal behaviors of a control agent is shown in Fig. 11. For a control agent the important internal behaviors are:

Agent Behaviors	Behavior Type
initializationSequentialBehaviour	SequentialBehavior
prepareAgentProcess	oneShotBehavior
registerServices	oneShotBehavior
connectToOpcServer	oneShotBehavior
createOpcGroups	oneShotBehavior
subscribeToDFForControlAgents	oneShotBehavior
SubscriptionInitiator	SubscriptionInitiator
subscribeToControlAgents	oneShotBehavior
readRemoteCrossVariables	TickerBehaviour
prepareNewSP	oneShotBehavior
TickerBehaviour (1000ms)	TickerBehaviour
handleDataChange	oneShotBehavior
higherLevelControl	oneShotBehavior
handleSubscriptionsForActuals	AchieveREResponder
handleSubscriptionsForCross	AchieveREResponder
handleOperatorActionRequests	AchieveREResponder
handleOperatorSetpoints	oneShotBehavior
manageOperatorSetpoints	FSMBehaviour
checkVariableValidity	oneShotBehavior
discardSP	oneShotBehavior
forwardSP	oneShotBehavior
notifyOperator	oneShotBehavior

Fig. 11 Complete behavior list for a control agent

1. *handleDataChange*: it is a one-shot behavior executed periodically to read process data and to check if there is a process data change to send the changed process data to the concerned remote operator agents and also to send the changed cross variables to other control agents. This behavior invokes another behavior for providing complex higher level control algorithms (i.e., interpolation, global synchronization, etc.), which require higher-capability resources, and can't be provided by the legacy limited-resources control systems.

2. *higherLevelControl*: this behavior is initiated by the *handleDataChange* behavior if there is any process data change. It is a one-shot behavior that contains a number of algorithms for processing the process variables. In other words this behavior realizes the dependency relations among control processes variables. For instance, the setpoint of a process variable may depend on the actual value of another process variable and the former may be calculated from the later through an interpolation algorithm, which needs higher computing power than that provided by the legacy PLC. Figure 12 presents an illustrative example; Var5.SP is calculated from Var4.PV through an interpolation algorithm executed by the control agent. Many other complex algorithms can be added to this behavior as required, i.e., PID controller algorithms. By this way, control agents provide an

Fig. 12 The process
variables dependency
relations

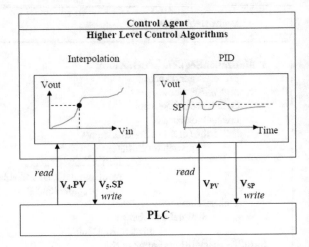

extra computational power for the underlying legacy control systems with limited resources.

3. *prepareNewSP*: after a control agent sends a subscription request to another control agent for getting cross reference process data, it continuously receives this cross process data and forwards it to this behavior for the processing and calculation of new setpoints for some specified local process variables.

4. *manageOperatorSetpoints*: it is a JADE FMS behavior implements a defined finite state machine. Figure 13 presents the designed FSM that is implemented by this behavior. This behavior is executed just after the control agent receives a new setpoint request from a remote agent. Further, this behavior includes four child behaviors each one of them extends the Jade *OneShotBehaviour*. The validity of operator setpoints can be evaluated based on the process variable allowable range (i.e., min and max).

The CA agents, after created and configured by the organization LS, register their assigned services to the organization local yellow page service (DF in JADE). Therefore, any operator agent situated in the organization will be able to find the required CA by searching the local DF. In this application, if a CA is selected to be shared through the overlapping relationships between system organizations, it will be requested to register its services to a remote DF, so other foreign remote operator agents can find and start interacting with it to get access to its assigned control process.

3.3.4 Remote Operator Agent (RA)

RAs are also a type of dynamic agents but in this case study they are created manually by system operators any where and at any time. Not all of the system agents need to be provided by a graphical user interface (GUI) except the GS static role for dynam-

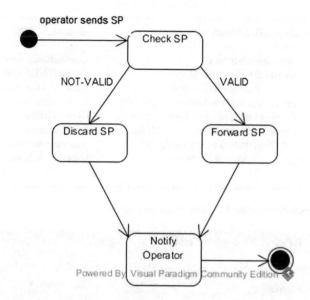

Fig. 13 The validation of operator setpoints

ically reconfiguring the organization by the system administrator (i.e. updating the organization address book), and the remote operator agent which is the operator inter-face to the system and he uses it to supervise and control physical control processes, they are monitored by the organization LS too. Figure 14 presents the internal JADE behaviors of a RA agent. A remote operator agent (RA) after starting (manually by the operator) searches for a certain required service (PLC) through the organization's local DF which then delivers the required service provider (CA) contact information to it if available. The dynamic control agents are created by the LS but the dynamic remote operator agents are created manually by the human operators as needed. In this prototype both of them are supervised by the LS. A simple GUI for remote operator agents is shown in Fig. 15.

The remote operator agent only needs to use the local yellow page service to find the provider dynamic agent of the required service or control process. But in case the required control process is located in a far environment and is under the direct supervision of another organization (control center) then an inter-organization dynamic reorganization is required, the organization which hosts the remote operator agent will try to interact with the organization which has the required control process to share it through the overlap dynamic behavior between the two organizations. If the overlap process is successful, the remote operator should be notified and be given the required information to be able to access the required control process and pursue the supervisory and control activities.

Agent Behaviors		Behavior Type
searchForService		OneShotBehaviour
subscribeToControlAgent		OneShotBehaviour
AchieveREInitiator		AchieveREInitiator
receiveActualVariables	(1000ms)	TickerBehaviour
receiveSetpointVariables	(1000ms)	TickerBehaviour
receiveNotifications	(1000ms)	TickerBehaviour
sendSetpointsToControlAgent		OneShotBehaviour
AchieveREInitiator		AchieveREInitiator

Fig. 14 Complete behavior list for a remote operator agent

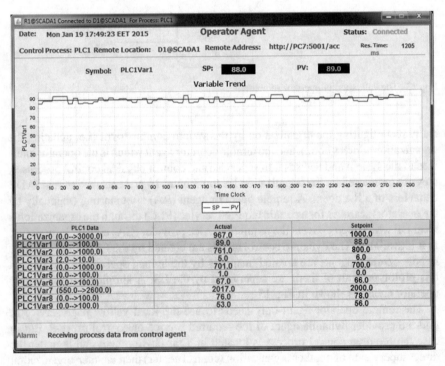

Fig. 15 The GUI designed for a remote operator agent

4 Performance Evaluation

Initially and just after starting up the system-to-be, only the GS is created by the administrator, then it reads the system configuration from an XML file which contains the control processes configurations such as process name, connection interface and the process variables required to be monitored. Another configuration XML-file

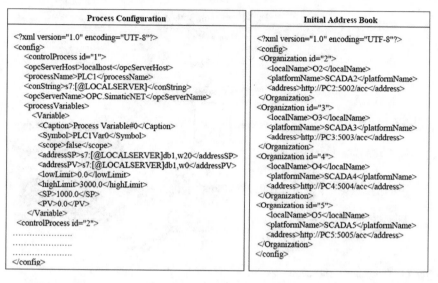

Process Configuration	Initial Address Book
| ```xml
<?xml version="1.0" encoding="UTF-8"?>
<config>
  <controlProcess id="1">
    <opcServerHost>localhost</opcServerHost>
    <processName>PLC1</processName>
    <conString>s7:[@LOCALSERVER]</conString>
    <opcServerName>OPC.SimaticNET</opcServerName>
    <processVariables>
      <Variable>
        <Caption>Process Variable#0</Caption>
        <Symbol>PLC1Var0</Symbol>
        <scope>false</scope>
        <addressSP>s7:[@LOCALSERVER]db1,w20</addressSP>
        <addressPV>s7:[@LOCALSERVER]db1,w0</addressPV>
        <lowLimit>0.0</lowLimit>
        <highLimit>3000.0</highLimit>
        <SP>1000.0</SP>
        <PV>0.0</PV>
      </Variable>
    <controlProcess id="2">
    .....................
    .....................
    .....................
</config>
``` | ```xml
<?xml version="1.0" encoding="UTF-8"?>
<config>
  <Organization id="2">
    <localName>O2</localName>
    <platformName>SCADA2</platformName>
    <address>http://PC2:5002/acc</address>
  </Organization>
  <Organization id="3">
    <localName>O3</localName>
    <platformName>SCADA3</platformName>
    <address>http://PC3:5003/acc</address>
  </Organization>
  <Organization id="4">
    <localName>O4</localName>
    <platformName>SCADA4</platformName>
    <address>http://PC4:5004/acc</address>
  </Organization>
  <Organization id="5">
    <localName>O5</localName>
    <platformName>SCADA5</platformName>
    <address>http://PC5:5005/acc</address>
  </Organization>
</config>
``` |

Fig. 16 The configuration files of control processes and acquaintances

is provided to the GS includes the initial information about its acquaintance organizations such as their names, addressees, etc. Figure 16 presents example segments of these XML configuration files. After reading the configuration XML-files, GS processes these configurations and creates the local supervisors (this happens by the interaction with the platform agent management system or AMS service), then it sends the configuration data to the created LS through a request message. The LS processes the configuration data and then creates the dynamic agents (specifically control agents), which will connect to the control processes using the OPC protocol. Then the LS agent assigns to each control agent its control process to supervise, in this prototype it is assumed that one control process is assigned to each control agent, but it is possible to assign more than one control process to a dynamic agent. After that, each control agent registers its assigned service to the local DF to tell other dynamic agents what services it provides.

Typically it is difficult for academic researchers to test large-scale industrial applications with physical work environments, and usually they have to create a simulation environment. To evaluate the performance of the proposed LCCIs architecture a simulation environment needs to be designed and developed. The architecture of the simulation environment is shown in Fig. 17. As demonstrated in the figure, the physical control systems (PLCs) are replaced with a software application developed in Visual Basic 6.0. The simulation environment consists of four main components described as follows. PLC is a type of control systems widely used in industry. The PLC connects to all the electrical sensors, devices, instruments in the industrial process and according to their states; it changes the output states to modify the current state of the industrial process according to predefined algorithms. OPC initially meant

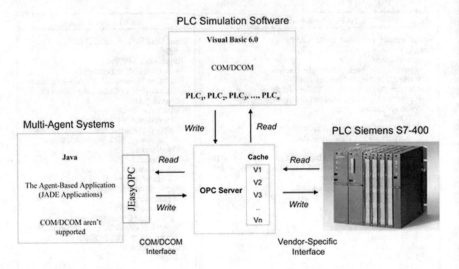

Fig. 17 Simulation architecture

Ole for Process Control, but after it becomes familiar for achieving interoperability among control systems, it was redirected to mean Open Process Control. OPC is open connectivity in industrial automation and the enterprise systems that support the industry. OPC is a popular process communication protocol that currently supported by a lot of PLC vendors. The OPC server uses a data pulling mechanism to continuously read process data from the PLC and store this data into its memory as it contains an internal memory to store the concerned process variables (V1, V2, ..., Vn) read from the physical PLC at adjustable rate (i.e., 400 ms). The software application was developed using Visual Basic 6.0, which is a familiar programming language developed by Microsoft, so it can access the OPC COM server directly without software adaptors. The simulation software application designed to contain n number of PLCs {PLC1, PLC2, PLC3, ..., PLCn}. Each PLC in the simulation software is designed to contain 10 process variables; each of them has a setpoint (SP) and a process value (PV). the proposed simulation software was adjusted to keep difference between a process variable SP and its PV in a predefined range (i.e., ±5.0), and the rate at which the simulation software changes the PV of the simulated OPC process variables is set to one second or 1000 ms.

The Agent-Based Java Applications component represents the proposed large-scale SCADA system which is developed using JADE agent development platform. Most MAS platforms are based on Java programming language, therefore to establish a connection between a Java agent and an OPC server, a Java-COM bridge or adapter is required. JEasyOPC (a Java-OPC adapter) was selected to interface a Java agent to OPC servers and of course the other available bridges can be used in a similar way. JEasyOPC is a Java OPC client that is now greatly enhanced. It uses a JNI (Java Native Interface) layer coded in Delphi. The current version supports both OPC DA 2.0 and OPC DA 3.0. JEasyOPC is free source and can be downloaded easily from

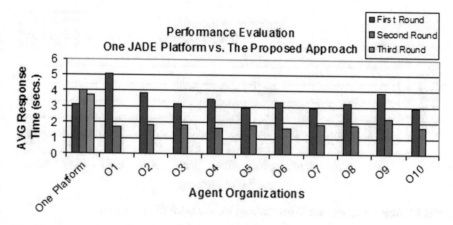

Fig. 18 One JADE platform versus multiple platforms aggregated based the proposed approach

the Internet. As demonstrated in the provided performance experiments, with the traditional JADE approaches there was no any improvement in system performance as time goes, and in contrast the system performance is getting worse as more agents are added. That is considered as a proof of concept for the adaptive behavior realized by the proposed approach. The realized paradigms with the adoption of the proposed NOSHAPE organizational model such as modularity, loose coupling, separation of concerns, decomposition, decentralization, adaptivity, flexibility are the key reasons of the success to provide the demonstrated improved performance compared to the traditional JADE approaches.

Figure 18 shows the average performance achieved with the adoption of the proposed approach compared to the traditional ACMAS approach with one JADE platform. With the traditional ACMAS approach (600 agents are running within one JADE platform) the average performance of the first, second, and third rounds are not stable and are getting worse with the increase of agents number. On the other hand, with the proposed NOSHAPE MAS approach the average performance of the system-to-be which is decomposed into 10 organizations are getting better (the second evaluation round gives a lower average response time compared to the first evaluation round for each agent organization). To explain this claim in numbers, the average response time of all system organizations in the first evaluation round is calculated to be 3.48376 s and that of the second round is calculated to be 1.81968 s, so a performance improvement of about 47% is achieved.

To compare, Fig. 19 shows the average performance when using the DFs federation provided by JADE platform. As shown in the figure, the second evaluation round average response time is getting worse for all system platforms. The reason is related to the long search time in the hierarchy formed by the JADE DFs federation. The average performance of the first round is 2.42052 s and the average performance of the second evaluation round is 2.75583 s, thus there is no improvement in the performance and in contrast it is getting worse. Table 8 summaries the comparative

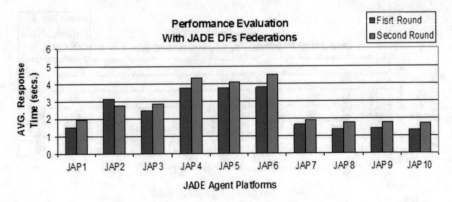

Fig. 19 Aggregating multiple JADE platforms based o JADE DFs Federation

Table 8 Comparative evaluation results

NOSHAPE MAS versus Present JADE approaches performance comparison

Development approach	First round (s)	Second round (s)	Performance improvement
10 NOSHAPE MAS organizations	3.48376	1.81968	47%
One JADE platform	3.1572	3.9864	No improvement and is getting worse
10 JADE platforms with DFs federation	2.42052	2.75583	No improvement and is getting worse

evaluation results obtained from experiment #1 (with one JADE platform), experiment #2 (with the proposed NOSHAPE MAS approach), and experiment #3 (with DFs federation).

5 Conclusions

Smart cities will not be possible without new innovative engineering approaches because of the expected complexity of the information, computation, and communication systems required for architecting them. LCCIs represent a challenging issues in terms of management, coordination, sustainability, resiliency, reliability, and autonomy aspects within future smart cities. Future smart cities should be designed to self-adaptive systems able to adapt uncertain dynamic work environments. Self-organizing multi-agent systems represent a novel paradigm for building large-scale complex systems with self-adaptive nature. This chapter presented a new approach for engineering the LCCIs of future smart cities based on self-organizing MAS able to self-organize themselves to adapt unforeseen situations based on a promising organizational model called NOSHAPE designed to enable large-scale MAS to dynamically reorganizing (restructuring) themselves as a response to changes in their objectives and/or work environments. Based on a simulation environment, a performance evaluation has been conducted that shows how the proposed approach is able to deal with the dynamic unanticipated work environment behaviors. Up to 47% of performance improvement has been achieved with the proposed approach compared to conventional organizational techniques in MAS (i.e. federation). Further future research is still required to address important issues of future smart cities and their associated LCCIs such as security, and resiliency.

References

1. Abbas HA, Shaheen S (2017) Towards a hybrid MAS organizational model: combining the ACMAS and OCMAS viewpoints. Int J Organ Collective Intell (IJOCI) 7(4):18–50
2. Abbas HA (2015) Realizing the NOSHAPE MAS organizational model: an operational view. Int J Agent Technol Syst (IJATS) 7(2):75–104
3. Esposito C, Di Martino C, Cinque M, Cotroneo D (2010) Effective data dissemination for large-scale complex critical infrastructures. In: 2010 Third international conference on dependability (DEPEND). IEEE, pp 64–69
4. Elhoseny H, Elhoseny M, Abdelrazek S, Bakry H, Riad A (2016) Utilizing service oriented architecture (SOA) in smart cities. Int J Advancements Comput Technol (IJACT) 8(3):77–84
5. Abbas H, Shaheen S, Elhoseny M, Singh Ak, Alkhambashi M (2018) Systems thinking for developing sustainable complex smart cities based on self-regulated agent systems and fog computing. In: Sustainable computing informatics and systems, (SUSCOM). Elsevier, Available online 17 May 2018
6. Elhoseny H, Elhoseny M, Riad AM, Hassanien AE (2018) A framework for big data analysis in smart cities. In: Hassanien A, Tolba M, Elhoseny M, Mostafa M (eds) The international conference on advanced machine learning technologies and applications (AMLTA2018). AMLTA 2018. Advances in intelligent systems and computing, vol 723. Springer, Cham
7. Darwish A, Hassanien AE, Elhoseny M et al (2017) J Ambient Intell Human Comput. https://doi.org/10.1007/s12652-017-0659-1
8. Boardman J, Sauser B (2006, April) System of Systems-the meaning of of. In: 2006 IEEE/SMC International conference on system of systems engineering. IEEE, p 6
9. Wu FF, Moslehi K, Bose A (2005) Power system control centers: past, present, and future. Proc IEEE 93(11):1890–1908
10. Jennings NR, Wooldridge M (2000) Agent-oriented software engineering. In: Bradshaw J (ed) Handbook of Agent Technology. AAAI/MIT Press, Cambridge
11. Abbas HA (2014) Exploiting the overlapping of higher order: entities within multi-agent systems. Int J Agent Technol Syst (IJATS) 6(3):32–57
12. Abdelaziza A, Elhoseny M, Salama AS, Riad AM (2018) A machine learning model for improving healthcare services on cloud computing environment. Measurement 119(2018):117–128
13. Darwish A, Hassanien AE, Elhoseny M, Sangaiah AK, Muhammad K (2017) The impact of the hybrid platform of internet of things and cloud computing on healthcare systems: opportunities, challenges, and open problems. J Ambient Intell Humanized Comput (Springer)
14. Elhoseny M, Salama AS, Abdelaziz A, Riad A (2017) Intelligent systems based on cloud computing for healthcare services: a survey. Int J Comput Intell Stud 6(2/3):157–188 (Inderscience)
15. Elhoseny M, Hosny A, Hassanien AE, Muhammad K, Sangaiah AK (2017) Secure automated forensic investigation for sustainable critical infrastructures compliant with green computing requirements. IEEE Trans Sustain Comput PP(99):1

16. Abbas HA, Shaheen SI, Amin MH (2018) Providing a transparent dynamic organization technique for efficient aggregation of multiple JADE agent platforms. In: Accepted to be presented in Aswan IEEE international conference on innovative trends in computer engineering (ITCE'18), Aswan, Egypt, 19–21 Feb 2018
17. Bellifemine F, Caire G, Greenwood D (2007) Developing multi-agent system with JADE. Wiley, New York
18. OPC Foundation (2010) OPC DA 3.0 Specification [DB/OL], 4 Mar 2010
19. OMG (2015) UML Specification Version 1.3. Object Management Group, Inc
20. Protégé, domain ontology editor, website: http://protege.stanford.edu/. Accessed 5/2018
21. Jade-Bean-Generator,http://protege.cim3.net/cgi-bin/wiki.pl?OntologyBeanGenerator. Accessed 5/2018
22. JEasyOPC, http://sourceforge.net/projects/jeasyopc/. Accessed 5/2018

Security Challenges in IoT Cyber World

Chintan Patel and Nishant Doshi

Abstract Internet of Things (IoT) has created revolutionary impact in the world of technology and social life of billions of people. "Things" in the IoT refer to the real world objects which gets cognitive and communicative capabilities with the help of smart sensors, cameras and other devices. Indeed, IoT is implanting its footprint in each domain (i.e. health care, transportation, electric grid, agriculture, retail, manufacturing) to make it smart. Like, smart health care uses smart wearable devices equipped with smart sensors and tracking systems to provide live connectivity between patient, doctor and hospitals. Similarly, intelligent transport provides quick response in the accident or other hazardous situations. Smart grid provides complete information about usage of electricity and controlling of power supply. Agriculture monitoring and irrigation of waters and fertilizer using smart agriculture can save lots of time and energy of the farmers. Smart retail and transport can help to track quality of food, to generate easy billing and to provide product recommendation to the customers based on their past shopping habits. Thus, IoT creates big comfort for the businesses, government and peoples in their work of day-to-day life. With the advancement in the comfort, IoT come up with many serious technological challenges. Reliable and secure implementation of IoT application is most important aspect for the long time adaption of IoT. Availability of internet connected tracking devices and environment capturing sensors keep track of personal life of people at the same time it transfer it via internet to the cloud. So to assure security triangle CIA (Confidentiality, Integrity and Availability) to the people is major challenge for the researchers and developers. Recent attacks using concept of ransomware, in which attackers was seeking bitcoins to enable blocked services has created big financial

This work was completed with the support of our Family, Friends and Guide.

C. Patel · N. Doshi (✉)
Department of Computer Science and Engineering,
Pandit Deendayal Petroleum University, 2nd Floor cubical area, E-Block,
Gandhinagar 382421, Gujarat, India
e-mail: doshinikki2004@gmail.com

C. Patel
e-mail: chintan.p592@gmail.com

© Springer Nature Switzerland AG 2019
A. E. Hassanien et al. (eds.), *Security in Smart Cities: Models, Applications, and Challenges*, Lecture Notes in Intelligent Transportation and Infrastructure,
https://doi.org/10.1007/978-3-030-01560-2_8

damage to many people. In this chapter, we have discussed major security challenges for the IoT, major cyber threats, attacks and remedies on various IoT parts and applications. Importance of role and requirement of light weight cryptography in the IoT is also discussed in this chapter.

Keywords IoT · Cyber security · Ransomware · Confidentiality · Integrity Machine learning · Industry 4.0 · Smart grid · Smart agriculture

1 An IoT Overview

In last decade, rapid evolution of internet has changed the life of people from ancient world to technology geek. Internet has changed the complete meaning of communication, From the pigeons to telephone and telephone to whatsapp. Mobile is the next revolution preceded by internet where People have started to communicate "from any where, "at any time" "with any one". Adoption of 2G, 3G, 4G, 5G, GPS, GPRS, Radio signals, satellite has changed the complete way of communication and also a cost of communication. Later on in 1999, Kevin Ashton in auto-id laboratory of Massachusetts institute of technology was working on the radio frequency identification (RFID) technology to prepare a communication application for proctor and gamble supply chain. As per mention in his article in the RFID journal, he has first time use the word "IoT" [1]. Later on many researchers have started to do think on this idea. CISCO says IoT as a internet of everything. There are many different way an IoT can define: "IoT is the network of interconnected objects which are uniquely addressable based on certain set of rules" [2]. International telecommunication unit defines Internet of Things (IoT) in its executive summary report as a "Any time connection (on the move, night, day, outdoor and indoor), Any place connections (at the pc, on the move, outdoors), Any thing connection (between pc, human to human, human to thing, thing to thing) [3]. IoT deployment involves many different entities. As a devices there will be smart sensors, actuators, gateways and various things and as a communication medium standardization, protocol and globally accepted architecture [4]. IoT can also be defined as a self configuring capable infrastructure based on certain standards and protocols, so "INTERNET" and "THINGS" can simply conclude IoT in to "Interconnected things".

After the internet deployment, connected number of devices had grown very fast. As per the predictions by statista, by 2025, total number of connected devices will be 75.41 billion [6]. Various sensors which are part of devices provides cognitive capabilities to the smart devices. Smoke sensor identify the smoke and send the signal to the controllers for further action, Temperature sensor senses the temperature. Sensor enables smart devices to hear, see, think, and perform task by using smart actuators and micro-controllers [7]. In the IoT, sensors are base of futuristic infrastructure. Capability expansions of sensor characteristics like range, detection, sensitivity, accuracy, linearity and precision creates strong foundation for IoT world. In the IoT, sensors are mostly connected with micro controllers or micro processors.

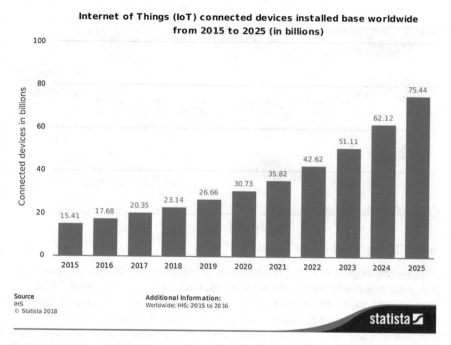

Fig. 1 IoT device statistics [5]

This controller or processors receives the data from thousands of connected sensors and forward that data to the fog devices (data mining enabled routers). Fog devices do the data processing and convert the unstructured data in to structured information. This information fog devices forwards to the cloud for the futuristic usage like decision makings and recommendation settings. Data stored in the cloud will be processed and knowledge generation will happen (Fig. 1).

1.1 An IoT Applications

IoT opens the door for smart applications development any many different areas and industries [8]. As per shown in Fig. 2, Different IoT application domains are city, transportation, retail, industry, agriculture, university, healthcare, grid logistics, mining, oil and gas industry and many more possible unexpanded domains. Use of machine learning in the in the industry 4.0 has created opportunities for the industries to expand globally by applying data mining, artificial intelligence in the machine functionalities.

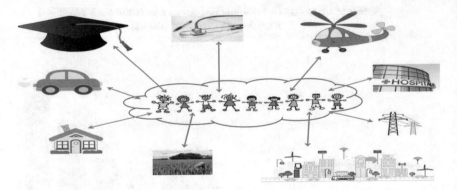

Fig. 2 IoT application opportunities

Smart Transportation and Logistics:

Intelligent transport system [9] defines as a "Millions of vehicles are connected with each other, with traffic control stations, with hospitals, with hotels, with fuel stations, with road network and mobiles to fulfill the major challenge of transport systems like enhancing the quality and comfort of travel, reduction of traffic, safety improvement and faster on road services by using various sensors and controllers connected with each other via sensors". ITS collects the information from the road, railway, air system and vehicles. This information are collected using sensors like fog sensors, motion sensors, air quality sensor, GPS location sensor and devices like CCTV. Radio frequency identification signals on road and vehicles also helps in smooth and light weight wireless communication. Collected data continuously forwarded via on road gateway to the traffic control system, if any accident happens than hospital and ambulance system will also get the information. People can get the advantage to track the bus or vehicles, to identify the traffic on particular route. ITS also helps to the first time traveler of City by providing information like near by hotel, hospital, places for visiting and fun, transportation medium available in city. Based on their traveling habits, IoT can provide the suggestions for best traveling facility available for them. As per mentioned in [9], probable number of death by road accidents by 2030 will be 3.54 millions, so ITS can help in the reduction of road death by its faster and efficient services. Major work in the ITS requires to focus on reliability of service and information or cyber security aspects of ITS. If hackers get the access of any service of ITS, than it can cause major damage for the traffic and safety of people. Big data and data analytics can play a major role in the decision making for the ITS [10]. Another application of IoT is logistics services, Major challenges faced by logistic industry is to monitor the quality of products, specially in the food and dairy products. RFID and NFC (Near field communication) plays important role in logistic monitoring. It plays role in row material purchasing and monitoring, manufacturing, packaging, data collection, package transportation, package delivery, selling and after sell services. Various sensors temperature, smell, humidity available in the container of product do the monitoring as well as maintain the required temperature and humidity in side the container for the very long distance delivery. These all the sensors will be connected with on road gateways. A sender can do the remote monitoring of quality of product, location of product and so on. So IoT can play the major role in transportation and logistics service enhancements [2].

Smart Healthcare:

Health care industry came up as a fastest adopting industry for IoT services. We can understand requirement of smart health care by the report of world health organization which indicates that Out of 1,00,000 people, due to ischaemic heart disease 119, due to Stroke 85, due to lower respiratory infections 43, due to chronic obstructive pulmonary disease 43, due to Trachea/bronchus/lung cancers 23, due to diabetes mellitus 22, Alzheimer's disease and other dementias 21, due to diarrhoeal diseases 19, due to tuberculosis 19, and due to road traffic accidents 10 deaths occurs. Major issues in these all the deaths are late diseases detection and delay in health care services. Patients which gets major stroke may not aware that they already have minor strokes earlier. So on time treatment is the major challenge in the health care. Smart health monitoring systems [11] can have various functionalities like patient monitoring system, medical inventory monitoring system, ambulance availability and ambulance facility monitoring, doctor and medical staff availability tracking, patient data collections and storage, identification and access control of patient data [2]. Other functionalities of smart hospital [12] have staff tele monitoring and collaboration and laboratory maintenance. Major sensors deployed in hospital, with patient body, ambulance vehicles are ECG, EEG, blood pressure, temperature, heart bit rate, feet. These all the sensors will collect the data from patient body, hospital and staff and forward this data to the central health data base. Remote patient monitoring will enable doctors to use this data base for to track the health of patient. Doctors can have continuous updates on major changes in the body of patient. So this can help to save the life of patient.

Smart Home:

Home becomes smart home when it is equipped with 3 major technology [13],

1. Internal network of wired and wireless devices.
2. Intelligent routes and processors to enable remote access and manage featured systems.
3. Home automation using home devices connected with local and remote devices.

Smart home can have smart lighting which changes the frequency of light based on the availability of sun light and person requirement. A smart coffee maker connected with the alarm of person will prepare the coffee for the person before he wake up or as per need [2]. Smart air condition equipped in home will maintain the room temperature. It is also connected with the person mobile or car of the person. Whenever car comes in to certain range, AC will start functioning. Motion sensor based alarming system will send the regular updates to the person mobile about the home situations. Smart door lock enables person to access, monitor and manage the door locking from remote place. Smart refrigerator will do the monitoring of inventory quantity and do the communication with the inventory suppler. Smart fire detection system will detect the fire and send the signals on the mobile of the home owner. Objective of smart home research is to increasing the comfort of daily life [13]. IoT based energy consumption plays major role in the reduction of energy consumption by controlling and monitoring unnecessary demands of energy. Context aware computing enables human activity identification and event automation. Context aware provides location, identity, activity and time to make home as a smart home [14].

Smart grid:

Grid means a electricity system. Basically grid have 4 different operations, operations are electricity generation, electricity transmission, electricity distribution and electricity control. Smart grid will have capabilities of efficient power delivery and response for wide range of conditions and events. Any where an event will be generated at any time. Smart grid will provide instance solution for that. Advanced capabilities which smart grid will have compare to existing grids are, digital and two way communication, distributed power generation, maximum use of sensors, self monitoring and self healing, pervasive control and customer choices. In the smart grid, energy transmission will be two way, by means we can say that power generation will not be performed only by government but also by user. User may generate extra energy using solar plate and sell to the government. Smart information subsystem provides information metering and measurement using smart meter, sensor, phasor measurement unit, and information management using data modeling and information analysis, integration and optimization [15]. Another challenges in the smart grid are [16] active and run time power regulations, voltage regulations, dynamic topology optimization and dynamic infrastructure. Smart meter provides connectivity between smart grid and smart home [17]. Smart home equipment will be directly connected with smart meter. Smart meter enables controlling of heating, ventilation and air condition inside home. Smart meter is intermediatory between electricity station and user home. Smart meter enables energy consumption data analysis, also deals with data acquisition, data transmission, data processing, and interpretation. Smart meter with water meter, heat meter and gas meter provides energy saving and services. It provides more timely and precise billing, cost reduction, avoiding electricity fraud and power peak control [18].

Smart agriculture

It is another major application of IoT which enables remote farm monitoring, live land record maintainable, land quality management, fertilizer requirement monitoring, smart irrigation system, crop recommendation system. A successful deployment of smart agriculture will increase the income of farmer, farmers will be able to sell products directly to the consumers. It will also help government to distribute loan, fertilizer and other benefits based on live land monitoring. It help easy calculations for loss happened due to disasters.

Smart retail enables retail industry to improve their businesses. Smart billing system will reduce lots of efforts of employee and time of customer. Using RFID and NFC, retail industry can easily track inventory. Internet enabled tags learns customer purchase habits and recommend the location of product inside the big mall. Customers may get live and recent feedback about the quality of product. IoT improve the quality of product and production process by internet enabled decentralized monitoring and control algorithms in the **smart industry**.

Effective secure and reliable data acquisition in the IoT improves the demand supply management. Data generated in the cloud will be big data and analytics of big data improves quality of decision making and understanding of complete business [19]. Sensor devices deployed in the ground creates foundations for the data generation and IoT system. Major challenges, in which research communities are facing major issues are sensor capability enhancement, congestion and collision free reliable communication protocols, interlinking of applications and ground level standardization and most important issues is the **to provide security** to the complete IoT world from the free handed technology giants **Cyber Hackers**.

1.2 An IoT Architecture

Till today many different organization like ITU, IETF, IEEE and different researchers have proposed tentative reference model for as an architecture of IoT, but still none of the proposed architecture got world wide acceptance to become reference model for IOT, In 2014, CISCO had given a 7 layer architecture. Seven layer architecture of CISCO was consisting edge nodes, communication, edge computing, data accumulation, data abstraction, application, and user and centers [20]. Many other researchers has given 3 layer architectures and 5 layer architectures, 3 layers architecture are proposed in [2, 21, 22], 4 layers architecture discussed in [8] while 5 layers architectures are proposed in [7, 23]. In [24], author has done well mannered survey about all the architectures for IoT proposed till date of publications. Over here we will discuss 4 layer architecture 3 by the view points of basic functionalities, devices, protocols used and possible security threats for that layer (Fig. 3).

1. **Perception or physical layer**:

   - **Functionalities**:
     - Basic functionality of this layer is to deploy the sensors, actuators, RFID readers and sensor gateway on the field.
     - It also focus on collecting environment data like location, temperature, pressure, humidity, motion, air pressure, pollution level and also collecting data from humans and machines for certain application [7, 8, 23].
     - Assigning universally accepted unique identification number to each devices.
     - Focus on capability enhancement of various IoT sensors in terms of cost, size, energy consumption, resource, communication and security.
     - Handling scalability and heterogeneity of devices and data, Reliability of communication.

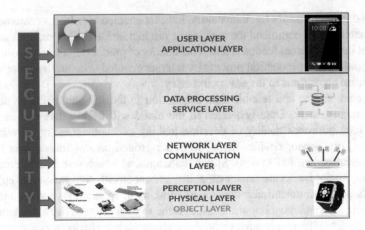

Fig. 3 IoT layered architecture

- **Devices used**:
  - Sensors, actuators, RFID readers, Mobile devices, Micro controllers, Micro processors [7].
- **Protocols used**:
  - RFID, NFC (Near field communication), ZigBee, Bluetooth low energy, Z-Wave, Infrared, and Wireless fidelity, IEEE 802.15.4 [7].
- **Security challenges**:
  - Physical security of devices deployed by animals, environments and some time humans [6]
  - Cyber security of device management, device behavior. Confidentiality and integrity of data transmission [6].
  - Authentication and access control of devices [6].

2. **Network layer**:

- **Functionalities**:
  - Data collection from physical or perception layer [4].
  - Data processing by using fog devices and forwarding data for further processing like decision making and high level services [4].
  - Forwarding control signals to the ground level sensors.
  - Network layer of IoT is having similar functionalities like OSI model, it performs long distance communication. additional functionality includes connecting smart things, network devices, and cloud servers [22, 23].
  - Taking care of routing, Discovering and mapping dynamically happening changing in network and devices topologies. Device role controlling, taking care of network energy and efficiency, handling quality of service in network [8].
- **Devices used**:
  - Fog devices, Routers and Gateways.

- **Protocols used**:
  - 2G, 3G, 4G LTE (Long term evolution), 5G, IPv4, IPv6, 6LoWPAN, Wireless fidelity, WI-Max [7].
- **Security challenges**:
  - Access Control of various networking devices like fog and routers, Accountability and Non-Repudiation of received and transmitted data, Authentication between sensor devices and routers [25].
  - Denial of service is the major challenge for this layer. If attackers successfully deploy flooding attack and results in to DoS than it can cause complete damage of sub system as well as eco system.

3. **Data processing layer**:

- **Functionalities**:
  - Major functionalities of this layer includes data mining and data processing of collected data in cloud. It's key enabler of applications and services in IoT [7, 8, 23].
  - Perform data analytics, graph generation, solution generation, recommendation generation.
  - Generating control signal, alerts, messages for the application layer users.
  - Perform the interlinking of application's common needs and provide services based on that [7].
  - this layer basically performs service discovery, service compositions, service trustworthiness and service API [8, 23].
  - Availing on demand services any time required by any one from the any place.
- **Devices used**:
  - Big data processors and servers, fiber optics, routers.
- **Protocols used**:
  - IPv6, TCP (Transmission control protocols)
- **Security challenges**:
  - Data Anonymity should be maintained and Data Privacy of users and devices should be taken care, Auditability of data by sensors and devices, Trust management, Confidentiality of stored data.

4. **User layer**:

- **Functionalities**:
  - This layer performs "User's ease of use IoT".
  - Preparing Many user oriented applications that collect the access based data from the data processing layer.
  - This layer is intermediator between user, IT and IoT.
  - Taking care of heterogenous private and public applications are running.
  - Domain oriented application designing Ex, Smart patient tracking system in Health care domains.
- **Devices used**:
  - Mobiles and computers.

- **Protocols used**:
  - MQTT (Message queuing telemetry transport), COAP (Constrained application protocol), XMPP (Extensible Messaging and Presence Protocol).
- **Security challenges**:
  - Confidentiality and Integrity of data and signal transmission, Authentication and Access Control between user application and cloud storage, DoS attack by hackers can lead user to lack of support. DoS attack in health care service application may cause death of person.

2 Cyber Security : An Overview

If you have the data, in the cyber world, you can do any thing. Most of the global companies believe that for them it is very easy to prepare the manufacturing plant but it is very very difficult to protect their systems from the cyber attackers. One side companies have global competitors who are increasing the production quantity and quality by adopting automation. Most of the owners of companies now can do the monitoring, and analytics of the companies production, employee management, customer management, row material purchasing and other task of plant using their mobile. Industry 4.0 [19, 26] has changed the world of manufacturing and industrial expansions. Most of the global companies are adopting IoT for their cost reduction, production quality and quantity improvement, ease of business management. People services and government functionaries are also focusing on technological adoption for the comfort of people. Smart Home [13, 14], intelligent transport system [9, 10], smart health [11], smart grid [15, 16] are the famous examples of this technological revolution. But at the same time cyber threats and challenges are increasing day by day. Major cyber attacks in the internet was able to affect infrastructure and financial costing. But in the IoT, cyber attacks can damage the life of peoples [27], can cause major threats in the functionaries of general public. So With the technology, Cyber attacks and cyber hackers are also expanding its presence in the technological world.

2.1 Security Parameters and Attack Terminologies

2.1.1 IoT Security Parameters

Confidentiality, integrity and availability are fundamental security parameters in the cyber world [28, 29]. While IoT security constrained can be divided in to 8 major parameters [25] like confidentiality, integrity, availability, accountability, auditability, trustworthiness, non-repudiation, privacy. We have define octagon 4 of 8 major parameters required for the complete security of the IoT.

1. **Confidentiality**:

   - Confidentiality can be defined in simple means as a "Hiding the information from the persons who are not authorized to read it". Confidentiality of private data like mobile numbers, passwords need to be secure in the traditional internet [30].

- In the IoT, devices collects the various sensitive information from the users, In the smart health care, medical devices will collect the personal information like bit rate, pulse information, body temperature [25]. In the smart home, person location, identity and other details like availability will be tracked. So all the information must be confidentially transmit from sensor devices to cloud storage.
- Confidentiality can be achieved using various symmetric and asymmetric encryption algorithms like RSA, DES. For the IoT, resource constrained devices are there, so various algorithms like Elliptic curve cryptography based encryption algorithms can be useful.

2. **Integrity**:

- Integrity can be defined in simple means as a "Preventing unauthorized user from modifying single bit of data in the communication".
- In the IoT, Integrity concerns becomes critical when it comes to modification of information in which human's insulin pumps and oxygen supplier type systems are going to function. Single bit change in the IoT can change the complete meaning from "True" to "False" and "Yes" to "No" [25, 30].
- Integrity in the financial transaction and industrial functioning also requires to ensure for the smooth functioning.
- Integrity in the security can be assured using various functions like check sum, one way hash function algorithms like SHA-1, SHA-2.

3. **Availability**:

- Availability can be defined in simple means as a "All the services running in the complete system will be available at any time to any authenticated entities [25]".
- In the IoT, Availability plays important role in terms of resources availability when ever it requires. Smart grid should control the power system in such a way that when ever user requires high power voltage. Smart home must ensure that automated air condition will cool the room when ever temperature will be above some threshold [30].
- Availability in automated payment system in the smart retails ensures the working of payment gateway when ever required.
- Availability in the security can be assured using parallel resource availability if any other fails.

4. **Trust and Privacy**:

- Trust management in IoT [31], can be defined in a simple means as a "Trust in IoT is ensuring that peoples and devices involved in IoT system accept the services and information with full faith and confidentiality [25]".

- In the IoT, Trust management involves reliable data collection, reliable data fusion and mining and enhanced user privacy. Successful trust creation in IoT ensures identity trust and Quality of IoT Services.
- Trust management involves behavior based trust control, reputation based trust control, Fuzzy approach to trust based access control [32].
- Some of the parameters that are used for the calculation of trust are frequency of answers, consistency of answer, physical proximity, common goals, common ecosystem, history of interaction, availability, common communities [32].
- Privacy in the IoT can be defined as "Not a single bit of information collected of person will be relived with any one without the permission of the person". Privacy preserving can be assured using various attribute based and anonymity based encryption algorithms.
- In the IoT, Most of the sensors which are part of public services will collect the many personal information. But to whom that personal information should be shared, must be decided by the person himself.

5. **Authentication and Access control**:

- Authentication means "Both the communicating parties are sure that they are communicating with the authenticate counter part". Successful authentication mechanism ensures the confidentiality, integrity and availability [30].
- Ensuring authentication in the IoT becomes critical due to heterogeneity of number of devices involved in the IoT [33]. Each device transmitting to data or want access of other device, each user who want an access of sensor via gateway need to authenticate himself.
- Authentication in the IoT requires low cost, low power schemes due to resource constraints. Light weight cryptography [34, 35] provides solution for this issue.
- Access control means "Only authenticate users, devices have access of other device or persons data and control".
- In IoT, Access control schemes can be divided based on arbitrary access control, mandatory access control, and role based access control [33]. Many researcher are working on Elliptic curve cryptography based schemes due to resource constrained devices and small key size of elliptic curve [34].

6. **Accountability**:

- Accountability in the IoT means "Who has generated data and who have processed the data will be clearly defined" [25].
- Accountability assurance can help in the heterogenous IoT environment that which device has generated which data and which device processed which data. It is an ability to hold users responsible for their actions [25].
- Accountability can be assured using proper identification mechanism for devices. Most of the devices in ground level will do communication using RFID technology or NFC technology, while edge level devices will get IPv6 based IP Address as a identification.

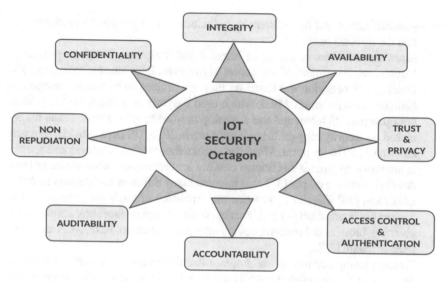

Fig. 4 IoT application opportunities

7. **Auditability**:

   - Accountability in the IoT means "We must be able to perform consistent monitoring of occurrence of each event in the system" [25, 30].
   - In the IoT, for the ground level device actions can be monitored by keeping various tools like wireshark at fog devices or data collecting devices.

8. **Non repudiation**:

   - Non repudiation in the IoT means "system must ensure weather the event was occurred or not".
   - Non repudiation can be assured using strict policy enforcement regarding event occurrence [25] (Fig. 4).

2.1.2 Cyber Attack Terminologies

1. **Attack vectors**: Attack vectors are the path that attacker choose for the attack. Major attack vectors that need to take care in the IoT system are backdoor, low encryption level, weak passwords.
2. **Viruses**: Viruses in the human body replicates it self inside the living cells and create the harm. Similarly viruses in the computer system viruses enters via infected program or infected host and later on replicates in side it. IoT is the network of devices which are developed on no common platform, no common

standardization and by different manufacturers. So there are more chances to find the infected host.

3. **Malware**: Malware is the set of harmful and dangerous computer programs. Viruses, worms, trojans all are subset of malware. Each subsets have some different set of behavior and based on they are divided in to various categories. Famous malware attack Mirai, which used busybox as a attack vector. It finds out open port 23 telnet and use a weak password to gain access. It has fist attacked on security blogger Brian Kreb's site in 2016, US DNS provider dyn was attacked in 2016 by mirai. Mirai has attacks thousands of cameras and routers in netwroks. Spambots and bricker bots are another recent malware attacks [36]. Another malware reaper [37] also infected many different IoT devices in 2017. CISCO in [38] identified most common malware grounds are potentially unwanted application (PUA) and binaries, trojan droppers, facebook scams, links, phishing, browser redirections, downloader, android trojan, FakeAvCN and HappyJS malware [38].

4. **Worms**: Computer worms are program that replicates it self within a network and exploit the complete network. It enters in to the system due to loopholes in the security policies of organizations. Hajime worm discovered in October 2017, which was developed on peer to peer network and more powerful than Mirai [39]. Another worm created by hacker janitor, Bricker bot has damaged millions of IoT devices as per claim mentioned in [40].

5. **Trojans**: Trojan is harmful computer program which enters in to system with the help of trusted applications. Major spreading of trojan happens via social engineering, backdoor, drive by download. For the spreading of Mirai attack, attackers have used social media, email based trust creation and downloads.

6. **Ransomware**: Ransomware are the attacks created with the intention of bribing money. Pattern in most of the ransomware are like attackers encrypt the files in the victim system and removes it self from the system. Later on it analyze the encrypted files. Out of which it identify the important files. Later on it sends a message with the website address to victim regarding procedure for decryption of files. Over there he compulsory have to pay money or crypto-currency for to get files back in plain text. Wannacry ransomware attack founded in may-2017 worldwide which encrypted files of many user and ask for $300–$600 via crypto-currency bitcoin. It was estimated that wannacry has damaged files of 300,000 computers across 150 nations [41]. Most of the systems which was affected by wannacry was running Microsoft windows operating system. Another ransomware Badrabbit founded in Russia and ukraine. it has asked user to pay 0.05 bitcoin or $276 in the exchange of file. Major motive of the ransom where attacks are money. In the IoT, there are lots of personal data collected at sensors and control devices. This data becomes critical point for system admin and open ground for the ransomware attackers [42]. In the IoT ransomware can attack the many internet enabled medical devices like pacemaker, insulin pump, nerve simulator. Cryptolocker and cryptowall are the famous family of ransom which can used to generate approximately $325 million. Ransomware will be primary threat for the IoT attacks [43].

7. **Adware**: Adware attacks in the cyber security used to create open ground for entering worms, trojan, or ransomware. Generally adware is not an attack but its advertising softwares or programs which generates parallel advertise when user click on particular link. In the IoT, use of social engineering sites like facebook and twitter can be used to add malware in IoT eco-system. Successful adware attack run on machine of owner can damage the working of complete factory. As per mentioned in [38], last year 75% organizations was infected by adware. Adware injects exploit kits, change browser and OS Settings, Gain full host controls, track users by location, capture personal data and credentials, gain full host control.

8. **Zero day vulnerabilities**: Zero day vulnerabilities are attacks which are founded first time in the system. Every week new zero day vulnerabilities was came in picture in the 2015 year. Adobe flash was the the biggest victim of Zero day vulnerabilities [43]. Zero day exploits can be used via malware ads. Watering hole attacks which uses compromised website to launch Z.D.V. when the particular attacks are found. Total 54 new vulnerabilities was founded in the year 2015 [44]. In the IoT, continuous behavior monitoring will be require to secure against Zero day vulnerabilities and credential attacks. Identifying zero day vulnerabilities in the IoT will be very critical due to heterogenous data and devices. It is predicted that every year there is 125% increase in the number of Zero day vulnerabilities [44].

9. **Stuxnet**: Stuxnet contains 3 interlinked attacks. First one is worm which executes routines, A linkfile which executes worm and a rootkit which hides all malicious activities. Stuxnet in 2010, was discovered by kaspersky lab. Stuxnet targets programmable logic control of the automated machines and targets the Supervisory control and data acquisition system. Stuxnet can damage the power plants, manufacturing industries, nuclear industries, gas industries, mining industries via attack on their SCADA system. Stuxnet attack on Iran nuclear program was the biggest stuxnet attack up to so far. In 2014, other attack on the German steel mill also warned about the security importance in the implementation of industry 4.0. Another trojan Lazoik which created with the aim to attack on energy sector [44]. Another stuxnet type malware Irongate was came in picture in 2016. which aimed to attack on industrial control system. Irongate in the system replaces valid dynamic link library (.dll) by malicious copy to allow particular malware to target on control system configuration [45]. In the IoT, Industry 4.0 mainly focus on the automation using internet connected PLC. So there will be no surprise if stuxnet will emerge as major threat for the IoT systems.

10. **Exploit toolkit**: Exploit kit are the freely available softwares which can be used by cyber criminals to attack on particular services like flash, Microsoft silver light, and java. Most of the exploit kits are run on the web servers. Exploit kit in the first step identify the security loopholes available in the software or website. Due to lack of system up-gradation and updating, 60% systems suffers from older exploit kits also. Exploit kit contains two parts, one is the control panel which allows attacker to generate faster services for the users, and services which allow attacker to monitor and infect the visitor or user of that service.

Ransomware and downloader was major payload for exploit kit [46]. Famous exploit kits are angler, nuclear, neutrino, RIG [38]. Exploit kit can be used for to implement zero day vulnerabilities [44]. In the IoT, Due to availability of many devices and controllers, to implement exploit kit becomes easy. Exploit kit can be used to launch malwares for activity analysis, location tracking, manufacturing monitoring, personal data collection and so on.

11. **Botnet**: Botnet are the remotely controlled infected systems to perform specific attack. A bot is a program that runs without human interaction in such a way that user can not understand it [29]. Mirai [39] and Reaper 63 are the IoT botnet which are responsible for the most effective denial of service attack of the entire attacking history. Reaper was having 9 most famous different vulnerabilities starting from loop hole detection to explosion. It has successfully infected netGear, linksys, GoAhead, and Avtech [47]. Mirai botnet support vector includes SYN-Flooding, UDP-Flooding, Valve source engine, query flooding, GRE-Flooding, ACK-Flooding, pseudo-random DNS label also known as DNS, Water Torture attacks, HTTP GET attacks, HTTP POST attacks, and HTTP HEAD attacks. Mirai supported access to TCP port 23,2323 and 103 [48].

12. **Man in the middle**: Man in the middle attack can be performed by both the type of attacker, active and passive. Passive attackers use man in the middle attack to monitor and gather the information between two parties with the motive of data collection or just for fun shake!!!. Active attacker do the modification in the communication bits. Motivation can be trust reduction, privacy leakage or any thing else. Now a days every communication are encrypted via secure algorithms and hash functions. To perform man in the middle attack becomes critical but due to big open ground in the IoT. Man in the middle attack also can play a role in collection of personal information like age, blood group, pulse rate, location and so on. Recently noticed man in the middle vulnerability in the chrysler corporation's connected jeeps and samsung's smart refrigerator opened the possibilities for man in the middle attack in IoT attack. An attacks mentioned in [49] using LightwaveRF and Belkin WeMo smart hub on connected home devices successfully shows that how man in the middle attack can be successfully deploy in the smart home devices using network sniffer tools like wireshark.

13. **phishing and spoofing**: Phishing attacks in which attacker aims to collect some confidential information via E-Mail spoofing or any other trusted entities. IoT Devices will be mostly automated and they will be working on request response based mechanism. It will be very difficult to provide intelligence context aware computing to filter the bad request and good request. Proper and strong authentication and identification mechanism will be required to implement at each and every potion of devices to secure from attacks like phishing. Cloud computing mechanism and device based communication in which credentials are dependent on the identity of the devices will major threat for the spoofing attacks.

14. **Side channel**: Side channel attacks are security threat in which attacker try to collect the information during the implementation of particular system. it collects the various measurement parameters and its value. Collected information will be applied in the cryptanalysis process. Side channel attack show how

cryptographic algorithm is implemented. It may not be bother about which cryptographic algorithms is implemented. and that is why side channel attacks are called as implementation attack. Side channel attacks takes benefits of co-relation between physical measurement taken for computation like time complexity, battery consumption, number of rotation occurred, frequency of signals generated. Involvement of many physical devices in the IoT which collects the different set of information and transmit it. Leakage of electro magnetic signal from the medical device can leak the important information about the health of patient. So side channel attack can help to get information about device and environment in which devices are working. Some of the famous side channel attacks are timing attack, fault attack, acoustic attack, visible light attack, error message attack, cache based attack [50].

15. **Denial of Service, distributed denial of service**: Denial of service attack is most famous attack in the cryptographic history. In DoS attack, attacker motivation is to reduce the service availability either partially or fully for the victim either for temporary timing or permanent. To implement DoS attack, attacker try to flood victim attack either by echo request or by sending unrequested files. Basic difference between DoS and distributed DoS attack is, in DoS attack, source of attack is mostly single machine while in DDoS attack source of attack is multiple machines. Telephone denial of service attack prevent the customers from calling in and out. Attack on BBC server in 2015 which blocked many famous service of BBC including iplayer considered one of the biggest DDOS attack. ICMP flood attack, TCP Syn flood attack, ping broadcast denial of service attack, Radio-frequency prowl denial of service attacks are one of the famous DDOS attacks. Botnet provide best services to implement DDOS attack. CCTV Cameras and sensors will be favorable victims of IoT attackers [44]. Some famous dos attack through which IoT devices can suffer are battery draining attack in which motive of an attacker to reduce the battery of IoT device. In the sleep deprivation attack an attacker send the request which looks legitimate so it consume the energy for node to look in the requests. Stuxnet attach which impacts the industrial attack is also one of the example of denial of service attack. Many devices in the IoT will do the communication using radio frequency channel, so there is big possibility where attacker try to jam the Radio frequency channel by using back doors [25]. As per mentioned in [51], in 2017, number of DDOS Attack increased 91%. **"The potential scale and power of IoT botnets has the ability to create Internet chaos and dire results for target victims said by stephensons"**. Figure 5 shows the general functioning of distributed denial of service attack. Ransom DDOS attack inspires the attacker to ask for money after performing attack to stop attack. As per reports of kaspersky lab, ransomware dos attack performed on south korean banks demanded $315, 000. On 21st October 2016, attack on DNS service provider Dyn was happened in which bogus traffic was transmitted on the server. Which impacted many other services including twitter because they all was using DNS service from Dyn. Similarly in IoT also if attacker able to attack on the control system which manages sensor devices. Than it can damage complete eco-system.

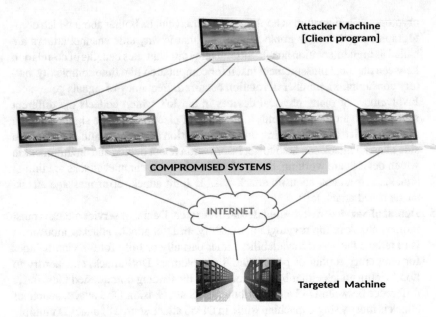

Fig. 5 DDOS attack [52]

16. **RFID based**: Radio frequency identification technology will be future for iden-
    tification of small billions of devices which are pillar of complete infrastructure
    for IoT. RFID Contains two components. One is RFID Tag and other is RFID
    Reader. Three variants of RFID available based on frequency range. short range,
    mid range and long range. Most famous attacks that are possible on RFID are:

    - Passive monitoring of data transmitted between sender and receiver.
    - Scanning the identity of sender and receiver via entering as a valid node and
      later on start behaving as a false node like node replication attack [25].
    - Jamming the RFID frequencies and perform denial of service attack.

    Inventorying RFID tag can provide valuable information. Ex. Electronic product
    code have to parts. First one is manufacturer code and product code. So intruder
    can easily obtain the information about the product. Another attack that is pos-
    sible in IoT is physical attack on devices and RFID tag. Physical attack can
    be possible due to environment or by mistake and also by specific intensions.
    Tag cloning attack in which attacker used the clone of valid tag to get access of
    financial transactions or valid systems. Eavesdropping of many different RFID
    tag can help to understand the types of devices, types of environment and types
    of incident which can help for the future analysis [25].

    So, in this chapter we have discussed about an overview of IoT, IoT architecture
    and cyber security terminologies and it's IoT point of view challenges.

Acknowledgements We are thankful to editors of this book who have taken their keen interest in our chapter proposal and support us in each stage. We also thankful to reviewers who given their comments for improvement of this chapter. Many thanks to all well wishers and supporters.

References

1. Kevin A (2009) That 'Internet of Things' Thing. RFID J. Eprint: http://www.rfidjournal.com/articles/pdf?4986
2. Atzori L, Iera A, Morabito G (2010) The Internet of Things: a Survey. Comput Netw 54(15):2787–2805. ISSN: 1389-1286. https://doi.org/10.1016/j.comnet.2010.05.010
3. Jamoussi B (2005) Executive summary on "The internet of things". eprint: https://www.itu.int/dms_pub/itu-s/opb/pol/S-POL-IR.IT-2005-SUM-PDF-E.pdf
4. Li S, Xu LD, Zhao S (2015) The internet of things: a survey. Inf Syst Frontiers 17(2):243–259. ISSN: 1387-3326. https://doi.org/10.1007/s10796-014-9492-7
5. Statista (2016) IOT Statistics by statista https://www.statista.com/statistics/471264/iot-number-of-connected-devices-worldwide/ (visited on 01/13/2018)
6. Carsten M (2017) Security and privacy in the internet of things. J Cyber Policy 2(2):155–184. eprint: https://doi.org/10.1080/23738871.2017.1366536
7. Al-Fuqaha A et al (2015) Internet of things: a survey on enabling technologies, Protocols, and Applications. IEEE Commun Surveys Tutorials 17(4):2347–2376. ISSn: 1553-877X. https://doi.org/10.1109/COMST.2015.2444095
8. Xu LD, He W, Li S (2014) Internet of things in industries: a survey. IEEE Trans Ind Informatics 10(4):2233–2243. ISSN: 1551-3203. https://doi.org/10.1109/TII.2014.2300753
9. Alam M, Ferreira J, Fonseca J (2016) Introduction to intelligent transportation systems. In: Alam M, Ferreira J, Fonseca J (eds) Intelligent transportation systems: dependable vehicular communications for improved road safety. Springer International Publishing, Cham, pp 1–17. ISBN 978-3-319-28183-4. https://doi.org/10.1007/978-3-319-28183-4_1
10. Zhang J et al (2011) Data-driven intelligent transportation systems: a Survey. IEEE Trans Intell Transport Syst 12(4):1624–1639. ISSN: 1524-9050. https://doi.org/10.1109/TITS.2011.2158001
11. Baig MM, Gholamhosseini H (2013) Smart Health Monitoring systems: an overview of design and modeling. J Med Syst 37(2):9898. ISSN: 1573-689X. https://doi.org/10.1007/s10916-012-9898-z
12. Smart hospitals security and resilience for smart health service and infrastructures. In: (2016). Eprint: https://doi.org/10.2824/28801
13. Alam MR, Reaz MBI, Ali MAM (2012) A Review of Smart homes past, present, and future. IEEE Trans Syst Man, and Cyber Part C (Appl and Rev) 42(6):1190–1203. ISSN: 1094-6977. https://doi.org/10.1109/TSMCC.2012.2189204
14. Khan M, Silva BN, Han K (2016) Internet of Things Based Energy Aware Smart Home Control System. IEEE Access 4:7556–7566. ISSN: 2169-3536. https://doi.org/10.1109/ACCESS.2016.2621752
15. Fang X et al (2012) Smart grid : the new and improved power grid: a survey. IEEE Commun Surveys Tutorials 14(4):944–980. ISSN: 1553-877X. https://doi.org/10.1109/SURV.2011.101911.00087
16. Mei S, Chen L (2013) Recent advances on smart grid technology and renewable energy integration. Sci China Technol Sci 56(12):3040–3048. ISSN: 1869-1900. https://doi.org/10.1007/s11431-013-5414-z
17. Benzi F et al (2011) Electricity smart meters interfacing the households. IEEE Trans Ind Electron 58(10):4487–4494. ISSN: 0278-0046. https://doi.org/10.1109/TIE.2011.2107713
18. Alahakoon D, Yu X (2016) Smart electricity meter data intelligence for future energy systems: a survey. IEEE Trans Ind Informatics 12(1):425–436. ISSN: 1551-3203. https://doi.org/10.1109/TII.2015.2414355

19. Haverkort BR, Zimmermann A (2017) Smart industry: how ICT will change the game. IEEE Internet Comput 21(1):8–10. ISSN: 1089-7801. https://doi.org/10.1109/MIC.2017.22

20. The internet of things reference model (2014) eprint: http://cdn.iotwf.com/resources/71/IoT_Reference_Model_White_Paper_June_4_2014.pdf

21. Khan R, Khan SU, Zaheer R, Khan S (2012) Future Internet: The internet of things architecture, possible applications and key challenges. In 2012 10th International Conference on Frontiers of Information Technology, pp 257–260. https://doi.org/10.1109/FIT.2012.53

22. Yang Z et al (2011) Study and application on the architecture and key technologies for IOT. In: 2011 International Conference on Multimedia Technology. pp. 747–751. https://doi.org/10.1109/ICMT.2011.6002149

23. Sethi P, Sarangi SR (2017) Internet of things: architectures, protocols, and applications. J Electr Comput Eng

24. Ray PP (2016) A survey on internet of things architectures. J King Saud University - Comput Inf Sci ISSN: 1319-1578. https://doi.org/10.1016/j.jksuci.2016.10.003. URL: http://www.sciencedirect.com/science/article/pii/S1319157816300799

25. Mosenia A, Jha NK (2017) A Comprehensive Study of Security of Internet-of-Things. IEEE Trans Emerging Topics in Comput 5(4):586–602. https://doi.org/10.1109/TETC.2016.2606384

26. Elmaghraby AS, Losavio MM (2014) Cyber security challenges in Smart Cities: safety, security and privacy. J Adv Res 5(4):491–497. ISSN: 2090-1232. https://doi.org/10.1016/j.jare.2014.02.006. URL: http://www.sciencedirect.com/science/article/pii/S2090123214000290

27. Zeadally S, Isaac JT, Baig Z (2016) Security attacks and solutions in electronic health (E-health) systems. J Med Syst 40(12):1–12. ISSN: 0148-5598. https://doi.org/10.1007/s10916-016-0597-z

28. Sfar AR et al (2017) A roadmap for security challenges in the internet of things. Digital Commun and Networks (2017). ISSN: 2352-8648. https://doi.org/10.1016/j.dcan.2017.04.003. URL: http://www.sciencedirect.com/science/article/pii/S2352864817300214

29. Rodosek GD, Golling M (2013) Cyber security: challenges and application areas. In: Essig M et al (eds) Supply chain safety management: security and robustness in logistics. Springer, Berlin Heidelberg, Berlin, Heidelberg, pp 179–197. ISBN 978-3-642-32021-7. https://doi.org/10.1007/978-3-642-32021-7_11

30. Alexander RD, Panguluri S (2017) Cybersecurity terminology and frameworks. In: Clark RM, Hakim S (eds) Cyber-physical security: protecting critical infrastructure at the state and local level. Springer International Publishing, Cham, pp 19–47. ISBN 978-3-319-32824-9. https://doi.org/10.1007/978-3-319-32824-9_2

31. Yan, Z, Zhang, P, Vasilakos AV (2014) A survey on trust management for Internet of Things. J Network Comput Appl 42:120–134. ISSN: 1084-8045. https://doi.org/10.1016/j.jnca.2014.01.014. URL: http://www.sciencedirect.com/science/article/pii/S1084804514000575

32. Sicari S et al (2015) Security, privacy and trust in Internet of Things: The road ahead. Comput Networks 76:146–164. ISSN: 1389-1286. https://doi.org/10.1016/j.comnet.2014.11.008. URL: http://www.sciencedirect.com/science/article/pii/S1389128614003971

33. Liu J, Xiao Y, Chen CLP (2012) Internet of Things' Authentication and Access Control. Int J Secur Netw 7(4):228–241. ISSN: 1747-8405. https://doi.org/10.1504/IJSN.2012.053461

34. Singh S et al (2017) Advanced lightweight encryption algorithms for IoT devices: survey, challenges and solutions. J Ambient Intell Humanized Comput. ISSN: 1868-5145. https://doi.org/10.1007/s12652-017-0494-4

35. Turan MS, Mouha N, McKay KA, Bassham L (2017) Report on lightweight cryptography. National institute of standards and technologies, department of US and commerce. https://doi.org/10.6028/NIST.IR.8114

36. Perry JS (2017) Anatomy of an IoT malware attack. URL: https://www.ibm.com/developerworks/library/iot-anatomy-iot-malware-attack/ (visited on 01/13/2018)

37. Greenberg A (2017) The reaper IoT botnet has already infected a million networks. URL: https://www.wired.com/story/reaper-iot-botnet-infected-million-networks/ (visited on 01/13/2018)

38. CISCO (2017) Annual cyber security report. URL: https://www.cisco.com/c/dam/m/digital/1198689/Cisco_2017_ACR_PDF.pdf (visited on 01/13/2018)
39. Grange W (2017) Hajime worm battles Mirai for control of the Internet of Things. https://www.symantec.com/connect/blogs/hajime-worm-battles-mirai-control-internet-things (visited on 01/13/2018)
40. Cimpanu C (2017) BrickerBot author claims he bricked two million devices. https://www.bleepingcomputer.com/news/security/brickerbot-author-claims-he-bricked-two-million-devices/ (visited on 01/13/2018)
41. woolaston V (2017) WannaCry ransomware: what is it and how to protect yourself http://www.wired.co.uk/article/wannacry-ransomware-virus-patch (visited on 01/13/2018)
42. Hatmaker T (2017) A new ransomware attack called bad rabbit looks related to notPetya. https://techcrunch.com/2017/10/24/badrabbit-notpetya-russia-ukraine-ransomware-malware/ (visited on 01/13/2018)
43. McAfee Labs (2017) Threat prediction for 2017. https://www.mcafee.com/in/resources/reports/rp-threats-predictions-2017.pdf (visited on 01/13/2018)
44. Symantec (2016) Internet security threat report. https://www.symantec.com/content/dam/symantec/docs/reports/istr-21-2016-en.pdf (visited on 01/13/2018)
45. Kumar M (2016) Irongate new stuxnet-like malware targets industrial control systems. https://thehackernews.com/2016/06/irongate-stuxnet-malware.html (visited on 01/13/2018)
46. williams F (2016) Understanding exploit kits: how they work and how to stop them. https://blog.barkly.com/how-exploit-kits-work (visited on 01/13/2018)
47. Fotiguard SE Team (2017) Reaper: the next evolution of IoT botnets. https://blog.fortinet.com/2017/11/16/reaper-the-next-evolution-of-iot-botnets (visited on 01/13/2018)
48. Dobbins R, Bjarnason S (2016) Mirai IoT Botnet Description and DDoS Attack Mitigation. https://www.arbornetworks.com/blog/asert/mirai-iot-botnet-description-ddos-attack-mitigation/ (visited on 01/13/2018)
49. Ballano M, Wueest C (2015) Insecurity in the internet of things. https://www.arbornetworks.com/blog/asert/mirai-iot-botnet-description-ddos-attack-mitigation/ (visited on 01/13/2018)
50. Zhou Y, Feng D (2005) Side-channel attacks: ten years after its publication and the impacts on cryptographic module security testing. zyb@is.iscas.ac.cn 13083 received 27 Oct 2005. http://eprint.iacr.org/2005/388
51. Rayome AD (2017) DDoS attacks increased 91 percentage in 2017 thanks to IoT. https://www.techrepublic.com/article/ddos-attacks-increased-91-in-2017-thanks-to-iot/ (visited on 01/13/2018)
52. Stacheldraht (2016) DDOS ATTACK. https://en.wikipedia.org/wiki/Denial-of-service_attack#/media/File:Stachledraht_DDos_Attack.svg (visited on 01/13/2018)

Blockchains for Smart Cities: A Survey

Ahmed G. Ghandour, Mohamed Elhoseny and Aboul Ella Hassanien

Abstract Blockchains and distributed ledger technologies (DLT) play significant role in the current modern applications in smart cities. By applying blockchain technologies, systems gain a full access to benefits for decentralized applications, and security of distributed ledger technologies which includes consensus algorithms and smart contracts. This paper presents a survey of blockchain technologies and their applications in smart cities. It shows how it increases security, performance, and reduce cost of systems in smart cities. It also provides a high-level explanation of what is blockchain, DLT, and what are consensus algorithms? And how they can be used? In this paper an evaluation also done for some selected researches. These researches were reviewed, summarized, and analyzed, then evaluated based on specific criteria. This paper also proposes a sample architecture for one example of how blockchain can be used in this example in smart cities.

Keywords Blockchain · Distributed ledger · Decentralization · Cryptography
Smart cities

A. G. Ghandour
Landmark Group, Dubai, United Arab Emirates

M. Elhoseny (✉)
Faculty of Computers and Information Sciences,
Mansoura University, Mansoura, Egypt
e-mail: mohamed_elhoseny@mans.edu.eg

M. Elhoseny
The Scientific Research Group in Egypt (SRGE), Cairo, Egypt

A. E. Hassanien
Faculty of Computers and Information, Cairo University, Giza, Egypt

© Springer Nature Switzerland AG 2019
A. E. Hassanien et al. (eds.), *Security in Smart Cities: Models, Applications, and Challenges*, Lecture Notes in Intelligent Transportation and Infrastructure, https://doi.org/10.1007/978-3-030-01560-2_9

1 Introduction

In the last decades, we noticed a huge increase into civilized cities due climate change, and lack of resources. We also started to hear the word smart cities in the last few years. The meaning of the word smart cities means that it's smart because it benefits from the latest technologies, and innovative applications.

Looking back in the last decade, one may observe the change in the trend between centralization and decentralization. A tremendous change in decentralization of computation infrastructure, power, storage, power, and code. Previously we heard of mainframe computers which are largely centralized. These mainframes host memory, data storage, computing power, and code. Accessing these mainframes done by terminals, which only take inputs and outputs, and these terminals do not store or process any data.

The distributed, and client-server architecture appeared with the rise of personal computers, private networks, and with advance in their capabilities and their computational power. This means that data could be replicated from server to server, and these data could be processed then synced back to the server.

Over time, Internet and cloud computing architectures enabled universal access from a variety of computing devices. In fact this 'cloud architecture' is decentralized in terms of hardware, but it's centralized in the application-level.

Currently, the world is witnessing the transition from centralized applications, computation, processing, and data storage to decentralized architectures and systems.

One of the key innovations which makes this shifting to decentralization possible is distributed ledger technology (DLT).

Distributed ledger is a data structure type which exists in multiple computer devices spreaded across multiple locations. Distributed ledger technology includes blockchains and smart contracts. The famous cryptocurrency "Bitcoin" and it's blockchain is one an example of permissionless blockchain. It uses all DLT technologies such as Peer-to-Peer (P2P) networks, cryptography, transactions timestamping, and shared computational power, along with a new consensus algorithm.

Distributed ledger technology generally consists of three main components. These components are: data model which captures the current state of the ledger, language of transactions which changes the ledger state, and protocol which used to build consensus between participants about which transactions shall be accepted, and in which order.

Blockchain is a decentralized technology which is a subset of distributed ledger technologies constructs a chronological chain of blocks, and consists of a peer-to-peer distributed ledger which is controlled by consensus, combined with a smart contracts system, and other technologies. Blockchain is well known currently as the new Internet. A block is a set of transactions combined together and added to the chain at the same time. Smart contracts are computer programs which execute predefined actions when certain conditions are met within the system. Consensus refers to a system which ensures that parties agree on a certain state of the system as true state [1].

Consensus algorithms includes many algorithms such as "Proof of Work" (PoW) which used in Bitcoin and it involves solving a computational challenging puzzle to create a new block. There are also some other consensus algorithms such as "Proof of Elapsed Time" (PoET) which uses a lottery function in which the node with the shortest wait time creates the next block. Other consensus algorithms also exist such as Proof of Stake (PoS), Simplified Byzantine Fault Tolerance (SBFT), Proof of Authority (PoA), etc.

A Peer-to-Peer Network is a network consists of computer systems. these computer systems are directly connected to each other via the Internet without a central server. Cryptocurrency is a digital asset used as a middle-ware in exchange. A cryptocurrency is exchanged using digital signatures to transfer ownership from one cryptographic key pair to another key pair. Since this digital asset has characteristics of money in terms of value and middle-ware in transactions, so it is referred to as currency. Bitcoin and Ethereum are two examples of these cryptocurrencies and they are different in their consensus algorithms.

Cryptography is a set of techniques used to allow secure communication between different parties. It's also used to ensure immutability and authenticity of data being communicated [1].

There are two types of blockchains one is permissionless such as Bitcoin or Ethereum where it's an open public blockchain and any one can join it. The other type of blockchain is permissioned or private blockchain which requires a pre-verification of the participating parties within the network, and these parties are usually known to each other. The selected type of permissionless or permissioned blockchain can be determined by the type of business or application where, when, and who will apply it.

Currently there is a huge demand for blockchain and there is a need to implement it in many fields of business and in smart cities.

Smart cities started to implement blockchains because of their benefits such as security, immutability, transparency, and cost saving, etc.

The smart city of Dubai for example started to implement blockchain technology in Dubai smart government. According to their website blockchain will save them the cost of 100 million documents every year, plus save up to 114 MTons CO_2 emissions from trip reductions, and 25.1 million hours of economic productivity in saved document processing time [2].

The interest in blockchain technology can be also shown from the price and capital market of their cryptocurrencies such as Bitcoin. For instance, according to the Coin Market Capitalizations website, Bitcoin price was 10,000 USD, and Bitcoin's market capitalization was nearly $178 billion USD as of 1st of December 2017. This price jumped to 16,300 USD, and market capitalization to nearly $272 billion USD on 8th of December 2017.

This paper has been arranged as follows: In Sect. 2 related work has been presented. Section 3 presents contributions of previous mentioned researches and summarize each research in a table format. Section 4 introduces evaluation criteria for these selected researches and studies. Section 5 presents architecture of a sample application for one of the blockchain Hyperledger framework which is Hyperledger

Indy on smart cities governments. Section 6 shows challenges, and limitations. Eventually, in Sect. 7 the conclusion and future work were presented.

2 Related Work

This literature review shows studies done on in the field of blockchain technologies and distributed ledgers. It also illustrates researches done in the previous mentioned fields and their applications in smart cities and current new business. Following are presentations for some researches related to the subject:

Sun et al. [3], introduced a conceptual framework to show how blockchain-based sharing services can contribute to smart cities. This conceptual framework contains three main factors. These factors are: human, technology, and organization. These factors where combined together to form in a peer-to-peer to form a blockchain in smart cities to be able to gain benefits from their sharing services. Research opinion, using this conceptual frame work to build a blockchain in smart cities can contribute in smart cities benefiting from the blockchain based sharing services.

Watanabe et al. [4], introduced a new mechanism for cryptographic blockchain. This mechanism for securing the blockchain was applied to smart contracts management. In this new mechanism, a new consensus algorithm were introduced that uses a credibility score to be used along with an existing consensus algorithm. This hybrid algorithm uses both proof-of-stake and the credibility score algorithm to prevent attackers from monopolizing resources and to keep blockchains secure. Researcher opinion, a new mechanism were introduced using a new hybrid consensus algorithm to provide more security layer for blockchain, and make it monopolization proof.

Zhao et al. [5], proposed opportunities of permissionless blockchains such as bitcoin for financial institutions in smart cities. And by doing so, how it can be more secure than the current systems, more cost and time saving, and more data sharing. Researcher opinion, shows open opportunities and reasons to use permissionless blockchain technologies in financial business. This could be an advantage in many fields such as currency trading field, or when data is extremely large to transform the network to blockchain, but for many other fields such as medical, governmental, financial, or enterprise transactions permissioned blockchain would be necessary.

Biswas and Muthukkumarasamy [6], proposed a security framework by implementing blockchain technology across smart devices to provide a secure communication channels and platforms in smart cities. Researcher opinion is that blockchain technology, and distributed ledger technologies plays an important role in smart cities security as it makes communication devices, channels, and protocols secure.

Watanabe et al. [7], made a proposal for recording contracts using blockchain technology. In this proposal, a new consensus protocol was described for smart contracts. This protocol proves that confirmation done by contractor consent, and archive this contract document in the blockchain. Researcher opinion is that the proposed protocol adds more security to blockchain, by confirming the agreement or the contract within the block, then archive the contract in the blockchain.

Gharaibeh et al. [8], presented a survey on data management, usability, and security in smart cities that has been collected using new enabled technologies such as Internet-of-Things (IOT). These new technologies have the ability to collect significant amount of data and information. In this survey they illustrated how to understand the value of this data, and ensure its security, and privacy. Moreover, managing these data through its correct life-cycle in smart cities were also presented. Researcher opinion that in this survey a data management methods were presented, along with data securing researches. Also, technologies were presented about methods for collecting data in smart cities. These data would be very useful if they were connected and distributed among all participated nodes, and all transaction were recorded in all these nodes.

Ibba et al. [9], introduced a solution for data immutability in smart cities by applying blockchains oriented smart cities. In this research they collected data from small mobile sensors using Internet-of-Things (IOT). In this research, they needed the data to be stable, and cannot be changed, and this is the reason they implemented the blockchain. One of blockchain characteristics is that it's immutable and cannot be changed, or extremely hard to be changed. To modify one node, all other nodes must be changed with same modification, which is impossible. Moreover if this modification happens it's easy to know what has been changed? who changed it? etc. as all committed actions done in the blockchain cannot be retrieved or rolledre any added line is shared across all the distributed ledgers. This is the reason they have chosen to apply blockchain technology in smart cities application for data collected by IOT sensors. Researcher opinion, blockchain adds many benefits specially in security, speed, data integrity, and orientation when applied in smart cities.

Guo et al. [10], presented smart cities issues, technologies, development, and solutions in China. They presented issues such as traffic jam, medical, and education problems. They also presented the development of these cities, and the rise of technologies, with some solutions to these problems. Researches opinion, all smart cities has problems, and technologies. These problems has to be addressed and analyzed to be able to select the correct solution which shall be applied with respect to cities technologies, in order to solve these problems.

Swan [11], presented in this book a detailed description about what is blockchain? And how it would contribute in shifting the currency global economy. The researcher presented a new model for connected world including blockchains technologies. He also presented how current technologies such as Internet-of-Things (IoT) sensors, tablets and smart phones, laptops, self-driving cars, smart cars, tracking devices, smart homes, and smart cities interact and would be very useful in within blockchain more than any other system or any other technology. Researcher opinion, blockchain technologies is leaving a blueprint in new world economy especially when combined with new other state-of-the-art technologies.

Sharma et al. [12], introduced a new vehicle network architecture based on blockchains in the smart cities, which can be a revolution in intelligent transport systems. Blockchain technologies can be used to build an intelligent, secure, distributed and autonomous transport system, and thus utilize intelligent transport systems resources and infrastructure better. This proposed architecture thus for builds a management system for distributed transportation. Researcher opinion, a new architecture has been proposed for vehicle networks in smart cities based on blockchain technologies.

Atzori [13], presented a survey about the architecture of three blockchain platforms for Internet-of-Things. Researcher also presented the blockchain as it is well known that it provides the disruptive solution for privacy and security issues of Internet-of-Things. Researcher also presented how the use of cryptocurrencies, and smart contracts which are available in blockchains with digital enhanced devices connected to the Internet will increase benefits, reduce cost, ensure of efficient usage of resources, and add extra security. Researcher opinion, implementing blockchain technologies adds many benefits to the usage of Internet-of-Things.

Dorri et al. [14], proposes a special type of blockchain for Internet-of-Things. It also represents privacy and security vulnerabilities related to the Internet-of-Things, and these issues can be overtaken by implementing blockchain technologies. They proposed a hierarchical distributed architecture consists of smart-homes, and many other nodes though the network. They have proved that using this decentralized topology increased the efficiency and effectiveness for IoT applications by providing and enhancing its security and privacy.

Talari et al. [15], published a review about smart cities, based on Internet-of-Things. They also reviewed their applications, benefits, and advantages. Researcher opinion, how Internet-of-Things affects smart cities, and benefits it. Applications for this technology in smart cities were also reviewed.

Li et al. [16], did a survey about the security of blockchain systems. In this paper, researchers conducted a systematic study on blockchain security threats, and they did a survey on real attacks by examining some of most popular blockchain systems. They also reviewed some enhancement solutions related to the security of blockchain systems. Researcher opinion, a survey has been done on blockchain systems security, which examined some popular blockchains and suggested some enhancements for blockchain security and privacy.

de la Rosa et al. [17], presented a survey about opportunities and open innovations which can be built depending on blockchain technologies. They used the distributed nature architecture and decentralization of the blockchain technology to set a definition for open innovation opportunities, and to build distributed reliable solutions. Researcher opinion, blockchain technologies sets a great opportunity to build smart solutions, and open innovations in smart cities.

Dustdar et al. [18], introduced an architecture and a new vision of smart cities. In this architecture they presented the coordinating activities, services, stakeholders and their devices. Instead of looking at smart cities from only the infrastructure perspective, they also considered the existing foundational technologies which used for controllability, provisioning, and coordination of activities with their alignment to fulfill their vision. Researcher opinion, a new visionary architecture was introduced for the interconnected nodes in smart cities.

Stanciu [19], proposed a novel model edge computing distributively controlled by blockchain system. As edge computing represented today in high-end connected resources in a network with many applications such as Internet-of-Things, health care, smart grids, smart-homes, smart cities, etc. The researcher used the distributed and hierarchical control architecture of the blockchain technology in this research. An application called Hyperledger Fabric which is an example of the permissioned blockchain network architecture with smart contracts has been implemented in this network. Hyperledger Fabric has allows components such as consensus to be plug-and-play which is called modular architecture. One of the major advantages of blockchain that it allows entities conducting transactions without passing information through a central authority. Researcher opinion, blockchain and their distributed control system architecture in nature benefits edge computing.

Liu et al. [20], introduced a new service framework for Internet-of-Things Data based on blockchain for data integrity, and immutability. As all current existing and available frameworks uses a third party auditor for authorization, and has many issues in data integrity, this lead researchers to use blockchain technologies, and distributed ledgers as it solves these issue. By implementing blockchains in a prototype, researchers proved that blockchain increased the performance, enhanced security, accountability, and dependability of IOT data. Researcher opinion, Implementation of blockchain technologies enhances data integrity for Internet-of-Things Data.

Xu et al. [21], introduced a new blockchain-based storage system called "Sapphire" which used for data analytics in Internet-of-Things in smart cities. As we are now witnessing an increase of the processing power, storage, and memory technologies in Internet-of-Things devices, researchers had the idea of using smart contracts in blockchain technologies to facilitate and enforce negotiation of contracts in IOT. Thus will help in parallel processing and map-reduce. It will also help in parallelism of making large-scale storage systems to be employed which will decrease data analytics tasks, and so execution time will be reduced. It will also reduce the risk of data to be centralized. Researchers also introduced a new smart contract protocol called "Object-based storage device 'OSD'-based smart contract (OSC)" applied in Sapphire, where IOT devices interact with blockchain. Data analytics applications done by IoT device processors which execute specific transactions. thus, only results returned to users instead of data files. Researcher opinion, a new system introduced called Sapphire based on blockchain technology on storage systems of the Internet of Things devices in smart cities can greatly help in data analytics and decrease the overhead.

Rabah [22], presented blockchain ecosystem, and blockchain technologies and how it can be used in the modern world as an engine leads to a new industrial revolution, and making our smart cities smarter, and powered by blockchain. Researcher here presented how blockchain can be used in writing smart contracts, voting, or even registering assets without the authority of a third party. Researcher also presented how it can be used in many other fields such as music, art industry, and commercial application benefiting from the distributed nature of the blockchain as it's a branch from distributed ledgers. Researcher opinion, blockchain represents the engine or the motor which leads to a new industrial revolution.

Ulieru [23], presented how blockchain technologies can be used to replace banking systems, and currencies in the modern world. Researcher presented the evolution of the decentralization, and how the new world of peer-to-peer, and crowd-funding powered by blockchain technologies in performing smart contracts fostered by consensus algorithms leads the world to change the current financial system. This can be seen by the rise of cryptocurrencies currently and the huge increase in their capitalization, and price, such as Bitcoin. Researcher opinion, blockchain technologies currently changing the financial system of the modern world.

Ruta et al. [24], proposed a novel Service-Oriented Architecture (SOA) to improve the scalability of Internet-of-Things (IOT) based on semantic blockchain. They benefited from blockchain technologies such as consensus, smart contracts, and distributed execution combined with Semantic Web of Things (SWoT) for operations such as discovery, selection, registration, and payment to overtake some of the well-known problems of Internet-of-Things such as computation and memory resources availability. Researcher opinion, semantic blockchain implementation helps in improve scalability in IOT.

McKee et al. [25], presented an overview of massive scale existing state-of-the-art automation, augmentation, and system integration across all domains in smart cities including autonomous driving cars, smart manufacturing, and smart energy efficiency. They introduced current open problems of Internet-of-Things (IoT) in smart cities, Internet of Simulation (IoS), and existing technologies such as edge computing, cloud computing, and fog computing. These problems were focused in complexity, centralization, integrity, scalability, and security. These issues can be solved with new technologies such as blockchain. Researcher opinion, new technologies can help in massive scale automation systems and solve their problems, and challenges.

Ruta et al. [26], proposed a semantic layer built on top the basic blockchain to form a service-oriented architecture (SOA) model to be used in many areas in smart cities. By doing so this model will increase the service or resources exploration, discovery, and validation by consensus algorithms. It will also increase the efficiency smart contract operations within the network. This model can be implemented in many applications in smart cities including trust less operations, material and resource management, markets, and enterprises. Researcher opinion. semantic blockchain service-oriented architecture adds many benefits in smart cities.

Smart cities face huge challenges, and problems in many fields including resource management and planning, industrial, medical, educational, cost, security, privacy, etc. This paper aims to provide an example sample solution by presenting one application sample of Hyperledger framework Indy in smart cities.

3 Contributions Summary

Table 1 presents the summary of all previous mentioned contributions and reviewed studies in 24 papers.

Table 1 Summary of contributions and previous reviewed studies

No.	Authors	Contribution
1	Sun et al. [3]	Introduced a framework about contribution of sharing services blockchain in smart cities
2	Watanabe et al. [4]	Proposed a new mechanism for securing blockchain which applied to smart contracts management
3	Zhao et al. [5]	Presented opportunities of blockchains which are permissionless such as Ethereum, or Bitcoin instead of the current financing systems in smart cities
4	Biswas and Muthukkumarasamy [6]	Introduced a new security framework by blockchain implementation across smart IOT devices to secure communications in smart cities
5	Watanabe et al. [7]	Proposed a new consensus protocol for recording smart contracts using blockchain technology
6	Gharaibeh et al. [8]	Presented a survey on data collected through technologies such as Internet-of-Things (IOT) and how to manage these data, make it usable, and secure in smart cities
7	Ibba et al. [9]	Introduced a solution called CitySense which applies blockchains for data immutability in smart cities using Internet-of-Things (IOT)
8	Guo et al. [10]	Presented smart cities in China with their issues, development, solutions, and technologies
9	Swan [11]	A book has been presented contains knowledge about blockchain, with its contribution in global economy shifting
10	Sharma et al. [12]	Introduced a revolution intelligent transport systems in the smart cities based on blockchains
11	Atzori [13]	Presented a survey about Internet-of-Things based blockchain platforms. Also benefits from cryptocurrencies, and smart contracts which are connected to IOT devices
12	Dorri et al. [14]	Proposed Internet-of-Things based on a special type of blockchain, and how it overtakes IOT related challenges
13	Talari et al. [15]	Published a review about Internet-of-Things and the effect of this technology in smart cities
14	Li et al. [16]	Presented a survey about blockchain systems security, and security threats. Researchers also reviewed some security enhancement solutions

(continued)

Table 1 (continued)

No.	Authors	Contribution
15	de la Rosa et al. [17]	Presented a survey about using blockchain technologies to open new opportunities for innovations
16	Dustdar et al. [18]	Introduced a visionary architecture with its coordinating activities, stakeholders, services, and devices in smart cities
17	Stanciu [19]	Proposed a novel model about distributed blockchain systems for edge computing
18	Liu et al. [20]	Introduced a service framework for using blockchain for data integrity, and immutability in Internet-of-Things applications
19	Xu et al. [21]	Introduced a new blockchain-based storage system for data analytics in Internet-of-Things
20	Rabah [22]	Presented that blockchain technologies, combined with blockchain ecosystem can be used as an engine for new industrial revolution
21	Ulieru [23]	Proved that current banking systems and the current money systems can be replaced by blockchain technologies. Researcher presented benefits and facts about blockchain technologies which qualifies it to lead in this field
22	Ruta et al. [24]	Proposed a semantic blockchain novel Service-Oriented Architecture (SOA) which can be used to improve the scalability of Internet-of-Things (IOT)
23	McKee et al. [25]	Presented a review about some new technologies in automation, augmentation, and system integration through all fields and domains in smart cities such as smart manufacturing, autonomous cars, and smart energy efficiency
24	Ruta et al. [26]	Introduced a new semantic service-oriented architecture (SOA) layer on top basic blockchain to form a model which can be used in many areas and fields in smart cities

4 Evaluation Criteria for Selected Researches

In this section, previous mentioned researches were reviewed and evaluated based on some evaluation criteria. These evaluation criteria are (applied method, scalability, usability, cost, blockchain, and if blockchain applied extra measures also considered such as application finality latency, and blockchain type ["permissioned"—"cryptocurrency based", "auditable", "modularity", "smart contracts", "immutability", "consensus protocol"]).

- Applied Method "AM": can be survey "Sur", framework "F", model "M", new technology "NT", system "Sys", Other "O".
- Scalability "S": is the ability to apply research to n number of nodes which can be High (H), Medium (M), or Low (L).
- Usability "U": refers to the ease to use the applied research after implementation, which can also be measured as High (H), Medium (M), or Low (L).
- Cost "C": is the expenses of applying and running suggested technique in the research and it can be also High (H), Medium (M), or Low (L).
- Blockchain "BC": determines whether blockchain technology has been applied in research "Y" or not "N".
- Latency "L": depends the consensus algorithm which will be used, e.g. POW algorithm consumes a lot of processing power, and time, therefor it affects in high latency of the final output. Latency can also be considered as High (H), Medium (M), or Low (L).
- Permissioned "P": is whether the blockchain type is permissioned "Y", permissionless "N", or semi permissioned "Semi".
- Cryptocurrency "CC": measures whether the blockchain based on cryptocurrency such as Bitcoin, or Ethereum "Y" or not "N".
- Auditable "A": determines whether this research blockchain based can be audited "Y" or not "N".
- Modularity "M": measures whether this blockchain type can be separated and recombined "Y" or not "N".
- Smart Contracts "SC": determines whether this blockchain based type applies smart contracts "Y" or not "N". .
- Immutability "IM": refers to whether this type of research technique is stable, and changing proof "Y" or not "N".
- Consensus Protocol "CP": determines what is the consensus algorithm which has been used in this blockchain. Consensus algorithm is mentioned, and if it's unavailable or inapplicable then Not Available "NA" has been placed.

Tables 2 and 3 contains the result for applying mentioned measurement criteria to previous mention researches. Table 2 contains evaluation criteria for (Applied Method "AM", Scalability "S", Usability "U", Cost "C", Blockchain "BC", Latency "L", and Permissioned "P").

Table 3 contains remaining evaluation criteria which were used to evaluate these researches. Evaluation criteria presented below are (Cryptocurrency "CC", Auditable "A", Modularity "M", Smart Contracts "SC", Immutability "IM", and Consensus Protocol "CP").

Table 2 Evaluation criteria for selected researches

No.	Authors	AM	S	U	C	BC	L	P
1	Sun et al. [3]	F	H	H	M	Y	H	N
2	Watanabe et al. [4]	NT	H	H	L	Y	L	Y
3	Zhao et al. [5]	O	H	H	M	Y	H	N
4	Biswas and Muthukkumarasamy [6]	F	H	H	L	Y	L	Y
5	Watanabe et al. [7]	M	H	H	L	Y	L	Y
6	Gharaibeh et al. [8]	Sur	H	M	H	N	H	N
7	Ibba et al. [9]	Sys	H	H	L	N	L	N
8	Guo et al. [10]	Sur	M	H	L	N	L	N
9	Swan [11]	O	H	H	L	Y	M	Semi
10	Sharma et al. [12]	Sys	H	H	M	Y	M	N
11	Atzori [13]	Sur	H	H	M	Y	L	N
12	Dorri et al. [14]	O	H	H	M	Y	L	N
13	Talari et al. [15]	O	M	M	H	N	H	N
14	Li et al. [16]	Sur	H	H	L	Y	L	Y
15	de la Rosa et al. [17]	Sur	H	H	L	Y	L	Y
16	Dustdar et al. [18]	NT	M	H	H	N	H	N
17	Stanciu [19]	M	H	H	L	Y	L	Y
18	Liu et al. [20]	F	H	H	L	Y	L	N
19	Xu et al. [21]	Sys	H	H	L	Y	L	Y
20	Rabah [22]	O	H	H	L	Y	L	N
21	Ulieru [23]	O	H	H	M	Y	H	N
22	Ruta et al. [24]	O	H	H	L	Y	L	Y
23	McKee et al. [25]	S	H	M	M	Y	L	Y
24	Ruta et al. [26]	O	H	H	L	Y	L	Y

5 Proposed Architecture for Using Hyperledger Indy for Identity Management in Smart Cities

This section describes Hyperledger, Hyperledger framework Indy, and how it can be applied in smart cities for smart identity management based on the distributed architecture of the blockchain technology. The presented framework consists of four layers which are application layer, data layer in blockchain "Hyperledger framework Indy", communication and consensus protocol layer, and physical hardware layer as shown in Fig. 1.

Hyperledger is a blockchain based technologies which includes many other technologies such as distributed ledger technologies (DLT), smart contracts, libraries set,

Table 3 Remaining evaluation criteria for selected researches

No.	Authors	CC	A	M	SC	IM	CP
1	Sun et al. [3]	Y	Y	N	N	Y	POW
2	Watanabe et al. [4]	N	Y	Y	Y	Y	PoET
3	Zhao et al. [5]	Y	Y	N	N	Y	Varies
4	Biswas and Muthukkumarasamy [6]	N	Y	Y	Y	Y	Varies
5	Watanabe et al. [7]	Y	Y	Y	Y	Y	POA
6	Gharaibeh et al. [8]	N	N	Y	N	N	NA
7	Ibba et al. [9]	N	Y	Y	Y	Y	PoET
8	Guo et al. [10]	N	Y	N	N	N	NA
9	Swan [11]	Y	Y	Y	Y	Y	Varies
10	Sharma et al. [12]	N	Y	Y	Y	Y	SBFT
11	Atzori [13]	Y	Y	Y	Y	Y	POS
12	Dorri et al. [14]	N	Y	Y	Y	Y	POC
13	Talari et al. [15]	N	Y	N	N	N	NA
14	Li et al. [16]	N	Y	Y	Y	Y	PBFT
15	de la Rosa et al. [17]	N	Y	Y	Y	Y	PoET
16	Dustdar et al. [18]	N	Y	Y	N	N	NA
17	Stanciu [19]	N	Y	Y	Y	Y	Varies
18	Liu et al. [20]	N	Y	Y	Y	Y	NA
19	Xu et al. [21]	N	Y	Y	Y	Y	OSC
20	Rabah [22]	N	Y	Y	Y	Y	NA
21	Ulieru [23]	Y	Y	N	N	Y	POW
22	Ruta et al. [24]	N	Y	Y	Y	Y	UTXO
23	McKee et al. [25]	N	Y	N	Y	Y	NA
24	Ruta et al. [26]	N	Y	Y	Y	Y	Varies

and some other applications samples. Hyperledger Indy is a DLT blockchain built for the purpose of decentralizing identity [1].

According to "Breach Level Index", almost 9.2 billion records of data were either stolen or lost since 2013, and only 4% from these data were encrypted. Hence Hyperledger Indy showed up to solve this issue of security and privacy for identity. Indy allows users to control, save, and share their digital identity by themselves. It also reduces data storage needed as it will be stored only in one connected node which is the user device, then validated, and presented only when needed. Such framework would increase the security, and build a public trust between parties. It also gains all the benefits of any other blockchain system of distributed architecture, security, interoperability, immutability, transparency, and cost reduction.

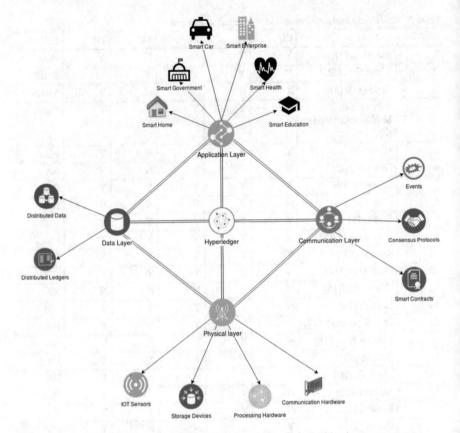

Fig. 1 Proposed architecture diagram for using Hyperledger Indy for identity management

The proposed architecture above in Fig. 1 proposes using Hyperledger Indy to benefit from its features in Identity management systems in smart cities. This architecture contains 4 layers, and they are as follows:

1. Application layer: which is where this blockchain Hyperledger can be implement in smart cities. It can be implemented in many application, such as in smart-homes, smart-governments, smart-cars, smart-enterprises, smart-medical-application, smart-education-system, etc. For example, user can share a reference to his/her identity to any of these mentioned applications when requested without sharing it actually with them, allowing more privacy, data saving, speed, and authentication. It will be authentic as all users have a reference in their ledger that has the latest update and any update in the user's identity if it's according to consensus protocols will add a new record to all connected nodes ledgers of blockchain that this user has updated his/her identity, and so it adds extra verification layer.

2. Data layer: in this layer all data will be distributed and saved in many different nodes, and any transaction in any of these node, will be added to something called ledger. This ledger is shared across all connected node, and it allows commit only, which means that no update, or delete is allowed in this ledger. It's also extremely hard to try to manipulate with this ledger, as all users have the same copy from this ledger. For example if someone hacked one node and changed any transaction, this hacker has to hack all other connected nodes in the same blockchain network to change other nodes with the same change at the same time, with the same previous actions, along with other actual actions happened at the same time in other connected blocks, which is impossible, and if someone did that, then it will be extremely easy to know who did it, when, and where because these data were stored in all distributed ledgers.

This considered an extremely powerful data layer compared to any other system such as cloud computing. In cloud computing it's distributed in applications but centralized in data architecture. In cloud computing for example, all data are stored in a central server, which can be easily hacked, and manipulated.

3. Physical layer: that contains all infrastructure and hardware devices used in any application to be connected through blockchain technology. These physical devices may include: Internet-of-Things (IOT) sensors, storage devices, communication hardware, or even processing hardware such as processors, and RAMs which may be used for data processing, analytics, or even for consensus algorithms verification like miners in Bitcoin.

4. Communication layer: which contains consensus protocols, and algorithms to be used in this communication, or even smart contract which will allow for event triggering based on the state of a block, or condition has been met in the blockchain.

6 Challenges, Limitations, and Open Problems

There are many challenges related to the implementation or adaptation of the blockchain technologies in smart cities, some of the key challenges or limitations to implement blockchain technologies are:

1. Lack of knowledge.

There is an exponential increase and high demand of the interest in blockchain technologies. Lack of available experts in this area is one of the major and most important challenges facing blockchains, and distributed ledger technologies.

2. Standards.

Blockchain is brand new technologies in its early days, and till now there are no standards or agreements related to development process, or business implementation. Standards is very important to ensure interoperability and to avoid taking risks.

A standardization organization has been established in 2016 called "International Organization for Standardization for Blockchain and Distributed Ledger Technologies".

3. Regulation.

There is a level of suspicion in all transactions in the blockchain system because of the lack of regulation. Each node is playing carefully while they are using blockchain technologies. there are no regularities for controlling smart contracts, which adds some concerns about it.

7 Conclusion and Future Work

Smart cities has a lot of issues to be solved nowadays and plenty of open opportunities. Blockchain also is one of the most important technologies these days due to its limitless benefits. This paper introduced a survey of the most recent researches done on using blockchain technologies in smart cities. Evaluation criteria, and comparison according to these evaluation criteria has been applied on these researches. Therefore, an architecture has been proposed in this paper about using Hyperledger Indy for managing identities in smart cities to help in solving smart cities issues. Some challenges were also presented in this paper such as regulations challenges, standards challenges, and lack of knowledge, which affect the implementation of blockchain technologies in smart cities.

Future work of this research may include using other Hyperledger frameworks such as Sawtooth, Iroha, etc. in smart cities, or even a hybrid combination of Hyperledger frameworks. Other future works is inventing or applying new consensus algorithms in different smart applications [27–36].

References

1. Cachin C (2016) Architecture of the Hyperledger blockchain fabric. In: Workshop on distributed cryptocurrencies and consensus ledgers
2. Smart Dubai Government (2016) Dubai blockchain strategy. Smart Dubai. www.smartdubai. ae/dubai_blockchain.php
3. Sun J, Yan J, Zhang KZK (2016) Blockchain-based sharing services: what blockchain technology can contribute to smart cities. Financ Innov 2(1):26
4. Watanabe H, Fujimura S, Nakadaira A, Miyazaki Y, Akutsu A, Kishigami J (2016) Blockchain contract: securing a blockchain applied to smart contracts. In: 2016 IEEE international conference on consumer electronics (ICCE). IEEE, pp 467–468
5. Zhao JL, Fan S, Yan J (2016) Overview of business innovations and research opportunities in blockchain and introduction to the special issue. Financ Innov 2(1):28
6. Biswas K, Muthukkumarasamy V (2016) Securing smart cities using blockchain technology. In: 2016 IEEE 18th international conference on high performance computing and communications;

IEEE 14th international conference on smart city; IEEE 2nd international conference on data science and systems (HPCC/SmartCity/DSS). IEEE, pp 1392–1393

7. Watanabe H, Fujimura S, Nakadaira A, Miyazaki Y, Akutsu A, Kishigami JJ (2015) Blockchain contract: a complete consensus using blockchain. In: 2015 IEEE 4th global conference on consumer electronics (GCCE). IEEE, pp 577–578

8. Gharaibeh A, Salahuddin MA, Hussini SJ, Khreishah A, Khalil I, Guizani M, Al-Fuqaha A (2017) Smart cities: a survey on data management, security, and enabling technologies. IEEE Commun Surv Tutorials 19(4):2456–2501

9. Ibba S, Pinna A, Seu M, Pani FE (2017) CitySense: blockchain-oriented smart cities. In: Proceedings of the XP2017 Scientific Workshops. ACM, p 12

10. Guo M, Liu Y, Yu H, Hu B, Sang Z (2016) An overview of smart city in China. China Commun 13(5):203–211

11. Swan M (2015) Blockchain: blueprint for a new economy. O'Reilly Media, Inc

12. Sharma PK, Moon SY, Park JH (2017) Block-VN: a distributed blockchain based vehicular network architecture in smart city. J Inf Process Syst 13(1):184–195

13. Atzori M (2016) Blockchain-based architectures for the Internet-of-Things: a survey

14. Dorri A, Kanhere SS, Jurdak R (2016) Blockchain in Internet-of-Things: challenges and solutions. arXiv preprint arXiv:1608.05187

15. Talari S, Shafie-khah M, Siano P, Loia V, Tommasetti A, Catalão JPS (2017) A review of smart cities based on the internet of things concept. Energies 10(4):421

16. Li X, Jiang P, Chen T, Luo X, Wen Q (2017) A survey on the security of blockchain systems. Future Gener Comput Syst

17. de la Rosa JL, Maicher L, Torres V, El-Fakdi A, Miralles F, Gibovic D (2017) A survey of blockchain technologies for open innovation. In: 4rd Annual World open innovation conference WOIC, pp 14–15

18. Dustdar S, Nastic S, Scekic O (2016) A novel vision of cyber-human smart city. In: 2016 Fourth IEEE workshop on hot topics in web systems and technologies (HotWeb). IEEE, pp 42–47

19. Stanciu A (2017) Blockchain based distributed control system for edge computing. In: 2017 21st International conference on control systems and computer science (CSCS). IEEE, pp 667–671

20. Liu B, Yu XL, Chen S, Xu X, Zhu L (2017) Blockchain based data integrity service framework for IoT Data. In: 2017 IEEE international conference on web services (ICWS). IEEE, pp 468–475

21. Xu Q, Aung KMM, Zhu Y, Yong KL (2017) A blockchain-based storage system for data analytics in the internet of things. In: New advances in the internet of things. Springer, Cham, pp 119–138

22. Rabah K (2017) Overview of blockchain as the engine of the 4th industrial revolution. Mara Res J Bus Manage-ISSN: 2519-1381 1(1):125–135

23. Ulieru M (2016) Blockchain 2.0 and beyond: adhocracies. In: Banking beyond banks and money. Springer International Publishing, pp 297–303

24. Ruta M, Scioscia F, Ieva S, Capurso G, Di Sciascio E (2017) Semantic blockchain to improve scalability in the internet of things. Open J Internet Things (OJIOT) 3(1):46–61

25. McKee DW, Clement S.J, Almutairi J, Xu J (2017) Massive-scale automation in cyber-physical systems: vision & challenges. In: 2017 IEEE 13th international symposium on autonomous decentralized system (ISADS). IEEE, pp 5–11

26. Ruta M, Scioscia F, Ieva S, Capurso G, Loseto G, Gramegna F, Pinto A, Di Sciascio E (2017) Semantic-enhanced blockchain technology for smart cities and communities. In: 3rd Italian conference on ICT

27. Elhoseny M, Hassanien AE (2019) An encryption model for data processing in WSN. Dynamic wireless sensor networks. Springer, Cham, pp 145–169

28. Elhoseny M, Hassanien AE (2019) Secure data transmission in WSN: an overview. Dynamic wireless sensor networks. Springer, Cham, pp 115–143

29. Elhoseny M, Hassanien AE (2019) Mobile object tracking in wide environments using WSNs. Dynamic wireless sensor networks. Springer, Cham, pp 3–28

30. Abbas H, Shaheen S, Elhoseny M, Singh AK, Alkhambashi M (2018) Systems thinking for developing sustainable complex smart cities based on self-regulated agent systems and fog computing. Sustain Comput Inf Syst

31. Elhoseny H, Elhoseny M, Riad AM, Hassanien AE (2018) A framework for big data analysis in smart cities. In: International conference on advanced machine learning technologies and applications. Springer, Cham, pp 405–414

32. Elhoseny H, Elhoseny M, Abdelrazek S, Riad AM, Hassanien AE (2017) Ubiquitous smart learning system for smart cities. In: 2017 Eighth international conference on intelligent computing and information systems (ICICIS). IEEE, pp 329–334

33. Elhoseny H, Elhoseny M, Abdelrazek S, Riad AM (2017) Evaluating learners' progress in smart learning environment. In: International conference on advanced intelligent systems and informatics. Springer, Cham, pp 734–744

34. Hassanien AE, Elhoseny M, Kacprzyk J (eds) (2018) Quantum computing: an environment for intelligent large scale real application. Springer, Berlin

35. Elhoseny H, Elhoseny M, Abdelrazek A, Bakry H, Riad A (2016) Utilizing service oriented architecture (SOA) in smart cities. Int. J. Adv. Comput. Technol. (IJACT) 8(3):77–84

36. Abdelaziz A, Elhoseny M, Salama AS, Riad AM (2018) A machine learning model for improving healthcare services on cloud computing environment. Measurement 119:117–128

Cryptanalysis of 4-Bit Crypto S-Boxes in Smart Applications

Sankhanil Dey and Ranjan Ghosh

Abstract 4-bit linear relations play an important role in cryptanalysis of 4-bit crypto S-boxes or S-boxes. 4-bit finite differences have also been a major part of cryptanalysis of 4-bit S-boxes. Existence of all 4-bit linear relations have been counted for all of 16 input and 16 output 4-bit bit patterns of 4-bit crypto S-boxes said as S-boxes has been reported in linear cryptanalysis of 4-bit S-boxes. Count of existing finite differences from each element of output S-boxes to distant output S-boxes have been noted in differential cryptanalysis of 4-bit S-boxes. In this paper a brief review of these two cryptanalytic methods for 4-bit S-boxes has been introduced in a very lucid and conceptual manner. Two new analysis techniques, one to search for the existing linear approximations among the input vectors (IPVs) and output Boolean functions (BFs) of a particular S-box has also been introduced in this paper. The search is limited to find the existing linear relations or approximations in the contrary to count the number of existent linear relations among all 16, 4-bit input and output bit patterns within all possible linear approximations. Another is to find number of balanced BFs in difference output S-boxes. Better the number of balanced BFs, better the security in smart applications.

Keywords Linear cryptanalysis · Differential cryptanalysis · Substitution boxes
S-boxes · Cryptography · Cryptanalysis

1 Introduction

Main goal of the smart object For intelligent application (SOFIA) project is to make "information" in the physical world available for smart services - connecting physical world with information world [1, 2, 3]. Full access to information present in the

S. Dey (✉) · R. Ghosh
Institute of Radio Physics and Electronics, University of Calcutta, Kolkata, India
e-mail: sdrpe_rs@caluniv.ac.in; sankhanil12009@gmail.com

R. Ghosh
e-mail: rghosh47@yahoo.co.in

© Springer Nature Switzerland AG 2019
A. E. Hassanien et al. (eds.), *Security in Smart Cities: Models, Applications, and Challenges*, Lecture Notes in Intelligent Transportation and Infrastructure,
https://doi.org/10.1007/978-3-030-01560-2_10

211

embedded computing devices has a potential for large impact on the daily lives of people living in this environment [4, 5, 6]. Sharing information safely has been of utmost importance in SOFIA. Modern block ciphers have been of utmost importance in doing so. The substitution boxes are the major part of ancient as well as modern block ciphers [7, 8].

Substitution boxes or S-boxes have been a part of block ciphers from the birth of commercial computer cryptography by Horst Feistel in IBM research [9]. A 4-bit S-box contains 16 elements [10]. If they are unique and distinct then the S-box has been termed as 4-bit crypto S-box. A 4-bit crypto S-box contains 16 unique and distinct elements vary from 0 to F in hex and index of each element which are unique and distinct also. They also vary from 0 to F in hex and follow a monotonically increasing sequential order. The elements of a crypto 4-bit S-box may be sequential or partly sequential or non-sequential in order [10]. The elements of index of S-box also construct an identity crypto S-box and termed as input S-box. A brief literature study of relevant literatures has been given in Sect. 2.

In differential cryptanalysis of 4-bit crypto S-boxes the 16 distant input S-Boxes have been obtained by xor operation with each of 16 input differences varies from 0 to F in hex to 16 elements of input S-box. The 16 distant S-boxes have been obtained by shuffling the elements of the original S-box in a certain order in which the elements of the input S-boxes have been shuffled in concerned distant input S-boxes. The 16 elements of each S-box and the elements in corresponding position of corresponding distant S-box has been xored to obtain the difference S-box. The difference S-Box may or may not be a crypto S-Box since it may not have all unique and distinct elements in it. The count of each element from 0 to F in Difference S-box have been noted and put in difference distribution table (DDT) for analysis of the S-box [11, 12]. The concept has been reviewed in Sect. 3.3 of Sect. 3.

In this paper a new algorithm using 4-bit BFs for differential cryptanalysis of 4-bit S-boxes have been introduced. An input S-box can be decomposed into four 4-bit input vectors (IPVs) with decimal equivalents 255 for 4th IPV, 3855 for 3rd IPV, 13,107 for 2nd IPV, and 21,845 for 1st IPV respectively. Now we complement all IPVs one, two, three and four at a time to obtain 16 4-bit distant input S-boxes. Each of four output BFs is shifted according to the shift of four IPVs of input S-boxes to form four IPVs of distant input S-boxes to obtain distant S-boxes. The four 4-bit output BFs of S-boxes are xored bitwise with four 4-bit BFs of distant S-boxes to obtain four 4-bit difference BFs. For 16 distant output S-boxes there are 64 difference BFs. Difference BFs are checked for balanced-ness i.e. for at most uncertainty. The Table in which the balanced-nesses of 64 difference BFs have been noted is called as differential analysis table (DAT). The Theory has been elaborated in Sect. 3.4 of Sect. 3.

In linear cryptanalysis of 4-bit S-boxes, every 4-bit linear relations have been tested for a particular 4-bit crypto S-box. The presence of each 4-bit unique linear relation is checked by satisfaction of each of them for all 16, 4-bit unique input bit patterns and corresponding 4-bit output bit patterns, generated from the index of each element and each element respectively of that particular crypto S-box. If they are satisfied 8 times out of 16 operations for all 4-bit unique input bit patterns and corresponding 4-bit output bit patterns, then the existence of the 4-bit linear equation is at a stake. The probability of presence and absence of a 4-bit linear relation both are $(=8/16)^{\frac{1}{2}}$. If a 4-bit linear equation is satisfied 0 times then it can be concluded that the given 4-bit linear relation is absent for that particular 4-bit crypto S-box. If a 4-bit linear equation is satisfied 16 times then it can also be concluded that the given 4-bit linear relation is present for that particular 4-bit crypto S-box. In both the cases full information is adverted to the cryptanalysts. The concept of probability bias was introduced to predict the randomization ability of that 4-bit S-box from the probability of presence or absence of unique 4-bit linear relations. The result is better for cryptanalysts if the probability of presence or absences of unique 4-bit linear equations are far away from ½ or near to 0 or 1. If the probabilities of presence or absence of all unique 4-bit linear relations are ½ or close to ½, then the 4-bit crypto S-box is said to be linear cryptanalysis immune, since the existence of maximum 4-bit linear relations for that 4-bit crypto S-box is hard to predict [11, 12]. Heys also introduced the concept of linear approximation table (LAT) in which the numbers of times, each 4-bit unique linear relation have been satisfied for all 16, unique 4-bit input bit patterns and corresponding 4-bit output bit patterns of a crypto S-box have been noted. The result is better for a cryptanalysts if the numbers of 8s in the table are less. If numbers of 8s are much more than the other numbers in the table then the 4-bit crypto S-box is said to be more linear cryptanalysis immune [11, 12].

In another look an input S-box can be decomposed into four 4-bit input vectors (IPVs) with decimal equivalents 255 for 4th IPV, 3855 for 3rd IPV, 13,107 for 2nd IPV, and 21,845 for 1st IPV respectively. The S-box can also be decomposed into 4, 4-bit output BFs (OPBFs). Each IPV can be denoted as a input variable of a linear relation and OPBF as a output variable and '+' as xor operation. Linear relations have been checked for satisfaction and 16-bit output variables (OPVs) due to linear relations have been checked for balanced-ness. Balanced OPVs indicates, out of 16 bits of IPVs and OPBFs, 8 bits satisfies the linear relation and 8 bits is out of satisfaction, i.e. best uncertainty. 256 4-bit linear relations have been operated on 4, 16-bit IPVs and 4, 16-bit OPBFs and 256 OPVs have been generated. The count of number of 1s in OPVs have been put in linear approximation table or LAT. Better the number of 8s in lAT, better the S-box security[HH96][HH02]. The concept has been reviewed in brief in Sect. 4.2 of Sect. 4.

In this paper a new technique to find the existing linear relations or linear approximations for a particular 4-bit S-Box has been introduced. If the nonlinear part of the ANF equation of a 4-bit output BF is absent or calculated to be 0 then the equation is termed as a linear relation or approximation. Searching for number of existing linear relations through this method is ended up with number of existing linear relations i.e. the goal to conclude the security of a 4-bit bijective S-box has been attended in a very lucid manner by this method. The method has been described in Sect. 4.3 of Sect. 4.

Result and analysis of all four algorithms have been given in Sect. 5. The conclusion have been made in Sect. 6 and acknowledgements have been made in last section respectively.

2 Literature Study

In this section an exhaustive relevant literature study or review work with their
specific references has been introduced to crypto literature. In Sect. 2.1 the relevant
topic has been cryptography and cryptology, in Sect. 2.2 the topic has been linear
cryptanalysis, in Sect. 2.3 the topic has been differential cryptanalysis, in Sect. 2.4
the topic has been cryptanalysis of stream ciphers and in Sect. 2.5 the relevant topic
has been Strict Avalanche Criterion (SAC) of substitution boxes. At last a literature
study on IPs and primitive polynomials have been given in Sect. 2.6.

2.1 Cryptography and Cryptology

In the end of twentieth century a bible of cryptography had been introduced [13].
The various concepts involved in cryptography and also some information on cryptanalysis had been provided to crypto-community in late nineties [14]. A simplified
version of DES. that has the architecture of DES but has much lesser rounds and
much lesser bits had also been proposed at the same time. The cipher has also been
better for educational purposes [15]. Later in early twenty first century an organized
pathway towards learning how to cryptanalyze had been charted [16]. Almost at the
same time a new cipher as a candidate for the new AES, main concepts and issues
involve in block cipher design and cryptanalysis had also been proposed [17] that
is also a measure of cipher strength. A vital preliminary introduction to cryptanalysis has also been introduced to cryptanalysts [18]. At the same time somewhat
similar notion as [18] but uses a more descriptive approach and focused on linear
cryptanalysis and differential cryptanalysis of a given SPN cipher had been elaborated [19]. Particularly, it discusses DES-like ciphers that had been extended with
it [20]. Comparison of modes of operations such as CBC, CFB, OFB and ECB had
also been elaborated [21]. A new cipher called 'Camelia' had been introduced with
its cryptanalysis technique to demonstrate the strength of the cipher [22]. History
of commercial computer cryptography and classical ciphers and the effect of cryptography on society had also been introduced in this queue [23]. The requirements
of a good cryptosystem and cryptanalysis had also been demonstrated later [24].
Description of Rijndael, the new AES provides good insight into many creative
cryptographic techniques that increases cipher strength had been included in literature. A bit Later a highly mathematical path to explain cryptologic concepts had also
been introduced [25]. Investigation of the security of Ron Rivest's DESX construction, a cheaper alternative to Triple DES had been elaborated [26]. A nice provision
to an encyclopedic look at the design, analysis and applications of cryptographic
techniques had been depicted later [27] and last but not the least a good explanation
on why cryptography has been hard and the issues which cryptographers have to
consider in designing ciphers had been elaborated [28]. Simplified Data Encryption
Standard or S-DES is an educational algorithm similar to Data Encryption Standard

(DES) but with much smaller Parameters [29, 30]. The technique to analyze S-DES using linear cryptanalysis and differential cryptanalysis has been of keen interest of crypto-community later [29, 30]. The encryption and decryption algorithm or cipher of twofish algorithm had been introduced to crypto community and a cryptanalysis of the said cipher had also been elaborated in subject to be a part of advance encryption algorithm proposals [31].

2.2 Some Old and Recent References on Linear Cryptanalysis

The cryptanalysis technique to 4-bit crypto S-boxes using linear relations among four, 4-bit input vectors (IPVs) and four, output 4-bit Boolean functions (OPBFs) of a 4-bit S-box have been termed as linear cryptanalysis of 4-bit crypto S-boxes [32, 33]. Another technique to analyze the security of a 4-bit crypto S-box using all possible finite differences had also been termed as differential cryptanalysis of 4-bit crypto S-boxes [32, 33]. The search for best characteristic in linear cryptanalysis and the maximal weight path in a directed graph and correspondence between them had also been elaborated with proper example [34]. It had also been proposed that the use of correlation matrix as a natural representation to understand and describe the mechanism of linear cryptanalysis [35]. It was also formalized the method described in [36] and showed that at the structural level, linear cryptanalysis has been very similar to differential cryptanalysis. It was also used for further exploration into linear cryptanalysis [37]. It had also been provided with a generalization of linear cryptanalysis and suggests that IDEA and SAFER K-64 have been secure against such generalization [38]. It had been surveyed to the use of multiple linear approximations in cryptanalysis to improve efficiency and to reduce the amount of data required for cryptanalysis in certain circumstances [39]. Cryptanalysis of DES cipher with linear relations [36] and the improved version of the said cryptanalysis [36] with 12 computers had also been reported later [40]. The description of an implementation of Matsui's linear cryptanalysis of DES with strong emphasis on efficiency had also been reported [41]. In early days of this century the cryptanalytic attack based on multiple linear approximations to AES candidate 'Serpent' had also been reported [42]. Later a Technique to prove security bounds against linear and differential cryptanalytic attack using Mixed-Integer Linear Programming (MILP) had also been elaborated [43]. Later to this on the strength of two variants of reduced round lightweight block cipher SIMON-32 and SIMON-48 had been tested against linear cryptanalysis and had been presented the optimum possible results [44]. Almost at the same time the strength of another light weight block ciphers SIMECK had been tested against linear cryptanalysis [45]. The fault analysis of light weight block cipher SPECK and linear cryptanalysis with zero statistical correlation among plaintext and respective cipher text of reduced round lightweight block cipher SIMON to test its strength had also been introduced in recent past [46].

2.3 Some Old and Recent References on Differential Cryptanalysis

The design of a Feistel cipher with at least 5 rounds that has been resistant to differential cryptanalysis had been reported to crypto community [47]. The exploration of the possibility of defeating differential cryptanalysis by designing S-boxes with equiprobable output XORs using bent functions had been reported once [48]. The description of some design criteria for creating good S-boxes that are immune to differential cryptanalysis and these criteria are based on information theoretic concepts had been reported later [49]. It had been Introduced that the differential cryptanalysis on a reduced round variant of DES [50] and broke a variety of ciphers, the fastest break being of two-pass Snefru [51] and also described the cryptanalysis of the full 16-round DES using an improved version [50, 52]. It had been shown that there have been DES-like iterated ciphers that does not yield to differential cryptanalysis [53] and also introduced the concept of Markov ciphers and explained its significance in differential cryptanalysis. It had also been Investigated that the security of iterated block ciphers shows how to and when an r-round cipher is not vulnerable to attacks [54]. It had also been proposed that eight round Twofish can be attacked and investigated the role of key dependent S-boxes in differential cryptanalysis [55]. It had been on the same line with [48] but proposed that the input variables be increased and that the S-box be balanced to increase resistance towards both differential and linear cryptanalysis [56]. Early in this century in previous decade estimation of probability of block ciphers against linear and differential cryptanalytic attack had been reported. Later a new algebraic and statistical technique of cryptanalysis against block cipher PRESENT-128 had been reported [57]. Almost 3 year later a new technique entitled Impossible Differential Cryptanalysis had also been reported [58]. A detailed comparative study of DES based on the strength of Data Encryption Standard (DES) against linear cryptanalysis and differential cryptanalysis had been reported later [59]. At last constraints of programming models of chosen key differential cryptanalysis had been reported to crypto community [60].

2.4 Linear Cryptanalysis and Differential Cryptanalysis of Stream Ciphers

In late 20th century A stepping stone of the Differential-Linear cryptanalysis method that is a very efficient method against DES had also been grounded [61]. The relationship between linear and differential cryptanalysis and present classes of ciphers which are resistant towards these attacks had also been elaborated [62]. Description of statistical cryptanalysis of DES, a combination and improvement of both linear and differential cryptanalysis with suggestion of the linearity of S-boxes have not been very important had been depicted [63]. Later in 21st century description of analysis with multiple expressions and differential-linear cryptanalysis with experimental results of an implementation of differential-linear cryptanalysis with multiple expressions applied to DES variants had also been proposed [64]. At the same time the attack on 7 and 8 round Rijndael using the square method with a related-key attack that can break 9 rounds Rijndael with 256 bit keys had been described [65]. In Late or almost end of 20th century the strength of stream ciphers have been tested against differential cryptanalytic attack [66]. Later the strength of them had also been tested against linear cryptanalytic attack [67]. A separate method of linear cryptanalytic attack had been reported once [68]. At least 6 years later the strength of stream cipher Helix had been tested against differential cryptanalytic attack [69]. Later the strength of stream ciphers Py, Py6, and Pypy had also been tested again differential cryptanalytic attack [70]. Recently the test of strength of stream cipher ZUC against differential cryptanalytic attack had also been reported to crypto community [71].

2.5 Strict Avalanche Criterion (SAC) of S-Boxes

In beginning Strict Avalanche Criterion of 4-bit Boolean Functions and Bit Independence Criterion of 4-bit S-boxes had been introduced [72] and design of good S-boxes based on these criteria had also been reported later [73]. In end of 20th century the construction of secured S-boxes to satisfy Strict Avalanche Criterion of S-boxes had been reported with ease [74]. The test of 4-bit Boolean functions to satisfy higher order Strict Avalanche Criterion (HOSAC) have had also been illustrated [75]. In early twenty first century the analysis methods to Strict Avalanche Criterion (SAC) had been reported. A new approach to test degree of suitability of S-boxes in modern block ciphers had been introduced to crypto-community [76]. 16!, 4-bit S-boxes had also been tested for optimum linear equivalent classes later [77]. The strength of several block ciphers against several cryptanalytic attacks had been tested and reported later [78]. Recently the key dependent S-boxes and simple algorithms to generate key dependent S-boxes had been reported [79]. An efficient cryptographic S-box design using soft computing algorithms have had also been reported [80]. In recent past the cellular automata had been used to construct good S-boxes [81].

2.6 Polynomials

In early twentieth century Radolf Church initiated the search for irreducible polynomials over Galois Field GF(p^q) for p = 2, 3, 5 and 7 and for p = 2, q = 1 through 11, for p = 3, q = 1 through 7, for p = 5, q = 1 through 4 and for p = 7, q = 1 through 3 respectively. A manual polynomial multiplication among respected EPs gives RPs in the said Galois field. All RPs have been cancelled from the list of BPs to give IPs over the said Galois field GF(p^q) [82]. Later the necessary condition for a BP to bean IPs had been generalized to even 2 characteristics. It had also been applied to RPs and gives irreducible factors mod 2 [83]. Next to it elementary techniques to compute over finite fields or Galois field $GF(p^q)$ had been descried with proper modifications [83]. In next the factorization of polynomials over Galois field $GF(p^q)$ had been elaborated [84]. Later appropriate coding techniques of polynomials over Galois field $GF(p^q)$ had been illustrated with example [85]. The previous idea of factorizing polynomials over Galois field $GF(p^q)$ [84] had also been extended to large value of P or large finite fields [86]. Later few probabilistic algorithms to find IPs over Galois Field $GF(p^q)$ for degree q had been elaborated with example [87]. Later factorization of multivariate polynomials over Galois fields $GF(p)$ had also been introduced to mathematics community [88]. With that the separation of irreducible factors of BPs [89] had also been introduced later [90]. Next to it the factorization of BPs with Generalized Reimann Hypothesis (GRH) had also been elaborated [91]. Later a probabilistic algorithm to find irreducible factors of basic bivariate Polynomials over Galois field $GF(p^q)$ had also been illustrated [92]. Later the conjectural deterministic algorithm to find primitive elements and relevant primitive polynomials over binary Galois field $GF(2)$ had been introduced [93]. Some new algorithms to find IPs over Galois field $GF(p)$ had also been introduced at the same time [94]. Another use of Generalized Reimann Hypothesis (GRH) to determine irreducible factors in a deterministic manner and also for multiplicative subgroups had been introduced later [95]. The table binary equivalents of binary primitive polynomials had been illustrated in literature [96]. The method to find roots of primitive polynomials over binary Galois field $GF(2)$ had been introduced to mathematical community [97]. A method to search for IPs in a random manner and factorization of BPs or to find irreducible factors of BPs in a random fashion had been introduced later [98]. After that a new variant of Rabin's algorithm [99] had been introduced with probabilistic analysis of BPs with no irreducible factors [100]. Later a factorization of univariate polynomials over Galois field $GF(p)$ in sub quadratic execution time had also been notified [101]. Later a deterministic algorithm to factorized IPs over one variable had also been introduced [102]. An algorithm to factorize bivariate polynomials over Galois field $GF(p)$ with hensel lifting had also been notified [103]. Next to it an algorithm had also been introduced to find factor of irreducible and almost primitive polynomials over Galois field $GF(2)$ [104]. Later a deterministic algorithm to factorize polynomials over Galois field $GF(p)$ to distinct degree factors had also been notified [105]. A detailed study of multiples and products of univariate primitive polynomials over binary Galois field $GF(2)$ had also been done [106]. Later algorithm to find optimal IPs over extended binary Galois field $GF(2^m)$ [107] and a deterministic algorithm to determine pascal polynomials over Galois field $GF(2)$ [108] had been added to literature. Later the search of IPs and primitive polynomials over binary Galois field $GF(2)$ had also been done successfully [109]. At the same time the square free polynomials had also been factorized [48] where a work on divisibility of trinomials by IPs over binary Galois field $GF(2)$ [110] had also been notified. Later a probabilistic algorithm to factor polynomials over finite fields had been introduced [111]. An explicit factorization to obtain irreducible factors to obtain for cyclotomic polynomials over Galois field $GF(p^q)$ had also been reported later [112]. A fast randomized algorithm to obtain IPs over a certain Galois field $GF(p^q)$ had been notified [113]. A deterministic algorithm to obtain factors of a polynomial over Galois field $GF(p^q)$ had also been notified at

Table 1 4-bit crypto S-box

Row	Column	1	2	3	4	5	6	7	8	9	A	B	C	D	E	F	G
1	Index	0	1	2	3	4	5	6	7	8	9	A	B	C	D	E	F
2	S-Box	E	4	D	1	2	F	B	8	3	A	6	C	5	9	0	7

the same time [114]. A review of construction of IPs over finite fields and algorithms to factor polynomials over finite fields had been reported to literature [115, 116]. An algorithm to search for primitive polynomials had also been notified at the same time [117]. The residue of division of BPs by IPs must be 1 and this reported to literature a bit later [118]. The IPs with several coefficients of different categories had been illustrated in literature a bit later [119]. The use of zeta function to factor polynomials over finite fields had been notified later on [120] At last Integer polynomials had also been described with examples [121].

3 A Brief Review of Differential Cryptanalysis of 4-Bit Crypto S-Boxes and a New Technique with Boolean Functions for Differential Cryptanalysis of 4-Bit S-Boxes

The given 4-bit crypto S-box has been described in Sect. 3.1. The relation between 4-bit crypto S-boxes and 4-bit BFs has been illustrated in Sect. 3.2. The differential cryptanalysis of 4-bit crypto S-boxes and DDT or differential distribution Table has been illustrated in Sect. 3.3. The differential cryptanalysis of 4-bit S-boxes with 4-bit BFs has been described in Sect. 3.4.

3.1 4-Bit Crypto S-Boxes

A 4-bit crypto S-box can be written as follows in Table 1, where the each element of the first row of Table 1, entitled as index, are the position of each element of the S-box within the given S-box and the elements of the 2nd row, entitled as S-box are the elements of the given substitution box. It can be concluded that the 1st row is fixed for all possible crypto S-boxes. The values of each element of the 1st row are distinct, unique and vary between 0 and F in hex. The values of the each element of the 2nd row of a crypto S-box are also distinct and unique and also vary between 0 and F in hex. The values of the elements of the fixed 1st row are sequential and monotonically increasing where for the 2nd row they can be sequential or partly sequential or non-sequential. Here the given substitution box is the 1st 4-bit S-box of the 1st S-Box out of 8 of Data Encryption Standard [122, 32, 33].

Table 2 Decomposition of 4-bit input S-box and given S-box (1st 4-bit S-box of 1st S-box out of 8 of DES) to 4-bit BFs

Row	Column	1	2	3	4	5	6	7	8	9	A	B	C	D	E	F	G	H. Decimal Equivalent
1	Index	0	1	2	3	4	5	6	7	8	9	A	B	C	D	E	F	
2	IPV4	0	0	0	0	0	0	0	0	1	1	1	1	1	1	1	1	00255
3	IPV3	0	0	0	0	1	1	1	1	0	0	0	0	1	1	1	1	03855
4	IPV2	0	0	1	1	0	0	1	1	0	0	1	1	0	0	1	1	13107
5	IPV1	0	1	0	1	0	1	0	1	0	1	0	1	0	1	0	1	21845
6	S-box	E	4	D	1	2	F	B	8	3	A	6	C	5	9	0	7	–
7	OPBF4	1	0	1	0	0	1	1	1	0	1	0	1	0	1	0	0	42836
8	OPBF3	1	1	1	0	0	1	0	0	0	0	1	1	1	0	0	1	58425
9	OPBF2	1	0	0	0	1	1	1	0	1	1	1	0	0	0	0	1	36577
A	OPBF1	0	0	1	1	0	1	1	0	1	0	0	0	1	1	0	1	13965

3.2 Relation Between 4-Bit S-Boxes and 4-Bit Boolean Functions (4-Bit BFs)

Index of each element of a 4-bit crypto S-box and the element itself is a hexadecimal number and that can be converted into a 4-bit bit sequence that are given in column 1 through G of row 1 and row 6 under row heading Index and S-box respectively. From row 2 through 5 and row 7 through A of each column from 1 through G of Table 2 shows the 4-bit bit sequences of the corresponding hexadecimal numbers of the index of each element of the given crypto S-box and each element of the crypto S-box itself. Each row from 2 through 5 and 7 through A from column 1 through G constitutes a 16 bit, bit sequence that is a 16 bit long input vectors (IPVs) and 4-bit output BFs (OPBFs) respectively. Column 1 through G of row 2 is termed as 4th IPV, row 3 is termed as 3rd IPV, row 4 is termed as 2nd IPV and row 5 is termed as 1st IPV whereas column 1 through G of Row 7 is termed as 4th OPBF, row 8 is termed as 3rd OPBF, row 9 is termed as 2nd OPBF and row A is termed as 1st OPBF [AT90]. The decimal equivalent of each IPV and OPBF are noted at column H of respective rows.

3.3 Review of Differential Cryptanalysis of 4-Bit Crypto S-Boxes [12, 123]

In differential cryptanalysis of 4-bit crypto S-boxes, elements of 4-bit input S-box (ISB) have been xored with a particular 4-bit input difference (ID) to obtain a distant input S-box (DISB). The distant S-boxes (DSB) have been obtained from original S-box (SB) by shuffling the elements of SB in such order in the way in which the elements of ISB have been shuffled to obtain DISB for a particular ID. Each element of difference S-box (DFSB) have been obtained by the xor operation of corresponding elements of SB and DSB. The count of each hexadecimal number from 0 to F have been put into the concerned cell of differential distribution table or DDT. As the number of 0s in DDT increases, information regarding concerned output difference (OD) increases so the S-box has been determined as weak S-box. The 4-bit Sequence of each element of ISB, ID, DISB, DSB, DFSB have been given in BIN ISB, BIN ID, BIN DISB, BIN DSB, BIN DFSB respectively.

The column 1 in Table 3 from row 1 through G shows the 16 elements of ISB in a monotonically increasing sequence or order. The ISB can also be concluded as an Identity 4-bit S-box. The elements of 1st 4-bit S-box, out of 4 of 1st S-Box of Data Encryption Standard (DES) out of 8, has been considered as S-box (SB), in column 7 from row 1 through G. The elements of ID, DISB, DSB, DFSB has been shown in row 1 through G of column 3, 5, 9 and C of Table 3 respectively. The 4-bit Binary equivalents of each elements of ISB, ID, DISB, SB, DSB, DFSB, has been shown in row 1 through G of column 2, 4, 6, 8, A and B of Table 3 respectively.

The review has been done in two different views; The S-box view has been described in Sect. 3.3.1 in which the concerned column of interest are row 1 through G of column 1, 3, 5, 7, 9 and C respectively. The 4-bit binary pattern view has also been described in Sect. 3.3.2 in which concerned column of interest are row 1 through G of column 2, 4, 6, 8, A and b respectively. The Pseudo Code of two algorithms with their time complexity comparison has been illustrated in Sect. 3.3.3.

3.3.1 S-Box View of Differential Cryptanalysis of 4-Bit Crypto S-Boxes

The S-box with a particular input difference or ID from 0 to F in which all elements have the same value 'B' in hex, is not a crypto S-box but an S-box and is shown in row 1 through G of column 3 of Table 3. The distant input S-box (DISB) is shown in row 1 through G of column 5 of the said table. In DISB each row element from row 1 through G is obtained by the xor operation of the elements in corresponding positions of each element of DISB from row 1 through G of column 1 (ISB) and column 3 (ID) respectively. In ISB for each row element from row 1 through G of column 1 just in corresponding position from row 1 through G of column 7, there is an element of SB. Now in DISB the elements of ISB have been shuffled in a particular order and In DSB the corresponding elements of SB has also been shuffled in that particular order. Each element of the Difference S-Box or DFSB from row 1 through G of column C has been obtained by xor operation of each element in corresponding positions from row 1 through G of column 7 and row 1 through G of column 9 respectively. The repetition of each existing elements in DSB have been counted and put into Difference Distribution Table or DDT. It is shown in Table 4 as follows,

The count of each existing elements in DFSB have been put into difference distribution table as follows, in row 2 of Table 5. For input difference (ID) = 'B' and output difference from 0 through F of row 1.

Table 3 Table of differential cryptanalysis of 1st 4-bit S-Box of 1st S-Box out of 8 of DES

COL	1	2	3	4	5	6	7	8	9	A	B	C
ROW	ISB	Bin IS4321	ID	Bin ID 4321	DISB	Bin DISB 4321	SB	Bin OSB 4321	DSB	Bin DSB 4321	Bin DFSB 4321	DFSB
1	0	0000	B	1011	B	1011	E	1110	C	1100	0010	2
2	1	0001	B	1011	A	1010	4	0100	6	0110	0010	2
3	2	0010	B	1011	9	1001	D	1101	A	1010	0001	7
4	3	0011	B	1011	8	1000	1	0001	3	0011	0010	2
5	4	0100	B	1011	F	1111	2	0010	7	0111	0101	5
6	5	0101	B	1011	E	1110	F	1111	0	0000	1111	F
7	6	0110	B	1011	D	1101	B	1011	9	1001	0010	2
8	7	0111	B	1011	C	1100	8	1000	5	0101	1101	D
9	8	1000	B	1011	3	0011	3	0011	1	0001	0010	2
A	9	1001	B	1011	2	0010	A	1010	D	1101	0001	7
B	A	1010	B	1011	1	0001	6	0110	4	0100	0010	2
C	B	1011	B	1011	0	0000	C	1100	E	1110	0010	2
D	C	1100	B	1011	7	0111	5	0101	8	1000	1101	D
E	D	1101	B	1011	6	0110	9	1001	B	1011	0010	2
F	E	1110	B	1011	5	0101	0	0000	F	1111	1111	F
G	F	1111	B	1011	4	0100	7	0111	2	0010	0101	5

Table 4 Count of repetition of each existing element in DSB

R|C	1	2	3	4	5	6	7	8	9	A	B	C	D	E	F	G	H
1	DSB el	0	1	2	3	4	5	6	7	8	9	A	B	C	D	E	F
2	Count	0	0	8	0	0	2	0	2	0	0	0	0	0	2	0	2

Table 5 The part of DDT with input difference 'B'

Row	1	Output difference														
1	Input difference	0	1	2	3	4	5	6	7	8	9	A	B	C	D	E	F
2	B	0	2	8	0	0	2	0	0	0	0	0	0	2	0	2

3.3.2 4-Bit Binary Pattern View of Differential Cryptanalysis of 4-Bit Crypto S-Boxes

The corresponding four bit bit patterns of input S-box elements (ISB) has been shown from row 1 through G of column 2 in Table 3 and termed as Bin ISB. The particular input difference '1101' is shown in each row from 1 through G of column 4 in Table 3. The Distant 4-bit input bit patterns are shown from row 1 through G of column 6 (Bin DISB) are obtained by the xor operation of the elements in corresponding positions of each element of BIN DISB from row 1 through G of column 2 (Bin ISB) and Column 4 (Bin ID) respectively. In Bin ISB for each element from row 1 through G of column 2 in corresponding position from row 1 through G of column 8, there is an element of Bin SB. Now in Bin DISB the elements of ISB have been shuffled in a particular order and in Bin DSB the corresponding elements of SB has also been shuffled in that particular order. Each element from row 1 through G of column 11 has been obtained by xor operation of each element in corresponding positions from row 1 through G of column 8 and row 1 through G of column 10 respectively. The repetition of each existing elements in Bin DFSB have been counted and put into difference distribution table or DDT. It is shown in Table 6 as follows,

The count of each existing elements in Bin DFSB have been put into the difference distribution table as follows, in row 2 of Table 7 for binary input difference (Bin ID) '1101' and output difference from 0 through F of row 1.

The Total DDT or difference distribution table for 16 IDs for the given S-box has been shown below in Table 8.

3.3.3 Pseudo Code for Differential Cryptanalysis of 4-Bit Crypto S-Boxes and Its Time Complexity Analysis

The pseudo code of algorithm for 4-bit binary pattern view with time complexity has been depicted in Sect. 3.3.3.1. The pseudo code of algorithm for S-box view with time complexity has been depicted in Sect. 3.3.3.2 and the comparison of time complexity of two algos has been given in Sect. 3.3.3.3.

Table 6 Count of repetition of each existing element in Bin DSB

RIC	1	2	3	4	5	6	7	8	9	A	B	C	D	E	F	G	H
1	DFSB el	0000	0001	0010	0011	0100	0101	0110	0111	1000	1001	1010	1011	1100	1101	1110	1111
2	Count	0	2	8	0	0	2	0	0	0	0	0	0	0	2	0	2

Table 7 The part of DDT with input difference '1101'

Row	1	Output difference (in Hex)														
1	Input difference	0	1	2	3	4	5	6	7	8	9	A	B	C	D	E	F
2	1101	0	2	8	0	0	2	0	0	0	0	0	0	0	2	0	2

Pseudo Code of Algorithm of Differential Cryptanalysis 4-Bit Binary Pattern View

The pseudo code has been given as follows,

Start. // Start of pseudo code
// Variable declarations, two dimensional array ISB[4][16] is for 4-bit bit patterns for Input S-box, IDIFF[4][16] is for 4-bit bit patterns of input difference, three dimensional Array ODIFF[4][16][16] is for all 4-bit bit patterns of output difference for 16 IDIFFs.
Step 0A: int ISB[4][16]; int IDIFF[4][16]; int ODIFF[4][16][16];
// Variable declarations, ISB'[4][16][16] is for 4-bit bit patterns of all elements of 16 distant ISBs. OSB[4][16] is for 4-bit bit patterns of the given S-box or output S-box, OSB'[4][16][16] is for 4-bit bit patterns of all elements of 16 distant OSBs, DDT[16][16] is for difference distribution table, and Count[16] is for count of each element in ODIFF for 16 OSBs.
Step 0B: int ISB'[4][16][16]; int OSB[4][16]; int OSB'[4][16][16]; int DDT[16][16]; int Count[16];
// Differential cryptanalysis block.
Step 01: For I = 1:16; For J = 1:16; For K = 1:4; // Start of for loop I, J, K respectively

$$ISB'[K][I][J] = ISB[K][J]^\wedge IDIFF[K][I];$$
$$OSB'[K][I][J] = OSB[ISB'[K][I][J]];$$
$$ODIFF[K][I][J] = OSB[K][J]^\wedge OSB'[K][I][J];$$

End For K. End For J. End For I.// End of for loop K, J, I respectively
// Generation of DDT.
Step 02: For I = 1:16 For J = 1:16 For K = 1:4 // Start of For Loop I, J, K respectively

$$DDT[I][J] = Count[ISB[K][J]];$$

End For K. End For J. End For I. // End of for loop K, J, I respectively
Stop. // End of pseudo code
Time complexity of the given algorithm. Since differential cryptanalysis block contains 3 nested loops so the time complexity of the algorithm has been $O(n^3)$.

Table 8 Difference distribution table or DDT of the given S-box

Table 7b DDT

Input Difference	Output difference															
	0	1	2	3	4	5	6	7	8	9	A	B	C	D	E	F
0	16	0	0	0	0	0	0	0	0	0	0	0	0	0	0	0
1	0	0	0	2	0	0	0	2	0	2	4	0	4	2	0	0
2	0	0	0	2	0	6	2	2	0	2	0	0	0	0	2	0
3	0	0	2	0	2	0	0	0	0	4	2	0	2	0	0	4
4	0	0	0	2	0	0	6	0	0	2	0	4	2	0	0	0
5	0	4	0	0	0	2	2	0	0	0	4	0	2	0	0	2
6	0	0	0	4	0	4	0	0	0	0	0	0	2	2	2	2
7	0	0	2	2	2	0	2	2	0	2	2	0	0	0	0	4
8	0	0	0	0	0	0	2	4	2	0	0	4	0	4	2	2
9	0	2	0	0	2	0	0	0	6	0	2	2	2	0	0	0
A	0	2	2	0	0	0	0	2	0	0	0	2	0	0	4	0
B	0	0	8	0	0	2	0	0	0	0	0	0	0	2	0	2
C	0	2	0	0	2	2	2	4	2	0	0	2	0	6	0	0
D	0	4	0	0	0	0	0	0	6	0	2	0	2	0	2	0
E	0	0	2	4	2	0	0	0	0	0	0	0	0	0	2	0
F	0	2	0	0	6	0	0	0	0	4	0	2	0	0	2	0

Table 9 Time complexity comparison of two algorithms

View	4-bit BP view	S-box view
Time complexity	$O(n^3)$	$O(n^2)$

Pseudo Code of Algorithm of Differential Cryptanalysis S-Box View

The pseudo code has been given as follows,

Start. // Start of pseudo code
// Variable declarations, one dimensional array ISB[16] is for input S-box in Hex, IDIFF[16] is for input difference in Hex, three dimensional array ODIFF[16][16] is for all output difference in Hex for 16 IDIFFs.
Step 0A: int ISB[16]; int IDIFF[16]; int ODIFF[16][16];
// Variable declarations, ISB'[16][16] is for all elements in Hex of 16 distant ISBs. OSB[16] is for elements in Hex of the given S-box or output S-box, OSB'[16][16] is for all elements in Hex of 16 distant OSBs, DDT[16][16] is for DDT, and Count[16] is for count of each element in ODIFF for 16 OSBs.
Step 0B: int ISB'[16][16]; int OSB[16]; int OSB'[16][16]; int DDT[16][16]; int Count[16].
// Differential cryptanalysis block
Step 01: For I = 1:16; For J = 1:16; // For loop I and J respectively.

$$ISB'[I][J] = ISB[J]^\wedge IDIFF[I];$$

$$OSB'[I][J] = OSB[ISB'[I][J]];$$

$$ODIFF[I][J] = OSB[J]^\wedge OSB'[I][J];$$

End For J. End For I.// End of for loop J and I respectively.
Step 02: For I = 1:16; For J = 1:16 // For loop I and J respectively.

$$DDT[I][J] = Count[ISB[J]];$$

End For J. End For I. // End of for loop J and I respectively.
Stop. // End of pseudo code
Time complexity of the given algorithm. Since differential cryptanalysis block contains 2 nested loops so the time complexity of the algorithm has been $O(n^2)$.

Comparison of Time Complexity of Two Views of Differential Cryptanalysis of 4-Bit Crypto S-Boxes

The comparison of time complexity of two algos has been given in Table 9 as follows,
     It can be concluded from the comparison that the execution time reduces in S-box view than the 4-bit Binary Pattern view. So in can be concluded from above review

Table 10 Complement of IPVs due to a particular ID

ID	1	0	1	1
Complement	C	N	C	C

work that the execution time of differential cryptanalysis depends upon the view of the algorithm and the S-box view has been proved to be a better algorithm than 4-bit binary pattern view algorithm.

3.4 Differential Cryptanalysis of 4-Bit Crypto S-Boxes Using 4-Bit BFs

The procedure to obtain four input vectors (IPVs) and four output BFs (OPBFs) from the elements of a particular 4-bit crypto S-box has been described in Sect. 2.1. The procedure to obtain distant input vectors (DIPVs) and distant output BFs (DOPBFs) for a particular input difference (ID) of the said S-box has been described with example in Sect. 3.4.1. Generation of difference 4-bit BFs, analysis of algorithm and generation of difference analysis algorithm in Sects. 3.4.2, 3.4.3 and 3.4.4 respectively. The differential analysis table of the given S-box, Pseudo code of algorithm with time complexity and comparison of time complexity of three Algos have been given in Sects. 3.4.5, 3.4.6 and 3.4.7 respectively.

3.4.1 Distant Input BFs (DIBFs) and Distant Output BFs (DOBFs) Generation from IBFs and OBFs for a Specific ID

Within 4 bits of binary input difference (Bin ID), 1 in position p means do complement of pth IPV and 0 means no operation on pth IPV. Similarly in the given example 1 in position 4 of Bin ID, as in position 4 from row 1 through G of column 4 of Table 3 indicates do complement of 4-bit IPV, IPV4 i.e. CIPV4 and 0 in position 3 as in position 3 from row 1 through G of column 4 of Table 3 means no operation on 4-bit IPV, IPV3 (CIPV3) or CIPV3 = IPV3. Similarly 1 in respective positions 2 and 1 as in positions 2 and 1 from row 1 through G of column 4 of Table 3 means do complement 4-bit IPV, IPV2 (CIPV2) and do complement of 4-bit IPV, IPV1 (CIPV1) respectively. CIPV4, CIPV3, CIPV2 and CIPV1 for Input S-box (ISB) and Input Difference (ID) have been shown from row 1 through G of column 1 and column 3 of Table 3 respectively (Table 10).

Here the 4th OPBF has been taken as an example of OPBF and termed as OPBF. Since complement of 4th IPV means interchanging each 8 bit halves of 16 bit long 4th IPV so The 2, 8 bit halves of OPBF have been interchanged due to complement of 4th IPV. The resultant OPBF has been shown from column 1 through G of row 6 in Table 11. Again no operation on 3rd IPV means CIPV3 = IPV3 so resultant

Table 11 Construction of DIBFs and DOBFs

Row|Col		1	2	3	4	5	6	7	8	9	A	B	C	D	E	F	G
1	CIPV4	1	1	1	1	1	1	1	1	0	0	0	0	0	0	0	0
2	CIPV3	1	1	1	1	0	0	0	0	1	1	1	1	0	0	0	0
3	CIPV2	0	0	1	1	0	0	1	1	0	0	1	1	0	0	1	1
4	CIPV1	1	0	1	0	1	0	1	0	1	0	1	0	1	0	1	0
5	OPBF	1	0	1	0	0	1	1	1	0	1	0	1	0	1	0	0
6	STEP1	0	1	0	1	0	1	0	0	1	0	1	0	0	1	1	1
7	STEP2	0	1	0	1	0	1	0	0	1	0	1	0	0	1	1	1
8	STEP3	0	1	0	1	0	0	0	1	1	0	1	0	1	1	0	1
9	STEP4	1	0	1	0	0	0	1	0	0	1	0	1	1	1	1	0
A	COPBF	1	0	1	0	0	0	1	0	0	1	0	1	1	1	1	0

Table 12 DBF generation

R|C		1	2	3	4	5	6	7	8	9	A	B	C	D	E	F	G
1	OPBF	1	0	1	0	0	1	1	1	0	1	0	1	0	1	0	0
2	COPBF	1	0	1	0	0	0	1	0	0	1	0	1	1	1	1	0
3	DIFF	0	0	0	0	0	1	0	1	0	0	0	0	1	0	1	0

OPBF is as same as STEP1 and has been shown from column 1 through G of row 7 in Table 11. Next to it, the complement of 2nd IPV means interchanging each 2 bit halves of each 4 bit halves of each 8 bit halves of resultant OPBF. The resultant OPBF has been shown from column 1 through G of row 8 in Table 11. Again the complement of 1st IPV means interchanging each bit of each 2 bit halves of each 4 bit halves of each 8 bit halves of resultant OPBF, The resultant OPBF after operation has been shown in column 1 through G of row 9 in Table 11. The complemented OPBF has been the resultant OPBF of STEP4 and has been shown from column 1 through G of row A in Table 11.

3.4.2 Generation of Difference Boolean Functions or DBFs for a Certain ID

The DBFs of each OPBF have been generated by bitwise xor of OPBFs and the corresponding COPBFs. The corresponding DBFs of OBPF4, OBPF3, OBPF2, OPBF1 are denoted as DIFF4, DIFF3, DIFF2, DIFF1 respectively. Generation of 4th DBF of ID '1011' has been shown in column 1 through G of row 3 of Table 12.

3.4.3 Analysis

If the DBFs are balanced then the number of bits changed and remains unchanged among corresponding bits of OPBFs and COPBFs is maximum. So uncertainty of

Table 13 Balanced-ness of DBFs

R|C	1	2
1	Difference BF	Total number of 1s
2	DIFF	4

determining a particular change in bits is maximum. As the number of balanced DBFs are increased among 64 ($=2^4 \times 4$) possible DBFs then the security will increase. The number of 1s or balanced-ness of the above DBF shown from row 1 through G of row 3 of Table 12 has been shown in column 2 of row 2 of Table 13.

3.4.4 DBFs Generation and Derivation of a Particular Row of Differential Analysis Table (DAT) for a Certain ID

Four IPVs in the order IPV4, IPV3, IPV2 and IPV1 for the S-box given in Table 1 and four CIPVs, CIPV4, CIPV3, CIPV2 and CIPV1 for a certain ID '1011' have been shown from column 1 through G of row 1, 2, 3, 4, 5, 6, 7 and 8 respectively in Table 14. Four OPBFs in the order OPBF4, OPBF3, OPBF2 and OPBF1 for the S-box given in Table 1 and four COPBFs COPBF4, COPBF3, COPBF2 and COPBF1 for a certain ID '1011' have been shown from column 1 through G of row 9, A, B, C, D, E, F and G respectively in Table 14. The resultant DBFs, DIFF4, DIFF3, DIFF2, DIFF1, have been shown in column 1 through G of row H, I, J, K of Table 14. The number of 1s or Balanced-ness of four DBFs have been shown in row from column 2 through 5 of row 1 in Table 15.

3.4.5 Differential Analysis Table or DAT

The balanced-ness of four DBFs for each ID have been shown from column 2 through 5 of row 2 through H of DAT or Table 16.

3.4.6 Pseudo Code for Differential Cryptanalysis of 4-Bit Crypto S-Boxes and Its Time Complexity Analysis

The pseudo code has been given as follows,

Start. // Start of pseudo code
// Variable declarations, one dimensional array ISB[16] is for input S-box in Hex, IDIFF[16] is for input difference in Hex, three dimensional array ODIFF[16][16] is for all output difference in Hex for 16 IDIFFs. Bin_ODIFF[4][16][16] is for all 4-bit bit patterns of output difference for 16 IDIFFs.
Step 0A: int ISB[16]; int IDIFF[16]; int ODIFF[16][16];
// Variable declarations, ISB'[16][16] is for all elements in Hex of 16 distant ISBs. OSB[16] is for elements in Hex of the given S-box or output S-box, OSB'[16][16]

Table 14 Generation of a particular row of differential analysis table (DAT)

Row|Col		1	2	3	4	5	6	7	8	9	A	B	C	D	E	F	G
1	IBF4	0	0	0	0	0	0	0	0	1	1	1	1	1	1	1	1
2	IBF3	0	0	0	0	1	1	1	1	0	0	0	0	1	1	1	1
3	IBF2	0	0	1	1	0	0	1	1	0	0	1	1	0	0	1	1
4	IBF1	0	1	0	1	0	1	0	1	0	1	0	1	0	1	0	1
5	CIBF4	1	1	1	1	1	1	1	1	0	0	0	0	0	0	0	0
6	CIBF3	0	0	0	0	1	1	1	1	0	0	0	0	1	1	1	1
7	CIBF2	1	1	0	0	1	1	0	0	1	1	0	0	1	1	0	0
8	CIBF1	1	0	1	0	1	0	1	0	1	0	1	0	1	0	1	0
9	OBF4	1	0	1	0	0	1	1	1	0	1	0	1	0	1	0	0
A	OBF3	1	1	1	0	0	1	0	0	0	0	1	1	1	0	0	1
B	OBF2	1	0	0	0	1	1	1	0	1	1	1	0	0	0	0	1
C	OBF1	0	0	1	1	0	1	1	0	1	0	0	0	1	1	0	1
D	COBF4	1	0	1	0	0	0	1	0	0	1	0	1	1	1	1	0
E	COBF3	1	1	0	0	1	0	0	1	0	1	1	1	0	0	1	0
F	COBF2	0	1	1	1	1	0	0	0	0	0	0	1	0	1	1	1
G	COBF1	0	0	0	1	1	0	1	1	1	1	0	0	0	1	1	0
H	DIFF4	0	0	0	0	0	1	0	1	0	0	0	0	1	0	1	0
I	DIFF3	0	0	1	0	1	1	0	1	0	1	0	0	1	0	1	1
J	DIFF2	1	1	1	1	0	1	1	0	1	1	1	1	0	1	1	0
K	DIFF1	0	0	1	0	1	1	0	1	0	1	0	0	1	0	1	1

Table 15 Balanced-ness of four DBFs

R|C	1	2	3	4	5
	Difference BFs	DIFF4	DIFF3	DIFF2	DIFF1
1	No. of ones	4	8	C	8

is for all elements in Hex of 16 distant OSBs, DAT[16][16] is for difference analysis table, and Count[16] is for count of each element in ODIFF for 16 OSBs.

Step 0B: int ISB'[16][16]; int OSB[16]; int OSB'[16][16]; int DAT[4][16]; int Count[16].

// Differential cryptanalysis block

Step 01: For $I = 1:16$; For $J = 1:16$; // For loop I and J respectively.

$$ISB'[I][J] = ISB[J]^\wedge IDIFF[I];$$

$$OSB'[I][J] = OSB\big[ISB'[I][J]\big];$$

$$ODIFF[I][J] = OSB[J]^\wedge OSB'[I][J];$$

For $K = 1:4$ Bin_ODIFF[K][I][J] = Hex to Binary(ODIFF[I][J])
End For J. End For I.// End of for loop J and I respectively.
Step 03: For $I = 1:4$; For $J = 1:16$; For $K = 1:16$ // For Loop I and J respectively.

Table 16 DAT for 1st 4-bit S-Box of 1st S-Box of DES

R\C	1	2	3	4	5
1	ID in Hex	DIFF1	DIFF2	DIFF3	DIFF4
2	0	0	0	0	0
3	1	8	8	8	C
4	2	C	8	C	4
5	3	8	8	8	C
6	4	8	C	8	8
7	5	8	8	8	8
8	6	C	8	C	8
9	7	8	C	8	8
A	8	C	C	C	C
B	9	8	8	8	8
C	10	4	8	4	C
D	11	8	C	8	4
E	12	C	4	C	8
F	13	8	8	8	8
G	14	4	8	4	8
H	15	8	4	8	8

Table 17 Time complexity comparison of three algos

View	4-bit BP view	S-box view	With 4-bit BFs
Time complexity	$O(n^3)$	$O(n^2)$	$O(n^2)$

$$DAT[I][J] = Count[Bin\_ODIFF[I][J][K]];$$

End For J. End For I. // End of for loop J and I respectively.
Stop. // End of pseudo code
Time complexity of the given algorithm. Since differential cryptanalysis block contains 2 nested loops so the time complexity of the algorithm has been $O(n^2)$.

3.4.7 Comparison of Time Complexity of Two Views of Differential Cryptanalysis of 4-Bit S-Boxes and Differential Cryptanalysis with 4-Bit BFs

The comparison of time complexity of three algos has been given in Table 17 as follows,

4 A Brief Review of Linear Cryptanalysis of 4-Bit Crypto S-Boxes and a New Technique with Boolean Functions for Linear Cryptanalysis of 4-Bit Crypto S-Boxes or Linear Approximation Analysis

The review of related relevant property of 4-bit BFs, algebraic normal form (ANF) of 4-bit BFs has been illustrated in Sect. 4.1. The review of linear cryptanalysis of 4-bit Crypto S-boxes has been described in brief in Sect. 4.2. At last the new technique to analyze 4-bit S-boxes by existing 4-bit linear approximations or linear approximation analysis has been described in brief in Sect. 4.3.

4.1 A Review of Boolean Functions (BF) and Its Algebraic Normal Form (ANF)

A 4-bit Boolean Function (BF) accepts 4 bits as input $\{x_1x_2x_3x_4\}$ having 16 combinations of decimal values varying between 0 and 15 and provides 1-bit output for each combination of input. The input-output relation is given in a Truth Table which provides 16-bit output vector corresponding to four 16-bit input vectors, each one attached to x_1, x_2, x_3 and x_4. The 4-bit BF is a mapping from $(0,1)^4$ to $(0,1)^1$ and its functional relation, $F(x)$ can be expressed in Algebraic Normal Form (ANF) with 16 coefficients as given in Eq. (1),

$$F(x) = a_0 + (a_1.x_1 + a_2.x_2 + a_3.x_3 + a_4.x_4) + (a_5.x_1.x_2 + a_6.x_1.x_3 + a_7.x_1.x_4$$
$$+ a_8.x_2.x_3 + a_9.x_2.x_4 + a_{10}.x_3.x_4) + (a_{11}.x_1.x_2.x_3 + a_{12}.x_1.x_2.x_4 + a_{13}.x_1.x_3.x_4$$
$$+ a_{14}.x_2.x_3.x_4) + a_{15}.x_1.x_2.x_3.x_4 \tag{1}$$

where x represents the decimal value or the hex value of 4 input bits represented by $\{x_1x_2x_3x_4\}$, BF assumes 1-bit output, '.' and '+' represent AND and XOR operations respectively. Here a_0 is a constant coefficient, (a_1-a_4) are 4 linear coefficients, and (a_5-a_{15}) are 11 nonlinear coefficients of which (a_5-a_{10}) are 6 non-linear coefficients of 6 terms with 2-AND-operated-input-bits, $(a_{11}-a_{14})$ are 4 nonlinear coefficients of 4 terms with 3-AND-operated-input-bits and a_{15} is a non-linear coefficient of one term with 4-AND-operated-input-bits. The 16 binary ANF coefficients, from a_0 to a_{15} are marked respectively as anf.bit0 to anf.bit15 in ANF representation and are evaluated from the 16-bit output vector of a BF designated as bf.bit0 to bf.bit15 using the following relations as given in Eq. (2),

$$anf.bit0 = bf.bit0;$$
$$anf.bit1 = anf.bit0 + bf.bit8;$$
$$anf.bit2 = anf.bit0 + bf.bit4;$$
$$anf.bit3 = anf.bit0 + bf.bit2;$$
$$anf.bit4 = anf.bit0 + bf.bit1;$$
$$anf.bit5 = anf.bit0 + anf.bit1 + anf.bit2 + bf.bit12;$$
$$anf.bit6 = anf.bit0 + anf.bit1 + anf.bit3 + bf.bit10;$$
$$anf.bit7 = anf.bit0 + anf.bit1 + anf.bit4 + bf.bit9;$$
$$anf.bit8 = anf.bit0 + anf.bit2 + anf.bit3 + bf.bit6;$$
$$anf.bit9 = anf.bit0 + anf.bit2 + anf.bit4 + bf.bit5;$$
$$anf.bit10 = anf.bit0 + anf.bit3 + anf.bit4 + bf.bit3;$$
$$anf.bit11 = anf.bit0 + anf.bit1 + anf.bit2 + anf.bit3 + anf.bit5$$
$$+ anf.bit6 + anf.bit8 + bf.bit14;$$
$$anf.bit12 = anf.bit0 + anf.bit1 + anf.bit2 + anf.bit4 + anf.bit5$$
$$+ anf.bit7 + anf.bit9 + bf.bit13;$$
$$anf.bit13 = anf.bit0 + anf.bit1 + anf.bit3 + anf.bit4 + anf.bit6$$
$$+ anf.bit7 + anf.bit10 + bf.bit11;$$
$$anf.bit14 = anf.bit0 + anf.bit2 + anf.bit3 + anf.bit4 + anf.bit8$$
$$+ anf.bit9 + anf.bit10 + bf.bit7;$$
$$anf.bit15 = anf.bit0 + anf.bit1 + anf.bit2 + anf.bit3 + anf.bit4$$
$$+ anf.bit5 + anf.bit6 + anf.bit7 + anf.bit8 + anf.bit9 + anf.bit10$$
$$+ anf.bit11 + anf.bit12 + anf.bit13 + anf.bit14 + bf.bit15 \qquad (2)$$

The DEBF (Decimal Equivalent of BF) varies from 0 through 65,535 and each decimal value is converted to a 16-bit binary output of the Boolean function from bf.bit0 through bf.bit15. Based on the binary output of a BF, the ANF coefficients from anf.bit0 through anf.bit15 are calculated sequentially using Eq. (2).

4.2 A Review on Linear Cryptanalysis of 4-Bit Crypto S-Boxes [HH96][HH02]

The given 4-bit crypto S-box has been described in sub-Sect. 4.2.1. The relations of 4-bit S-boxes with 4 bit BFs and with linear approximations are described in Sects. 4.2.2 and 4.2.3 respectively. LAT or Linear Approximation Table has also been illustrated in Sect. 4.2.4. Algorithm of linear cryptanalysis with time complexity analysis has been described in Sect. 4.2.5.

Table 18 4-bit crypto S-box

Row	Column	1	2	3	4	5	6	7	8	9	A	B	C	D	E	F	G
1	Index	0	1	2	3	4	5	6	7	8	9	A	B	C	D	E	F
2	S-Box	E	4	D	1	2	F	B	8	3	A	6	C	5	9	0	7

4.2.1 4-Bit Crypto S-Boxes

A 4-bit crypto S-box can be written as follows in Table 1, where the each element of the first row of Table 18, entitled as index, are the position of each element of the S-box within the given S-box and the elements of the 2nd row, entitled as S-box are the elements of the given substitution box. It can be concluded that the 1st row is fixed for all possible crypto S-boxes. The values of each element of the 1st row are distinct, unique and vary between 0 and F in hex. The values of the each element of the 2nd row of a crypto S-box are also distinct and unique and also vary between 0 and F in hex. The values of the elements of the fixed 1st row are sequential and monotonically increasing where for the 2nd row they can be sequential or partly sequential or non-sequential. Here the given substitution box is the 1st 4-bit S-box of the 1st S-Box out of 8 of Data Encryption Standard [122, 32, 33].

4.2.2 Relation Between 4-Bit S-Boxes and 4-Bit Boolean Functions (4-Bit BFs)

Index of each element of a 4-bit crypto S-box and the element itself is a hexadecimal number and that can be converted into a 4-bit bit sequence that are given in column 1 through G of row 1 and row 6 under row heading Index and S-box respectively. From row 2 through 5 and row 7 through A of each column from 1 through G of Table 19 shows the 4-bit bit sequences of the corresponding hexadecimal numbers of the index of each element of the given crypto S-box and each element of the crypto S-box itself. Each row from 2 through 5 and 7 through A from column 1 through G constitutes a 16 bit, bit sequence that is a 16 bit long input vectors (IPVs) and 4-bit output BFs (OPBFs) respectively. Column 1 through G of row 2 is termed as 4th IPV, row 3 is termed as 3rd IPV, row 4 is termed as 2nd IPV and row 5 is termed as 1st IPV whereas column 1 through G of Row 7 is termed as 4th OPBF, row 8 is termed as 3rd OPBF, row 9 is termed as 2nd OPBF and row A is termed as 1st OPBF [122]. The decimal equivalent of each IPV and OPBF are noted at column H of respective rows.

4.2.3 4-Bit Linear Relations

The elements of input S-box have been shown under column heading 'I' and the input vectors have been shown under field IPVs (Input vectors) and subsequently

Table 19 Decomposition of 4-bit input S-box and given S-box (1st 4-bit S-box of 1st S-box out of 8 of DES) to 4-bit BFs

Row	Column	1	2	3	4	5	6	7	8	9	A	B	C	D	E	F	G	H. Decimal Equivalent
1	Index	0	1	2	3	4	5	6	7	8	9	A	B	C	D	E	F		
2	IPV4	0	0	0	0	0	0	0	0	1	1	1	1	1	1	1	1		00255
3	IPV3	0	0	0	0	1	1	1	1	0	0	0	0	1	1	1	1		03855
4	IPV2	0	0	1	1	0	0	1	1	0	0	1	1	0	0	1	1		13107
5	IPV1	0	1	0	1	0	1	0	1	0	1	0	1	0	1	0	1		21845
6	S-box	E	4	D	1	2	F	B	8	3	A	6	C	5	9	0	7		
7	OPBF4	1	0	1	0	0	1	1	1	0	1	0	1	0	1	0	0		42836
8	OPBF3	1	1	1	0	0	1	0	0	0	0	1	1	1	0	0	1		58425
9	OPBF2	1	0	0	0	1	1	1	0	1	1	1	0	0	0	0	1		36577
A	OPBF1	0	0	1	1	0	1	1	0	1	0	0	0	1	1	0	1		13965

under column headings 1, 2, 3 and 4. The 4th input vector has been depicted under column heading '4', 3rd input vector has been depicted under column heading '3', 2nd input vector has been depicted under column heading '2' and 1st input vector has been depicted under column heading '1'. The elements of S-box have been shown under column heading 'SB' and the output 4-bit BFs are shown under field OPBFs (Output Boolean functions) and subsequently under column headings 1, 2, 3 and 4. The 4th output BF has been depicted under column heading '4', 3rd output BF has been depicted under column heading '3', 2nd output BF has been depicted under column heading '2' and 1st output BF has been depicted under column heading '1' (Table 20).

The IPEs or input equations are all possible xored terms that can be formed using four IPVs 4, 3, 2 and 1. On the other hand OPEs are possible xored terms that can be formed using four OPVs 4, 3, 2 and 1. All possible IPEs and OPEs are listed under the column and also row heading (IPE = OPE) from row 2 through H and column 1 through G respectively. Each cell is a linear equation equating IPE to OPE. Such as $L_{1+2+4,2+3}$ is the linear equation formed by IPE '1 + 2 + 3' i.e. the xored combination of three IPVs 1, 2 and 4 and OPE '2 + 3' i.e. the xored combination of two OPBFs 2 and 3. The 256 possible 4-bit linear equations are shown in Table 21.

4.2.4 Linear Approximation Table (LAT) [14]

According to Heys each linear equation is tested for each of 16 4-bit patterns shown in each row under the field IPVs and subsequently under the column headings 1, 2, 3 and 4 and the corresponding 16 4-bit patterns under field OPBFs and subsequently under the column headings 1, 2, 3 and 4. If a linear equation satisfies 8 times out of 16 then the existence of the linear equation is highly unpredictable. That is the

Table 20 IPVs and OPBFs for given S-Box

I	IPVs				SB	OPBFs			
	4	3	2	1		4	3	2	1
0	0	0	0	0	E	1	1	1	0
1	0	0	0	1	4	0	1	0	0
2	0	0	1	0	D	1	1	0	1
3	0	0	1	1	1	0	0	0	1
4	0	1	0	0	5	0	1	0	1
5	0	1	0	1	9	1	0	0	1
6	0	1	1	0	0	0	0	0	0
7	0	1	1	1	7	0	1	1	1
8	1	0	0	0	2	0	0	1	0
9	1	0	0	1	F	1	1	1	1
A	1	0	1	0	B	1	0	1	1
B	1	0	1	1	8	1	0	0	0
C	1	1	0	0	3	0	0	1	1
D	1	1	0	1	A	1	0	1	0
E	1	1	1	0	6	0	1	1	0
F	1	1	1	1	C	1	1	0	0

probability is ½. If the numbers of satisfaction of each linear equation is noted in respective cells of Table 22 then it is called as Linear Approximation Table or LAT. The Linear Approximation Table for the given S-Box has been shown in Table 22.

4.2.5 Pseudo Code of Algorithm with Time Complexity Analysis of Linear Cryptanalysis of 4-Bit Crypto S-Boxes

The algorithm to execute the linear cryptanalysis for 4-bit Crypto S-boxes following Heys [12, 123] considers 4-bit Boolean variables A_i and B_j whose i and j are the

Table 21 256, 4-bit linear equations with input equations (IPE) and output equations (OPE)

Rows	Columns	C	D	E	F	G
1	IPE = OPE	$1+2+3$	$1+2+4$	$1+3+4$	$2+3+4$	$1+2+3+4$
2	0	$L_{0,1+2+3}$	$L_{0,1+2+4}$	$L_{0,1+3+4}$	$L_{0,2+3+4}$	$L_{0,1+2+3+4}$
3	1	$L_{1,1+2+3}$	$L_{1,1+2+4}$	$L_{1,1+3+4}$	$L_{1,2+3+4}$	$L_{1,1+2+3+4}$
4	2	$L_{2,1+2+3}$	$L_{2,1+2+4}$	$L_{2,1+3+4}$	$L_{2,2+3+4}$	$L_{2,1+2+3+4}$
5	3	$L_{3,1+2+3}$	$L_{3,1+2+4}$	$L_{3,1+3+4}$	$L_{3,2+3+4}$	$L_{3,1+2+3+4}$
6	4	$L_{4,1+2+3}$	$L_{4,1+2+4}$	$L_{4,1+3+4}$	$L_{4,2+3+4}$	$L_{4,1+2+3+4}$
7	$1+2$	$L_{1+2,1+2+3}$	$L_{1+2,1+2+4}$	$L_{1+2,1+3+4}$	$L_{1+2,2+3+4}$	$L_{1+2,1+2+3+4}$
8	$1+3$	$L_{1+3,1+2+3}$	$L_{1+3,1+2+4}$	$L_{1+3,1+3+4}$	$L_{1+3,2+3+4}$	$L_{1+3,1+2+3+4}$
9	$1+4$	$L_{1+4,1+2+3}$	$L_{1+4,1+2+4}$	$L_{1+4,1+3+4}$	$L_{1+4,2+3+4}$	$L_{1+4,1+2+3+4}$
A	$2+3$	$L_{2+3,1+2+3}$	$L_{2+3,1+2+4}$	$L_{2+3,1+3+4}$	$L_{2+3,2+3+4}$	$L_{2+3,1+2+3+4}$
B	$2+4$	$L_{2+4,1+2+3}$	$L_{2+4,1+2+4}$	$L_{2+4,1+3+4}$	$L_{2+4,2+3+4}$	$L_{2+4,1+2+3+4}$
C	$3+4$	$L_{3+4,1+2+3}$	$L_{3+4,1+2+4}$	$L_{3+4,1+3+4}$	$L_{3+4,2+3+4}$	$L_{3+4,1+2+3+4}$
D	$1+2+3$	$L_{1+2+3,1+2+3}$	$L_{1+2+3,1+2+4}$	$L_{1+2+3,1+3+4}$	$L_{1+2+3,2+3+4}$	$L_{1+2+3,1+2+3+4}$
E	$1+2+4$	$L_{1+2+4,1+2+3}$	$L_{1+2+4,1+2+4}$	$L_{1+2+4,1+3+4}$	$L_{1+2+4,2+3+4}$	$L_{1+2+4,1+2+3+4}$
F	$1+3+4$	$L_{1+3+4,1+2+3}$	$L_{1+3+4,1+2+4}$	$L_{1+3+4,1+3+4}$	$L_{1+3+4,2+3+4}$	$L_{1+3+4,1+2+3+4}$
G	$2+3+4$	$L_{2+3+4,1+2+3}$	$L_{2+3+4,1+2+4}$	$L_{2+3+4,1+3+4}$	$L_{2+3+4,2+3+4}$	$L_{2+3+4,1+2+3+4}$
H	$1+2+3+4$	$L_{1+2+3+4,1+2+3}$	$L_{1+2+3+4,1+2+4}$	$L_{1+2+3+4,1+3+4}$	$L_{1+2+3+4,2+3+4}$	$L_{1+2+3+4,1+2+3+4}$

(continued)

Table 21 (continued)

Rows	Columns	1	2	3	4	5	6	7	8	9	A	B
1	IPE=OPE	0	1	2	3	4	1+2	1+3	1+4	2+3	2+4	3+4
2	0	$L_{0,0}$	$L_{0,1}$	$L_{0,2}$	$L_{0,3}$	$L_{0,4}$	$L_{0,1+2}$	$L_{0,1+3}$	$L_{0,1+4}$	$L_{0,2+3}$	$L_{0,2+4}$	$L_{0,3+4}$
3	1	$L_{1,0}$	$L_{1,1}$	$L_{1,2}$	$L_{1,3}$	$L_{1,4}$	$L_{1,1+2}$	$L_{1,1+3}$	$L_{1,1+4}$	$L_{1,2+3}$	$L_{1,2+4}$	$L_{1,3+4}$
4	2	$L_{2,0}$	$L_{2,1}$	$L_{2,2}$	$L_{2,3}$	$L_{2,4}$	$L_{2,1+2}$	$L_{2,1+3}$	$L_{2,1+4}$	$L_{2,2+3}$	$L_{2,2+4}$	$L_{2,3+4}$
5	3	$L_{3,0}$	$L_{3,1}$	$L_{3,2}$	$L_{3,3}$	$L_{3,4}$	$L_{3,1+2}$	$L_{3,1+3}$	$L_{3,1+4}$	$L_{3,2+3}$	$L_{3,2+4}$	$L_{3,3+4}$
6	4	$L_{4,0}$	$L_{4,1}$	$L_{4,2}$	$L_{4,3}$	$L_{4,4}$	$L_{4,1+2}$	$L_{4,1+3}$	$L_{4,1+4}$	$L_{4,2+3}$	$L_{4,2+4}$	$L_{4,3+4}$
7	1+2	$L_{1+2,0}$	$L_{1+2,1}$	$L_{1+2,2}$	$L_{1+2,3}$	$L_{1+2,4}$	$L_{1+2,1+2}$	$L_{1+2,1+3}$	$L_{1+2,1+4}$	$L_{1+2,2+3}$	$L_{1+2,2+4}$	$L_{1+2,3+4}$
8	1+3	$L_{1+3,0}$	$L_{1+3,1}$	$L_{1+3,2}$	$L_{1+3,3}$	$L_{1+3,4}$	$L_{1+3,1+2}$	$L_{1+3,1+3}$	$L_{1+3,1+4}$	$L_{1+3,2+3}$	$L_{1+3,2+4}$	$L_{1+3,3+4}$
9	1+4	$L_{1+4,0}$	$L_{1+4,1}$	$L_{1+4,2}$	$L_{1+4,3}$	$L_{1+4,4}$	$L_{1+4,1+2}$	$L_{1+4,1+3}$	$L_{1+4,1+4}$	$L_{1+4,2+3}$	$L_{1+4,2+4}$	$L_{1+4,3+4}$
A	2+3	$L_{2+3,0}$	$L_{2+3,1}$	$L_{2+3,2}$	$L_{2+3,3}$	$L_{2+3,4}$	$L_{2+3,1+2}$	$L_{2+3,1+3}$	$L_{2+3,1+4}$	$L_{2+3,2+3}$	$L_{2+3,2+4}$	$L_{2+3,3+4}$
B	2+4	$L_{2+4,0}$	$L_{2+4,1}$	$L_{2+4,2}$	$L_{2+4,3}$	$L_{2+4,4}$	$L_{2+4,1+2}$	$L_{2+4,1+3}$	$L_{2+4,1+4}$	$L_{2+4,2+3}$	$L_{2+4,2+4}$	$L_{2+4,3+4}$
C	3+4	$L_{3+4,0}$	$L_{3+4,1}$	$L_{3+4,2}$	$L_{3+4,3}$	$L_{3+4,4}$	$L_{3+4,1+2}$	$L_{3+4,1+3}$	$L_{3+4,1+4}$	$L_{3+4,2+3}$	$L_{3+4,2+4}$	$L_{3+4,3+4}$
D	1+2+3	$L_{1+2+3,0}$	$L_{1+2+3,1}$	$L_{1+2+3,2}$	$L_{1+2+3,3}$	$L_{1+2+3,4}$	$L_{1+2+3,1+2}$	$L_{1+2+3,1+3}$	$L_{1+2+3,1+4}$	$L_{1+2+3,2+3}$	$L_{1+2+3,2+4}$	$L_{1+2+3,3+4}$
E	1+2+4	$L_{1+2+4,0}$	$L_{1+2+4,1}$	$L_{1+2+4,2}$	$L_{1+2+4,3}$	$L_{1+2+4,4}$	$L_{1+2+4,1+2}$	$L_{1+2+4,1+3}$	$L_{1+2+4,1+4}$	$L_{1+2+4,2+3}$	$L_{1+2+4,2+4}$	$L_{1+2+4,3+4}$
F	1+3+4	$L_{1+3+4,0}$	$L_{1+3+4,1}$	$L_{1+3+4,2}$	$L_{1+3+4,3}$	$L_{1+3+4,4}$	$L_{1+3+4,1+2}$	$L_{1+3+4,1+3}$	$L_{1+3+4,1+4}$	$L_{1+3+4,2+3}$	$L_{1+3+4,2+4}$	$L_{1+3+4,3+4}$
G	2+3+4	$L_{2+3+4,0}$	$L_{2+3+4,1}$	$L_{2+3+4,2}$	$L_{2+3+4,3}$	$L_{2+3+4,4}$	$L_{2+3+4,1+2}$	$L_{2+3+4,1+3}$	$L_{2+3+4,1+4}$	$L_{2+3+4,2+3}$	$L_{2+3+4,2+4}$	$L_{2+3+4,3+4}$
H	1+2+3+4	$L_{1+2+3+4,0}$	$L_{1+2+3+4,1}$	$L_{1+2+3+4,2}$	$L_{1+2+3+4,3}$	$L_{1+2+3+4,4}$	$L_{1+2+3+4,1+2}$	$L_{1+2+3+4,1+3}$	$L_{1+2+3+4,1+4}$	$L_{1+2+3+4,2+3}$	$L_{1+2+3+4,2+4}$	$L_{1+2+3+4,3+4}$

Table 22 Linear Approximation Table (LAT) for given S-Box

Input sum	Output sum															
	0	1	2	3	4	5	6	7	8	9	A	B	C	D	E	F
0	+8	0	0	0	0	0	0	0	0	0	0	0	0	0	0	0
1	0	0	−2	−2	0	0	−2	+6	+2	+2	0	0	+2	+2	0	0
2	0	0	−2	−2	0	0	−2	−2	0	0	+2	+2	0	0	−6	+2
3	0	0	0	0	0	0	0	0	+2	−6	−2	−2	+22	+2	−2	−2
4	0	+2	0	−2	−2	−4	−2	0	0	−2	0	+2	+2	−4	+2	0
5	0	−2	−2	0	−2	0	+4	+2	−2	0	−4	+2	0	−2	−2	0
6	0	+2	−2	+4	+2	0	0	+2	0	−2	+2	+4	−2	0	0	−2
7	0	−2	0	+2	+2	−4	+2	0	−2	0	+2	0	+4	+2	0	+2
8	0	0	0	0	0	0	0	0	−2	+2	+2	−2	+2	−2	−2	−6
9	0	0	−2	−2	0	0	−2	−2	−4	0	−2	+2	0	+4	+2	−2
A	0	+4	−2	+2	−4	0	+2	−2	+2	+2	0	0	+2	+2	0	0
B	0	+4	0	−4	+4	0	+4	0	0	0	0	0	0	0	0	0
C	0	−2	+4	−2	−2	0	+2	0	+2	0	+2	+4	0	+2	0	−2
D	0	+2	+2	0	−2	+4	0	+2	−4	−2	+2	0	+2	0	0	+2
E	0	+2	+2	0	−2	−4	0	+2	−2	0	0	−2	−4	+2	−2	0
F	0	−2	−4	−2	−2	0	+2	0	0	−2	+4	−2	−2	0	+2	0

decimal indices varying from 0 to 15 and Ai and Bj are taking corresponding bit values from [0000] to [1111]. The algorithm to fill the (16×16) elements of the LAT is,

```
for(i = 0;i < 16;i ++){
        A = 0;
        for(k = 0;k < 16;k ++)  A = A+(Ai0.Xk0 + Ai1.Xk1 + Ai2.Xk2 +
Ai3.Xk3)%2;
        for(j = 0;j < 16;j ++){
                B = 0;
                for(k = 0;k < 16;k ++)B =  B + (Bj0.Yk0 + Bj1.Yk1 +
Bj2.Yk2 + Bj3.Yk3)%2;
                Sij   = (A + B)%2;
                if (Sij ==0) Cij ++; Nij   = Cij − 8;
        }
}
```

Time complexity of the given algorithm. Since the pseudo code contains two nested loops so the time complexity of the given algorithm has been $O(n^2)$.

4.3 Linear Approximation Analysis

A crypto 4-bit S-box (1st 4-bit S-box out of 32 4-bit S-boxes of DES) has been described in Sect. 4.2.1. The table for four input vectors, output 4-bit BFs and cor-

ISB	IPVs	OSB	OPBFs	ANFs
	4321		4321	4321
0	0000	E	1110	1110
1	0001	4	0100	1010
2	0010	D	1101	0011
3	0011	1	0001	1100
4	0100	2	0010	1101
5	0101	F	1111	0110
6	0110	B	1011	0111
7	0111	8	1000	0011
8	1000	3	0011	1010
9	1001	A	1010	0110
A	1010	6	0110	1010
B	1011	C	1100	1000
C	1100	5	0101	0101
D	1101	9	1001	0010
E	1110	0	0000	1010
F	1111	7	0111	0000

Table 23 Input and output Boolean functions with corresponding ANF coefficients of the given S-Box

responding ANFs has been depicted in Sect. 4.3.2. The analysis has been described in Sect. 4.3.3. The result of Analysis has been given in Sect. 4.3.4.

4.3.1 Input Vectors (IPVs)-Output BFs (OPBFs)-Algebraic Normal Forms (ANFs)

The elements of input S-box have been shown under column heading 'ISB' and the Input Vectors have been shown under the field IPVs (input vectors) and subsequently under column headings 1, 2, 3 and 4. The 4th input vector has been depicted under column heading '4', 3rd input vector has been depicted under column heading '3', 2nd input vector has been depicted under column heading '2' and 1st input vector has been depicted under column heading '1'. The elements of S-box have been shown under column heading 'OSB' and the Output 4-bit BFs have been shown under field OPBFs (Output Boolean Functions) and subsequently under column headings 1, 2, 3 and 4. The 4th output BF has been depicted under column heading '4', 3rd output BF has been depicted under column heading '3', 2nd output BF has been depicted under column heading '2' and 1st output BF has been depicted under column heading '1'. The corresponding ANFs for 4 OPBFs, OPBF-4th, OPBF-3rd, OPBF-2nd, OPBF-1st, are depicted under field 'ANFs' subsequently under column heading 4, 3, 2 and 1 respectively of Table 23.

4.3.2 Linear Approximation Analysis (LAA)

An algebraic normal form or ANF equation is termed as Linear Equation or Linear Approximation if the Nonlinear Part or NP (i.e. The xored value of all product terms of Eq. (2) for corresponding 4 bit values of IPVs, with column heading 4, 3, 2, 1) is 0 and The linear part or LP for corresponding 4 bit values of IPVs, with column heading 4, 3, 2, 1 is equal to corresponding BF bit values. The corresponding ANF coefficients of output BFs $F(4)$, $F(3)$, $F(2)$, and $F(1)$ are given under row heading ANF(F4), ANF(F3), ANF(F2) and ANF(F1) respectively from row 2 through 5 and column 4 through J. In which Column 4 of row 2 through 5 gives the value of constant coefficient (a_0 according to Eq. 2.) of ANF(F4), ANF(F3), ANF(F2) and ANF(F1) respectively. Column 5 through 8 of row 2 through 5 gives the value of respective linear coefficients more specifically a_1, a_2, a_3, a_4 (according to Eq. 2.) of ANF(F4), ANF(F3), ANF(F2) and ANF(F1). They together termed as LP or linear part of the respective ANF equation. Column 9 through J of row 2 through 5 gives the value of respective non-linear coefficients more specifically a_5–a_{15} (according to Eq. 2.) of ANF(F4), ANF(F3), ANF(F2) and ANF(F1). They together termed as NP or non-linear part of the respective ANF equation.

The 4th, 3rd, 2nd, 1st IPV for the given S-box have been noted in the field 'IPVs' under column heading 4, 3, 2, 1 respectively from row 8 through M of Table 24. The 4 output BFs F4, F3, F2, F1 are noted at column 4, 8, C, G from row 8 through M respectively. The corresponding LP, NP, Satisfaction (SF) values (LP=BF) are noted at column 5 through 7, 9 through B, C through F and H through J from row 8 through M respectively of Table 24.

4.3.3 Results

Total Number of Existing Linear Approximations: 21.

4.3.4 Pseudo Code with Time Complexity Analysis of the Linear Approximation Analysis Algorithm

The Nonlinear Part for the given analysis has been termed as NP. The ANF coefficients are illustrated through array anf [24]. IPVs are termed as x_1, x_2, x_3, x_4 for IPV1, IPV2, IPV3, IPV4 respectively. The pseudo code of the algorithm of the above analysis is given below,

Start.
Step 1. NP = (anf[5].$\&x_1\&x_2$)^(anf[6]$\&x_1$ $\&x_3$)+(anf[7]$\&x_1$ $\&x_4$)+(anf[8] $\&x_2$ $\&x_3$)+(anf[9]$\&x_2$ $\&x_4$)+(anf[10]$\&x_3$ $\&x_4$)(anf[11]$\&x_1$ $\&x_2$ $\&x_3$)+(anf[12]$\&x_1$ $\&x_2$ $\&x_4$)+(anf[13]$\&x_1$ $\&x_3$ $\&x_4$) +(anf[14] $\&x_2$ $\&x_3$ $\&x_4$)+(anf[15]$\&x_1$ $\&x_2$ $\&x_3$ $\&x_4$))

Table 24 Linear approximation analysis

RIC	1	2	3	4	5	6	7	8	9	A	B	C	D	E	F	G	H	I	J
1	Co-Effs			C	LP				NP										
2	ANF(F4)			1	1	0	1	1	0	0	0	1	0	1	1	0	0	1	0
3	ANF(F3)			1	0	0	1	1	1	1	0	0	1	0	0	1	0	0	0
4	ANF(F2)			1	1	1	0	0	1	1	1	1	1	1	0	0	1	1	0
5	ANF(F1)			0	0	1	0	1	0	1	1	0	0	0	0	1	0	0	0
6	I D	IPVs	S B	F	L P	N P	S F	F	L P	N P	S F	F	L P	N P	S F	F	L P	N P	S F
7		4321		4				3				2				1			
8	0	0000	E	1	0	0	1	1	0	0	1	1	1	0	0	0	1	0	1
9	1	0001	4	0	0	0	0	1	0	0	1	0	1	1	0	0	0	0	0
A	2	0010	D	1	1	0	0	1	1	0	0	0	1	0	1	1	0	0	1
B	3	0011	1	0	1	1	1	0	1	0	1	0	1	1	1	1	1	0	0
C	4	0100	2	0	1	0	1	0	1	0	1	1	0	0	1	0	1	0	1
D	5	0101	F	1	0	0	1	1	0	1	1	1	0	1	1	1	1	0	0
E	6	0110	B	1	1	1	1	0	1	0	1	1	0	1	1	1	1	0	0
F	7	0111	8	1	0	1	0	0	0	1	1	0	0	0	0	0	1	0	1
G	8	1000	3	0	0	0	0	0	1	0	1	1	0	0	1	1	0	0	1
H	9	1001	A	1	1	0	0	0	0	0	0	1	0	1	1	0	0	1	1
RIC	I D	IPVs 4321	S B	F 4	L P	N P	S F	F 3	L P	N P	S F	F 2	L P	N P	S F	F 1	L P	N P	S F
I	A	1010	6	0	0	0	0	1	1	1	1	1	0	0	1	1	0	0	1
J	B	1011	C	1	1	1	1	1	0	1	1	0	0	1	0	0	0	0	0
K	C	1100	5	0	0	0	0	1	1	1	1	0	1	1	1	0	1	0	0
L	D	1101	9	1	1	0	0	0	0	1	1	0	1	1	1	1	1	1	0
M	E	1110	0	0	0	0	0	0	1	0	1	1	0	0	1	0	1	1	1
N	F	1111	7	0	1	0	1	1	0	0	1	1	1	0	0	1	1	1	1

No. of LA with BF1	No. of LA with BF2	No. of LA with BF3	No. of LA with BF4
7	4	2	8

Table 25 Time complexity comparison of two algos

View	4-bit LC	4-bit LA
Time complexity	$O(n^2)$	$O(n)$

Step   2.   LP= anf[0] ^(anf[1].&x_1)^ (anf[2].&x_2)^ (anf[3].&x_3)^ (anf[4].&x_4).

Step   3.   if(NP==0&& BF($x_1x_2x_3x_4$) == LP) then Linear equation.

else Nonlinear equation.

Stop.

Time complexity. Since the analysis contains no loops so the time complexity of the algorithm has been $O(n)$.

4.3.5 Comparison of Execution Time Complexity of Linear Cryptanalysis of 4-Bit Crypto S-Boxes and Linear Approximation Analysis of 4-Bit S-Boxes

The Comparison of time complexity of two algorithms has been given in Table 25 as follows,

It can be concluded from the comparison that the execution time reduces in linear approximation analysis than the linear cryptanalysis of 4-bit crypto S-boxes. So in can be concluded from above review work that the execution time of 4-bit LA algorithm is much less that 4-bit LC algorithm so 4-bit LA algorithm has been proved to be much better algorithm.

5   Result and Analysis and Security Criterion for All Four Algorithms of 4-Bit Crypto S-Boxes

In this section the analysis criterion of differential cryptanalysis of 4-bit S-boxes has been described in Sect. 5.1 and the analysis criterion of differential cryptanalysis with 4-bit BFs of 4-bit S-boxes has been described in Sect. 5.2. The same of linear cryptanalysis of 4-bit S-boxes has been illustrated in Sect. 5.3 and at last the analysis criterion of linear approximation analysis of 4-bit S-boxes has been depicted in Sect. 5.4. The result and analysis of results of four Algorithms on 32 DES 4-bit S-boxes has been described in brief in Sect. 5.5.

5.1 The Analysis Criterion of Differential Cryptanalysis of 4-Bit Crypto S-Boxes

In DDT there have been 256 cells, i.e. 16 rows and 16 columns. Each row has been for each input difference varies from 0 to F. Each column in each row represents each output difference varies from 0 to F for each input difference. 0 in any cell indicates absence of that output difference for subsequent input difference. Such as 0 in 2nd cell of Table 8 of relevant DDT means for input difference 0 the corresponding output difference o is absent. If number of 0 is too low or too high it supplies more information regarding concerned output difference. So an S-box is said to be immune to this cryptanalytic attack if number of 0s in DDT is close to 128 or half of total cells or 256. In the said example of 1st DES 4-bit S-box total number of 0s in DDT are 168. That is close to 128. So the S-box has been said to be almost secure from this attack.

5.2 The Analysis Criterion of Differential Cryptanalysis with 4-Bit BFs

As total number of balanced 4-bit BFs increases in DAT the security of S-box increases since balanced 4-bit BFs supplies at most uncertainty. Since Number of 0s and 1s in balanced 4-bit BFs are equal i.e. they are same in number means determination of each bit has been at most uncertainty. In the said example of 1st DES 4-bit S-box total number of 8s in DAT are 36. That is close to 32 half of total 64 cells. So the S-box has been said to be almost less secure from this attack.

5.3 The Analysis Criterion of Linear Cryptanalysis of 4-Bit Crypto S-Boxes

In LAT there are 256 cells for 256 possible 4-bit linear relations. The count of 16 4-bit binary conditions to satisfy for any given linear relation has been put into the concerned cell. 8 in a cell indicate that the particular linear relation has been satisfied for 8, 4-bit binary conditions and remain unsatisfied for 8, 4-bit binary conditions. That is at most uncertainty. In the said example of 1st DES 4-bit S-box total number of 8s in LAT have been 143. That is close to 128. So the S-box has been said to be less secure from this attack.

5.4 The Analysis Criterion of Linear Approximation Analysis of 4-Bit Crypto S-Boxes

The value of ^nC_r has been maximum when the value of r is ½ of the value of n (when n is even). Here the maximum number of linear approximations is 64. So if the total satisfaction of linear equation is 32 out of 64 then the number of possible sets of 32 linear equations has been the largest. Means if the total satisfaction is 32 out of 64 then the number of possible sets of 32 possible linear equations is $^{64}C_{32}$. That is maximum number of possible sets of linear equations. If the value of total No of linear approximations is closed to 32 then it is more cryptanalysis immune. Since the number of possible sets of linear equations are too large to calculate. As the value goes close to 0 or 64 it reduces the sets of possible linear equations to search, that reduces the effort to search for the linear equations present in a particular 4-bit S-box. In this example total satisfaction is 21 out of 64. Which means the given 4-bit S-Box is not a good 4 bit S-Box or not a good Crypt analytically immune S-Box.

If the values of total number of existing linear equations for a 4-bit crypto S-Box are 24–32, then the lowest numbers of sets of linear equations are 250649105469666120. This is a very large number to investigate. So the 4-bit S-Box is declared as a good 4-bit S-Box or 4-bit S-Box with good security. If it is between 16 and 23 then the lowest numbers of sets of linear equations are 488526937079580. This not a small number to investigate in today's computing scenario so the S-boxes are declared as medium S-Box or S-Box with medium security. The 4-bit crypto S-Boxes having existing linear equations less than 16 are declared as Poor 4-bit S-Box or vulnerable to cryptanalytic attack.

5.5 The Result and Analysis of Results of Four Algorithms on 32 DES 4-Bit S-Boxes

The four algorithms have been operated on 32 DES 4-bit S-boxes as shown Table 26. No-ELR stands for Number of existing linear relations, N8-LAT stands for number of 8 s in LAT, N0-DDT stands for number of 0s in DDT and N8-DAT stands for number of balanced (8-8) 4-bit BFs in DAT The discussion has been given below,

Table 26 Analysis of 32 DES S-boxes by four cryptanalytic attacks discussion

DES 4-bit S-boxes	NO-ELR	N8-LAT	N0-DDT	N8-DAT
e4d12fb83a6c5907	21	143	168	36
0f74e2d1a6cb9538	29	143	168	36
41e8d62bfc973a50	23	138	168	36
fc8249175b3ea06d	25	154	166	42
f18e6b34972dc05a	24	132	162	30
3d47f28ec01a69b5	21	143	166	30
0e7ba4d158c6932f	31	143	166	21
d8a13f42b67c05e9	20	126	168	36
a09e63f51dc7b428	17	133	162	30
d709346a285ecbf1	22	133	168	30
d6498f30b12c5ae7	23	151	166	21
1ad069874fe3b52c	28	158	174	30
7de3069a1285bc4f	22	136	168	36
d8b56f03472c1ae9	22	136	168	36
a690cb7df13e5284	20	136	168	36
3f06a1d8945bc72e	22	136	168	36
2c417ab6853fd0e9	25	137	162	30
eb2c47d150fa3986	20	143	166	36
421bad78f9c5630e	30	130	160	27
b8c71e2d6f09a453	21	134	166	18
c1af92680d34e75b	30	141	159	36
af427c9561de0b38	29	127	164	36
9ef528c3704a1db6	24	127	168	18
432c95fabe17608d	24	130	162	30
4b2ef08d3c975a61	26	134	168	30
d0b7491ae35c2f86	27	145	166	30
14bdc37eaf680592	28	137	168	36
6bd814a7950fe23c	25	135	173	0
d2846fb1a93e50c7	23	144	161	30
1fd8a374c56b0e92	20	147	174	27
7b419ce206adf358	27	132	166	18
21e74a8dfc90356b	28	138	168	39

Out of 32 DES S-boxes 1 have 17, 3 have 21, 4 have 22, 1 have 23, 3 have 24, 3 have 25, 1 have 26, 2 have 27, 3 have 28, 2 have 29, 2 have 30 and 1 have 31 Existing Linear Relations i.e. 24 S-boxes out of 32 have been less secure from this attack and 8 out of 32 have been immune to this attack. Again out of 32 DES S-boxes 1 have 126, 2 have 127, 2 have 130, 1 have 132, 2 have 133, 2 have 134, 1 have 135, 4 have 136, 2 have 137, 2 have 138, 1 have 141, 5 have 143, 1 have 144, 1 have 145, 1 have 147, 1 have 151, 1 have 154 and 1 have 158 8s in LAT. That is All S-boxes are less immune to this attack. Again out of 32 DES S-boxes 1 have 159, 1 have 160, 1 have 161, 4 have 162, 1 have 164, 8 have 166, 13 have 168, 1 have 173 and 2 have 174 0s in DDT. That is all S-boxes have been secured from this attack. At last out of 32 DES S-boxes 1 have 0, 3 have 18, 2 have 21, 2 have 27, 10 have 30, 12 have 36, 1 have 39 and 1 have 42 8s in DAT i.e. they have been less secure to this attack. The comparative analysis has proved that Linear Approximation analysis has been the most time efficient cryptanalytic algorithm for 4-bit S-boxes

6 Conclusions

In this paper a detailed discussion on four cryptanalytic attacks on 4-bit Crypto S-boxes have been done. From them point of view of execution time of algorithms the new attack Linear Approximation Analysis have been the best. A detail Analysis of 32 DES 4-bit S-boxes has also been included in this paper and it has been proved that DES S-boxes have constructed with the knowledge of differential cryptanalysis of 4-bit crypto S-boxes. All S-boxes are unsecure to rest of three attacks. So we cannot consider DES S-boxes as cryptographically secure S-boxes. All analysis has been chosen to strengthen ciphers to use in smart applications. It has been claimed to be a landmark of development of secure smart devices.

Acknowledgements For This exhaustive work we want to acknowledge the continuous encouragement of Prof. (Dr.) Amlan Chakrabarti, Dean Faculty Council of Engineering and Technology, University of Calcutta and the infrastructure provided by Prof. (Dr.) Debatosh Guha, Head Dept. Department of Radio Physics and Electronics, University of Calcutta. We also acknowledge TEQIP-Phase-II, University of Calcutta for providing financial support up to 30th November 2016.

References

1. Keller PE, Kouzes RT, Kangas LJ, Hashem S (1995) Transmission of Olfactory information for telemedicine. Studies in health technology and informatics, vol 18, pp 168–172. Interactive Technology and the New Paradigm for Healthcare
2. Haghpanah V, Saeedi M (2013) Smart article: a scientific crosstalk. Front Physiol 4:161
3. Jusoh S, Alfawareh HM (2012) Techniques, applications and challenging issue in text mining. Int J Comput Sci 9:0814–1694
4. Akhras G (2000) Smart materials and smart systems for the future. Can Mil J 1
5. Varadan VK (2005) Handbook of smart systems and materials. Institute of Physics Pub, London
6. Meyer G et al (2009) Advanced microsystems for automotive applications 2009 - Smart Systems for Safety, Sustainability and Comfort, Springer 2009
7. Shannon Claude (1949) Communication theory of secrecy systems (PDF). Bell Syst Tech J 28(4):656–715
8. Vaudenay S (2002) Security flaws induced by CBC padding applications to SSL, IPSEC, WTLS.... Advances in cryptology—EUROCRYPT 2002. In: Proceedings of international conference on the theory and applications of cryptographic techniques, vol 2332, pp 534–545. Springer
9. Data Encryption Standard (1977) Federal information processing standards publication (FIPS PUB) 46. National Bureau of Standards, Washington, DC
10. Data Encryption Standard (DES) (1999) Federal information processing standards publication (FIPS PUB) 46-3. National Institute of Standards and Technology, Gaithersburg, MD)
11. Feistel H (1971) Block cipher cryptographic system. US Patent 3798359 (Filed June 30, 1971)
12. Poonen B (2017) Using zeta functions to factor polynomials over finite fields. arXiv:1710.00970
13. Menezes A, van Oorschot P, Vanstone S (1996) Handbook of applied cryptography. CRC Press, Year
14. Schneier B (1996) Applied cryptography, 2nd edn. Wiley
15. Schaefer E (1996) A simplified data encryption standard algorithm. Cryptologia 20

16. Schneier B (2000) A self-study course in block-cipher cryptanalysis, Counterpane Internet Security
17. Schneier B et al (1999) The twofish encryption algorithm. Wiley, Year
18. Mirzan F (2000) Block ciphers and cryptanalysis. Department of Mathematics, Royal Holloway University of London, Egham
19. Heys HM (2000) A tutorial on linear and differential cryptanalysis. Memorial University of Newfoundland, Canada, Year
20. Schulzrinne H, Security Network (2000) Secret key cryptography. Columbia University, New York, Year
21. Pierson LG (2000) Comparing cryptographic modes of operation using flow diagrams. Sandia National Labarotaries, U.S.A
22. Aoki K et al (2000) Camellia: A 128-bit block cipher suitable for multiple platforms. NTT Corporation and Mitsubishi Electric Corporation, Year
23. Singh S (2001) The science of secrecy. Fourth Estate Limited
24. Susan Landau Standing the Test of Time (2000) The data encryption standard, sun microsystems
25. Garrett P (2001) Making, breaking codes. Prentice Hall, U.S.A
26. Kilian J, Rogaway P (2001) How to protect DES against exhaustive key search. NEC Research Institute U.S.A, Year
27. Yeun CY (2000) Design, analysis and applications of cryptographic techniques. Department of Mathematics, Royal Holloway University of London, Egham
28. Schneier B (2000) Why cryptography is harder than it looks. Counterpane Internet Security
29. Schaefer EF (1996) A Simplified data encryption standard algorithm. Cryptologia 20
30. Ooi KS, Vito BC (2002) Cryptanalysis of S-DES. University of Sheffield Center, Taylor College
31. Schneier B et al (1999) The twofish encryption algorithm. Wiley
32. Heys HM, Tavares SE (1996) Substitution-permutation networks resistant to differential and linear cryptanalysis. J Cryptology 9:1–19
33. Heys HM (2002) A tutorial on linear and differential cryptanlysis. Cryptologia 26:189–221
34. Buttayan L, Vajda I (1995) Searching for best linear approximation on DES-like cryptosystems. Electron Lett 31(11):873–874
35. Daemen J, Govaerts R, Vandewalle J (1995) Correlation matrices. Fast Software encryption, Lecture Notes in Computer science(LNCS)1008, pp 2–21. Springer
36. Matsui M (1994) Linear cryptanalysis method for DES cipher. Eurocrypt no. 765:386–397
37. Biham E (1994) On Matsui's linear cryptnalysis. Technion, Israel Institute of Technology, Israel
38. Harpes C, Kramer G, Massey J (1995) A generation of linear cryptanalysis and the applicability of Matsui's pilling-up lemma. Advances in Cryptology–Eurocrypt'95, pp 24–38
39. Kaliski B, Robshaw M (1994) Linear cryptanalysis using multiple approximations. Advances in cryptology-CRYPTO'94, pp 26–39
40. Matsui M (1994) The first experimental cryptanalysis of data encryption standard. Advances in cryptology–CRYPTO'94, pp 1–11
41. Junod PA (1998) Linear cryptanalysis of DES. Eidgenssische Tenhcische Hochsschule, Zurich
42. Collard B, Standaert FX, Quisquater JJ (2008) Experiments on the multiple linear cryptanalysis of reduced round serpent. In: Nyberg K (ed) Fast software encryption. FSE 2008. Lecture Notes in Computer Science, vol 5086. Springer, Berlin, Heidelberg
43. Mouha N, Wang Q, Gu D, Preneel B (2012) Differential and linear cryptanalysis using mixed-integer linear programming. In: Wu CK, Yung M, Lin D (eds) Information security and cryptology. Inscrypt 2011. Lecture Notes in Computer Science, vol 7537. Springer, Berlin, Heidelberg
44. Abdelraheem MA, Alizadeh J, AlKhzaimi H, Aref MR, Bagheri N, Gauravaram P (2015) Improved linear cryptanalysis of reduced-round SIMON-32 and SIMON-48, Cryptology e-print archive, Report-2015/988

45. Bagheri N (2015) Linear cryptanalysis of reduced-round SIMECK Variants. In: Biryukov A., Goyal V. (eds) Progress in cryptology—INDOCRYPT 2015. Lecture Notes in Computer science, vol 9462. Springer, Cham
46. Yu XL, Wu WL, Shi ZQ et al (2015) Zero-correlation linear cryptanalysis of reduced-round SIMON. J Comput Sci Technol 30:1358. https://doi.org/10.1007/s11390-015-1603-5,Year
47. Canteaut A (1997) Differential cryptanalysis of Feistel ciphers and differentially d-uniform mappings. Domaine de Voluceau, France
48. Adams C (1992) On immunity against Biham and Shamir's differential cryptanalysis. Inf Process Lett 41:77–80
49. Dawson M, Tavares S (1991) An expanded set of S-box design criteria based on information theory and its relation to differential-like attacks. Advances in cryptology—EUROCRYPT '91, pp 353–367
50. Biham E, Shamir A (1990) Differential cryptanalysis of DES-like cryptosystems. Advances in cryptology—CRYPTO '90, pp 2–21. Springer
51. Biham E, Shamir A (1991) Differential cryptanalysis of Snefru, Khafre, REDOC-II, LOKI and Lucifer. Advances in cryptology—CRYPTO '91, pp 156–171. Springer
52. Biham E, Shamir A (1992) Differential cryptanalysis of the full 16-round DES. Advances in Cryptology—CRYPTO '92, pp 487–496. Springer
53. Nyberg K (1991) Perfect nonlinear S-boxes. Advances in cryptology—EUROCRYPT '91, pp 378–386
54. Lai X, Massey JL (1991) Markov Ciphers and differential cryptanalysis. Swiss Federal Institute of Technology, Royal Holloway University of London, Egham
55. Murphy S, Robshaw MJB (2000) Differential cryptanalysis, key-dependant, S-boxes, and Twofish
56. Selçuk AA (2008) On probability of success in linear and differential cryptanalysis. J Cryptology 21:131. https://doi.org/10.1007/s00145-007-9013-7
57. Albrecht M, Cid C (2009) Algebraic techniques in differential cryptanalysis. In: Dunkelman O (ed) Fast software encryption. Lecture Notes in Computer science, vol 5665. Springer, Berlin, Heidelberg
58. Bouillaguet C, Dunkelman O, Fouque PA, Leurent G (2012) New insights on impossible differential cryptanalysis. In: Miri A, Vaudenay S (eds) Selected areas in cryptography. SAC 2011. Lecture Notes in Computer science, vol 7118. Springer, Berlin, Heidelberg
59. Rajashekarappa KM, Soyjaudah S, Sumithra Devi KA (2013) Comparative study on data encryption standard using differential cryptanalysis and linear cryptanalysis. Int J Adv Eng Technol 6(1):158–164
60. Gerault D, Minier M, Solnon C (2016) Constraint Programming models for chosen key differential cryptanalysis. In: Rueher M (ed) Principles and practice of constraint programming. CP 2016. Lecture Notes in Computer science, vol 9892. Springer, Cham
61. Hellman M, Langford S (1994) Differential-linear cryptanalysis. Crypto '94, no. 839:26–39
62. Vaudenay S, Moriai S (1994) Comparison of the randomness provided by some AES candidates. Eurocrypt no. 950:386–397. Springer
63. Vaudenay S (1994) An experiment on DES statistical cryptanalysis. Ecole Normale Supérieure, France
64. Gorska A et al (2016) New experimental results in differential-linear cryptanalysis of reduced variant of DES. Polish Academy of Sciences, Warsaw
65. Ferguson N et al (2001) Improved cryptanalysis of Rijndael. Counterpane Internet Security, USA
66. Ding C (1993) The differential cryptanalysis and design of natural stream ciphers. Fast software encryption. Cambridge Security Workshop, LNCS 809
67. Golic JD (1994) Linear cryptanalysis of stream ciphers. In: Fast software encryption, Second international workshop, LNCS 1008
68. Tanaka M, Hamaide T, Hisamatsu K, Kaneko T (1998) Linear cryptanalysis by Linear Sieve method. IECE Trans Fundam Electron Commun Comput Sci E81-A(1):82–87

69. Muller F (2004) Differential attacks against the helix stream cipher. In: Roy B, Meier W (eds) Fast software encryption. FSE 2004. Lecture Notes in Computer science, vol 3017. Springer, Berlin, Heidelberg
70. Wu H, Preneel B (2007) Differential cryptanalysis of the stream ciphers Py, Py6 and Pypy. In: Naor M (eds) Advances in cryptology—EUROCRYPT 2007. Lecture Notes in Computer science, vol 4515. Springer, Berlin, Heidelberg
71. Wu H, Huang T, Nguyen PH, Wang H, Ling S (2012) Differential attacks against stream cipher ZUC. In: Wang X, Sako K (eds) Advances in Cryptology—ASIACRYPT 2012. Lecture Notes in Computer science, vol 7658. Springer, Berlin, Heidelberg
72. Webster AF, Tavares SE (1986) On the design of S-boxes. In: Williams HC (ed) Advances in cryptology—CRYPTO '85 proceedings. CRYPTO 1985. Lecture Notes in Computer science, vol 218. Springer, Berlin, Heidelberg
73. Adams C, Tavares SJ (1990) The structured design of cryptographically good S-boxes. J Cryptology 3:27. https://doi.org/10.1007/BF00203967
74. Kim K, Matsumoto T, Imai H (1990) A recursive construction method of S-boxes satisfying strict avalanche criterion. In: Menezes AJ, Vanstone SA (eds) Advances in cryptology-CRYPTO' 90. CRYPTO 1990. Lecture Notes in Computer science, vol 537. Springer, Berlin, Heidelberg
75. Cusick TW (1994) Boolean functions satisfying a higher order strict avalanche criterion. In: Helleseth T (ed) Advances in cryptology — EUROCRYPT '93. EUROCRYPT 1993. Lecture Notes in Computer science, vol 765. Springer, Berlin, Heidelberg
76. Lisiskaya IV, Melnychuk ED, Lisitskiy KE (2012) Importance of S-blocks in modern block ciphers. Int J Comput Netw Inf Secur 4(10):1–12
77. Saarinen MJO (2012) Cryptographic analysis of all 4 × 4-Bit S-boxes. In: Miri A, Vaudenay S (eds) Selected areas in cryptography. SAC 2011. Lecture Notes in Computer science, vol 7118.Springer, Berlin, Heidelberg
78. Alkhzaimi HA, Knudsen LR (2016) Cryptanalysis of selected block ciphers. Kgs. Lyngby: Technical University of Denmark(DTU). (DTU Compute PHD; No.360)
79. Kazlauskas K, Smailiukas R, Vaicekaus G (2016) A novel method to design S-boxes based on key-dependent permutation schemes and its quality analysis. Int J Adv Comput Sci Appl 7(4):93–99
80. Ahmad M, Mittal N, Garg P, Khan MM (2016) Efficient cryptographic substitution box design using travelling salesman problem and chaos. Perspect Sci 8:465–468. Online publication date: 1-Sep-2016
81. Mazurkov MI, Sokolov AV (2016) Radioelectron. Commun Syst 59:212. https://doi.org/10.3103/S0735272716050034
82. Church R (1935) Tables of irreducible polynomials for the first four prime moduli. Ann Math. 2nd Series 36(1):198–209. http://www.jstor.org/stable/1968675
83. Swan RG (1962) Factorization of polynomials over finite fields. Pacific J Math 12(3):1099–1106. https://projecteuclid.org/euclid.pjm/1103036322
84. Bartee TC, Schneider DI (1963) Computation with finite fields. Inf Control 6(2):79–98. https://doi.org/10.1016/S0019-9958(63)90129-3
85. Berlekamp ER (1967) Factoring polynomials over finite fields. Bell Syst Tech J (Blackwell Publishing Ltd), 46(8):1853–1859. http://dx.doi.org/10.1002/j.1538-7305.1967.tb03174.x, https://doi.org/10.1109/tit.1968.1054226
86. Kasami T, Lin SH, Peterson W (1968) Polynomial codes. IEEE Trans Inf Theor 14(6):807–814
87. Berlekamp ER (1970) Factoring polynomials over large finite fields. Math Comp 24:713–735. https://doi.org/10.1090/S0025-5718-1970-0276200-X
88. Rabin MO (1980) Rabin, probabilistic algorithms in finite fields. SIAM J Comput 9(2):273–280. https://doi.org/10.1137/0209024
89. Lenstra AK (1985) Factoring multivariate polynomials over finite fields. J Comput Syst Sci 30(2):235–248. https://doi.org/10.1016/0022-0000(85)90016-9
90. McEliece RJ (1987) Factoring polynomials over finite fields. Finite fields for computer scientists and engineers, pp 75–96. https://doi.org/10.1007/978-1-4613-1983-2_7

91. Rónyai L (1988) Factoring polynomials over finite fields. J Algorithms 9(3):391–400. https://doi.org/10.1016/0196-6774(88)90029-6
92. Da Wan Q (1990) Factoring multivariate polynomials over large finite fields. Math Comput 54:755–770. https://doi.org/10.1090/S0025-5718-1990-1011448-0
93. Rybowicz M (1990) Search of primitive polynomials over finite fields. J Pure Appl Algebra 65(2):139–151. https://doi.org/10.1016/0022-4049(90)90115-X
94. Shoup V (1990) New algorithms for finding irreducible polynomials over finite fields. Math Comput 54(189):435–447
95. Rónyai L (1992) Galois groups and factoring polynomials over finite fields. SIAM J Discrete Math 5(3). https://doi.org/10.1137/0405026
96. Zivkovic M (1994) Table of primitive binary polynomials. Math Comput 63(207):301–301. https://doi.org/10.1090/s0025-5718-1994-1240662-8
97. Shparlinski I (1996) On finding primitive roots in finite fields. Theor Comput Sci 157(2):273–275. https://doi.org/10.1016/0304-3975(95)00164-6
98. Flajolet P, Gourdon X, Panario D (1996) Random polynomials and polynomial factorization. In: Lecture Notes in Computer science, 1099, pp 232–243. Springer, New York/Berlin
99. Gao S, Panario D (1997) Tests and constructions of irreducible polynomials over finite fields. In: Cucker F, Shub M (eds) Foundations of computational mathematics. Springer, Berlin, Heidelberg
100. Kaltofen E, Shoup V (1998) Subquadratic-time factoring of polynomials over finite fields. Math Comput Am Math Soc 67(223):1179–1197
101. Bach E, von zur Gathen J, Lenstra Jr HW (2001) Factoring polynomials over special finite fields. Finite Fields Appl 7(1):5–28. https://doi.org/10.1006/ffta.2000.0306
102. Gao S, Lauder AGB (2002) Hensel lifting and bivariate polynomial factorisation over finite fields. Math Comput 71:1663–1676
103. Brent RP, Zimmermann P (2003) Algorithms for finding almost irreducible and almost primitive trinomials. In: Proceedings of Brent 2003 Algorithms FF
104. Saxena NR, McCluskey EJ (2004) Primitive polynomial generation algorithms implementation and performance analysis. Technical Report(CRC TR 04-03), Center for Reliable Computing
105. Gao S, Kaltofen E, Lauder AGB (2004) Deterministic distinct-degree factorization of polynomials over finite fields. J Symbolic Comput 38(6):1461–1470. https://doi.org/10.1016/j.jsc.2004.05.004
106. Maitraa S, Gupta KC, Venkateswar A (2005) Results on multiples of primitive polynomials and their products over GF(2). Theor Comput Sci 341(1–3):311–343. https://doi.org/10.1016/j.tcs.2005.04.011
107. Scott M (2007) Optimal irreducible polynomials for GF(2^m) arithmetic. IACR Cryptology ePrint Archive
108. Fernandez CK (2008) Pascal polynomials over GF(2), Master's thesis, Naval Postgraduate School Monterey CA Dept of Mathematics, Accession Number:ADA483773
109. Saha C (2008) A note on irreducible polynomials and identity testing
110. Ahmed A, Lbekkouri A (2009) Determination of irreducible and primitive polynomials over a binary finite field. In: Conference: Workshop sur les Technologies de l'Information et de la Communication, Agadir, Maroc, p 94
111. Richards C (2009) Algorithms for factoring square-free polynomials over finite fields
112. Kim R, Koepf W (2009) Divisibility of trinomials by irreducible polynomials over F2. Int J Algebra 3(4):189–197
113. Hanif S, Imran M (2011) Factorization algorithms for polynomials over finite fields. Degree Project, School of Computer Science, Physics and Mathematics, Linnaeus University, Sweden
114. Wang L, Wang Q (2012) On explicit factors of cyclotomic polynomials over finite fields. Des Codes Cryptogr 63:87. https://doi.org/10.1007/s10623-011-9537-6
115. Couveignes JM, Lercier R (2013) Fast construction of irreducible polynomials over finite fields. Isr J Math 194:77. https://doi.org/10.1007/s11856-012-0070-8

116. Marquis David (2014) Deterministic factorization of polynomials over finite fields, thesis: MS in pur mathematics. Carleton University, Ottawa, Canada
117. Hammarhjelm G (2014) Construction of irreducible polynomials over finite fields, U.U.D.M Project Report 2014:17, Uppasala Universitet
118. Cavanna N (2014) Polynomial factoring algorithms and their computational complexity (2014). Honors Scholar Theses 384. http://digitalcommons.uconn.edu/srhonors_theses/384
119. Jiantao W, Zheng D (2014) Simple method to find primitive polynomials of degree n over $GF(2)$ where $2^{n}-1$ is a Mersenne prime[OL]. http://www.paper.edu.cn/lwzx/en_releasepaper/content/4587059
120. Sadique JKM, Zaman UZ, Dey S, Ghosh R (2015) An algorithm to find the irreducible polynomials over galois field GF(p^m).Int J Comput Appl 109(15):24–29. https://doi.org/10.5120/19266-1012
121. Ha J (2016) Irreducible polynomials with several prescribed coefficients. Finite Fields Appl 40:10–25. https://doi.org/10.1016/j.ffa.2016.02.006
122. Daemen J, Rijmen V (2000) AES Proposal: Rijndael,http://csrc.nist.gov/encryption/aes/ Last Visited: 7 Feb 2001
123. Weisstein EW Integer Polynomial. From MathWorld–A Wolfram Web Resource. http://mathworld.wolfram.com/IntegerPolynomial.html

Part III
Secure Network Communication in Smart Cities Applications

A Robust Watermarking Scheme Using Machine Learning Transmitted Over High-Speed Network for Smart Cities

Ankur Rai and Harsh Vikram Singh

Abstract Medical images are more typical than any other ordinary images. In telemedicine applications, transmission of medical image via open channel, demands strong security and copyright protection. This paper discusses a safe and secure watermarking technique using a machine learning algorithm. In this paper, propagation of watermarked image is simulated over a 3GPP/LTE downlink physical layer. In our proposed robust watermarking model, a double layer security is introduced to ensure the robustness of embedded data and then a transform domain-based hybrid watermarking technique, embeds the scrambled data into the transform coefficients of the cover image. Support Vector Machine (SVM) is work as a classifier, which classify a medical image into two distinct areas i.e. Non-Region of Interest (NROI) and Region of Interest (ROI). The secure watermark information is embedded into the unimportant part of the medical image, using the proposed embedding algorithm. The objective of the projected model is to avoid any quality degradation to the medical image. The result achieved in this experiment reveal that 10^{-6} Bit Error Rate (BER) value is realizable for greater value of Signal to Noise Ratio (SNR) i.e. more than 10.4 dB of SNR. The Peak Signal to Noise Ratio (PSNR) of received cover image is more than 35 dB, which is acceptable for clinical applications.

Keywords Medical image · ROI & NROI · Transform domain
Discrete wavelet transform · Singular value decomposition · Image classification
Support vector machine · LTE · OFDM · PSNR · SSIM

A. Rai (✉) · H. V. Singh
Kamla Nehru Institute of Technology, Sultanpur, U.P., India
e-mail: ankur_rai@knit.ac.in

H. V. Singh
e-mail: harshvikram@gmail.com

© Springer Nature Switzerland AG 2019
A. E. Hassanien et al. (eds.), *Security in Smart Cities: Models, Applications, and Challenges*, Lecture Notes in Intelligent Transportation and Infrastructure,
https://doi.org/10.1007/978-3-030-01560-2_11

1 Introduction

In the early 1800s, people used naive methods for transferring information and messages from one place to another, but the transfer of these methods depend on the distance and length of the message. For example, special flags, smoke signs or short messages and other visual signals were used for small distance communication. But to send a message over a long distance, using couriers was a more feasible basic choice. With the span of time, a new era of communication has taken place, where a long message (audio, video or text) can be easily transmitted over a long distance with less latency [1]. Therefore, the demand for mobile internet grew exponentially. As per the opera's web browser the number of pages viewed has raised from about 23 billion in January 2010 to 177 billion pages in November 2013 [2].

Now-a-days data acquisition, transmission and exchange are become so simple tasks due to the internet and availability of printers, fax and computers. The transmission of digital data through open channel network creates opportunities to the third parties or attackers for accessing data without any permission [3]. These digital data can be manipulated and retransmitted to the receiver end. The advancement in steganography, cryptography and watermarking have supported a lot to the digital contents for its security [4–6]. In modern days, exchange of patient's information from one hospital to another is be-come an important part in medical field. However, such exchange of information requires three basic preventions i.e. only the authentic recipient must receive the patient's information and the secret information of the patient must not be altered by an unauthenticated recipient and the data should belong to the same patient to whom it is related. Furthermore, these techniques can solve the several problems like End-user Privacy, Security, and Copyright issues.

Over the past years, the significant change in usage of Smartphone, tablets, net books, and laptops with wireless as well as broadband connections have added as many tens of millions of users (Research savy 2011), [5–7]. These examples show the huge demand for higher transfer rate, better availability, higher speed for mobile internet Connections. The researchers and the TELECOM industry have been working hard to come with new ideas for network architecture at low cost as well as high speed broadband connection for mobile access. The generation of wireless network 1G, 2G, 2.5G, 3G and Worldwide Interoperability for Microwave Access (WiMAX) like network infrastructure has been installed for several applications such as text, voice, data transfer etc. This infrastructure is being used in several telemedicine applications. Though, the lower data rates and secure transmission were the main issues.

The 3rd generation of mobile communication standards is announced by 3rd Generation Partnership Project (3GPP) as Lang Term Evolution (LTE). In mobile communication, LTE is the modern technology. It is considered as a 3.9G mobile network, is being standardized by 3GPP [8]. The title LTE is taken from the Evolved Universal Terrestrial Radio Access Network [E-UTRAN] [9, 10]. It is constituted on all IP framework which is not limited by earlier design. In downlink, the data rate is enlarged to 100 Mbps for data transmission using OFDM, whereas uplink access is upgraded to 50 Mbps by means of Single Carrier Frequency Division Multiple Access (SC-FDMA) modulation technique. The cellular network bandwidth is ascendable from 1.25 to 20 MHz and improved infrastructure is employed using LTE [11, 12]. Different bandwidth allocations and spectrum availability of different network operator to provide deferent services, is accommodated by means of this technique. The LTE technique full utilizes the given spectral bandwidth for high data rate and voice services than 3G network carriers.

Therefore, these recent advancements in new emerging technologies for the transmission of data through wired or wire communication has opened the gate for the intruders. Now-a-days data acquisition, transmission and exchange are become so simple tasks due to the internet and availability of printers, fax and computers. This transmission of digital data through open channel network creates opportunities to the third parties or intruders for accessing data without any permission.

A novel watermarking technique is proposed in this paper, in addition with transmission of watermarked image over a simulated OFDM based LTE physical layer. In this watermarking technique, a machine learning based algorithm is applied to the cover medical image for classification of ROI and NROI. Since a small misclassification in selection of ROI & NROI can cause a big trouble in diagnosis. Considering the LTE/3GPP technology for telemedicine applications, BER and PSNR of received signal have been evaluated.

The result achieved in this experiment reveal that 10^{-6} BER value is realizable for greater value of SNR i.e. more than 10.4 dB of SNR. The PSNR of received cover image is more than 35 dB, which is acceptable for clinical applications. The heartening results of proposed embedding method can have potential applications in telemedicine.

2 Transmission Over LTE Network

2.1 LTE—An Overview

In cellular technology, LTE emerges as a new technology towards a fast data communication by utilizing the existing allocated frequency spectrum. It enhances all the services like data rate, voice and video communication for the end users.

Efficient transmission of data and control information between an Enhanced Base Station (eNode B) and mobile User Equipment (UE) is enlarged by means of LTE [13]. Some new advanced technology and services have been added to the cellular communication such as OFDM, SC-FDMA and Multiple Input Multiple Output (MIMO) [14]. In downlink, the data rate is enlarged using OFDM, whereas uplink access is upgraded by means of SC-FDMA modulation scheme in LTE [15]. In OFDMA, quantified number of symbols are permitted to or from several users on a subcarrier by subcarrier basis. The three vital components of LTE network are; E-UTRAN, UE, Evolved Packet Core (EPC), Fig. 1.

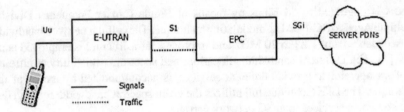

Fig. 1 System interface

The EPC connects data packet to the external world via internet or by other means of networking. Uu, S1 and SGi are shown as the interfaces between the different parts of the system.

Inter Symbol Interference (ISI) and Inter Carrier Interference (ICI) has no effects in OFDM. Since, higher data bit streams are divided into small parallel bit stream in OFDM and transmitted over a huge number of parallel narrow band sub-carriers in spite of a single wide band carrier [16, 17]. The orthogonality property of OFDM remove any sign of interference among the data streams. In addition, it is robust against narrow band interference and multi path problem.

OFDM based modulation scheme is having some demerits. It is sensitive to frequency errors and phase noise, since subcarriers are closely spaced. It is delicate to Doppler shift, causes interference between the subcarriers (ICI) [18]. Another demerit of OFDM is high peak-to-average signals, which introduces SC-FDMA in uplink, which recompense the disadvantage of OFDM. The Functional Commonality of SC-FDMA and OFDMA is shown in Fig. 2.

1. Constellation mapper: Depending on channel conditions (BPSK, QPSK, or 16QAM) it changes the arriving bit stream into a sole carrier symbols
2. Serial/parallel converter: As an input to FFT engine it setups time domain SC symbols into blocks

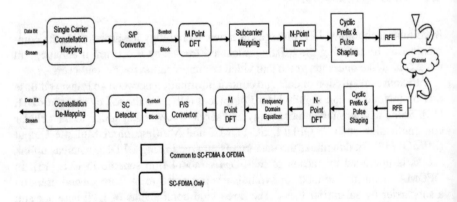

Fig. 2 Functional commonality of SC-FDMA and OFDMA

3. M-point DFT: time domain serial convertor symbol block is converted into M discrete tones
4. Sub-carrier mapping: DFT output tones are mapped to the definite subcarriers for broadcast. SC-FDMA schemes either use localized tones or distributed spaced tones. localized subcarrier mapping is used in LTE
5. N-point IDFT: The transmission took place by converting mapped subcarriers back into time domain
6. Pulse shaping and cyclic prefix: cyclic prefix is attached to SC-FDMA symbols to deliver multipath immunity. Pulse shaping is employed to prevent spectral regrowth, as in the case of OFDM
7. RFE: it performs the conversion of digital signal to analog signal and upconvert to RF for transmission.

The process is just reversed at the receiver side (Table 1).

Table 1 Technical specification of LTE network

Bandwidth	Scalable bandwidth 1.4, 3.0, 5.0, 10.0, 15.0 and 20.0 MHz
Multiple access technology	DL: OFDMA UL: SC-FDMA
Peak data rate	DL (2-channel MIMO): 100 Mb/s in 20 MHz channel UL (1-channel Tx): 50 Mb/s in 20 Mega Hearts channel
Supported antenna configurations	DL: 1×1, 4×2, 1×2, 2×2 UL: 1×1, 1×2
Spectral efficiency	5[bits/sec/Hz]
Mobility	Adjusted for lower speeds (<14 km/hr) Best functioning at speeds up to 130 km/hr Preserve link at rapidity up to 380 km/hr
Coverage	Best performance up to 4.8 km Minor deprivation 4.9–35 km Performance up to 110 km must not be excluded standard
Latency	~10 ms
Peak downlink speed 64QAM (Mbps)	331 (4×4 MIMO), 110 (SISO), 176 (2×2 MIMO)
Peak uplink speeds (Mbps)	52 (QPSK), 88 (64QAM), 58 (16QAM)

3 Overview of Data Hiding Scheme

Secure data embedding in a digital media can be described as the digital watermarking without introducing any perceptual distortion to the host signal. In recent years, data hiding has got much more attention from researcher as an option for communicating secure and robust communication [19].

Data hiding generally cover two areas: steganography and watermarking. Steganography deals with science of communicating information leaving the attacker without any clue that communication is occurring. The aim of steganography is to obscure a watermark or secret message into the host image in such a way, that it become undetectable for the intruders to identify the embedded data into the host image. The data embedding technique is generally classified into two categories, Spatial domain technique and transform domain technique.

3.1 *Spatial Domain Watermarking Techniques*

Embedding secure messages straight into the pixels of an image can be defined as the watermarking technique in spatial domain. The pixels of an 2D image are altered in spatial domain technique but there is trade-off between robustness and invisibility, since it cannot prevent image processing attacks and noise, some of the spatial domain techniques are discussed here:

Least Significant Bit (*LSB*): In spatial domain, LSB is a well-known technique for embedding, where watermark data is embedded in LSB of the host image. This method transforms the host image into 8 bit gray scale image. Since the pixels in an image are represented in 8 bits in which LSB is defined as the last bit or 0th bit in a pixel whereas 7th bit is considered as the most significant bit (MSB) in a pixel [20, 21]. This embedding technique does not impact severe transformation to the cover image and convenient to apply but any modification in the 7th bit plane affects the perceptibility of the cover image. This technique modifies some or all LSB pixel bits of cover image from the pixel bits of watermark image. It cannot be used in practical applications since it is very sensitive to noise and image processing attacks.

Spread Spectrum (*SS*): The watermark is comprised of pseudo random signal for the host image. The summation of a pseudo random signal to an 2D image is the main concept of spread spectrum watermarking algorithm that must be low of the perception threshold and cannot be noticed easily. This cannot be clear without knowing the constraints of watermarking algorithm [22]. SS basically involves two stages for image watermarking. In the first stage, the watermark is embedded in the host image. The bit sequence of watermark image is spread over the host image, then amplification of spread bit sequence takes place, this amplified bit sequence is then modulated by a pseudo noise sequence which act as a key for watermark signal. Now the host image is added to the modulated signal, resulting a watermarked image. In the second stage extraction of watermark from host image takes place. It involves the multiplication of key that was generated in embedding to the host image. Thresholding will be used to recover the watermark.

3.2 Frequency Domain Watermarking Techniques

In place of spatial domain methods, frequency domain modulation methods are commonly applied for watermarking. In frequency domain technique, the spectral coefficients can easily observe the characteristics of Human Visual System (HVS). To make an equilibrium among robustness and imperceptibility, utmost watermark algorithms choose to embed information in the mid-range frequencies. The most often used transform domain techniques are discussed here;

Discrete Cosine Transforms (DCT): The frequency domain-based methods are much robust of signal processing attacks than the spatial domain techniques [23]. In this type of transform domain data are represented in terms of frequency domain rather than spatial domain. Joint Photographic Experts Group (JPEG) uses DCT as a compression technique. This technique classifies the whole image data into three class of frequencies, known as low, middle and high frequency band [24]. This watermarking algorithm splits the host image into 8×8 of non-overlapping blocks for inserting the watermark. It also revels imperceptibility property of watermarking. DCT follows the lossy compression technique, in which the hidden data in higher frequency band is often discarded. Therefore, lower and middle frequency are best for embedding watermark in host image. Generally, the middle frequency band is chosen for embedding watermark, since this band has high capacity than lower and higher band [25].

Discrete Wavelet Transform: DWT is an advanced technique commonly used in various applications such as compression, image processing as well as in watermarking. In digital image watermarking techniques DWT has its own importance due to its favourable spatial localization and multiresolution properties. Its multiresolution property reveals several information of an image [26]. Wavelet function partitioned the data into distinct frequency components. Discontinuities and sharp spikes of a signal can be studied in DWT, which shows its advantage over traditional transform methods [27].

In one-dimensional DWT, the signal is fragmented into two parts i.e. low-level frequency and high-level frequency band using low pass filter as well as high pass filter respectively [28]. The high frequency band mostly included the edges of a signal whereas low level frequency band is furthermore separated into low frequency level and high-level frequency sub-bands. This process can be repeated till the desired number of partition level. Inverse DWT (IDWT) is used to reassemble the sub-bands to construct the novel image, shown in Fig. 3.

Singular Value Decomposition (SVD): SVD is one of numerical technique, which is used for diagonalization of input image [29]. This numerical technique (SVD) has several applications. An image I can be represented as a $M \times N$ non-negative matrix with scalar values. SVD decomposes the $M \times N$ matrix into a diagonal matrix S

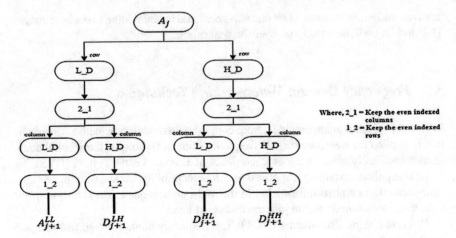

Fig. 3 Decomposition of 2D DWT

of singular values and two orthogonal matrices U, V of non-negative matrix I. The decomposition of matrix can be mathematically expressed by the given expression [30–32]:

$$I = U * S * V' \qquad (1)$$

The two major advantages of using SVD in image steganography are; first, it provides robustness against known signal processing attacks and second, a small variation or change in Singular Values (SVs) of image does not have any impact on the visual perceptual quality. In digital image processing, SVs of image have some key attributes, which are discussed here:

- Have a good stability when slight variations are embedded into an image, its singular values do not deviate expansively which is robust against several attacks.
- It Represent intrinsic algebraic attribute of an image.

Owing to these attributes, SVD matrix can be used to insert the watermark without large variation in the watermarked image which robust against several geometric attacks [20, 33].

Amit et al. [34] offered a DWT-SVD grounded watermarking scheme in which watermark is encrypted and encoded using ECC, which provide robustness and imperceptibility to the proposed model, however it increases the computational complexity to some extent. Colin et al. [35] have presented a model for watermarking scheme which comprises of lossy compression, scrambling and image moment theory to medical images. Image moment is used for embedding watermark near to the borders, which is vigorous against several attacks such as translation and scaling, but this model exposes degradation in watermarked image. Giakoumaki et al. [36] have proposed a watermarking scheme using multiple watermark, which observes

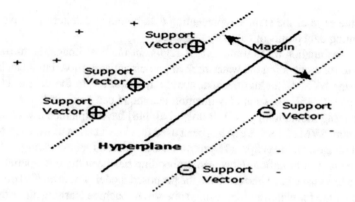

Fig. 4 Support vector classification

the strict provisions concerning health data handling by conserving their quality and diagnostic value. Hyung-Kyo Lee et al. (Hyung et al. [37]) proposed watermarking model applied 3-level DWT for embedding and central area of an image has been selected as ROI and watermark is embedded in the NROI region. Amit et al. [38] carried out a model based on SS and DWT to embed text as well as signature for watermarking. In the algorithm two level DWT is performed to insert the watermark in selective coefficients. A reversible watermaking scheme is applied by Chiang et al. [39] based on Difference Extension (DE) method, but it allows smaller hiding capacity for data embedding. Eswaraiah and Reddy [40] have proposed a water-marking technique based on fragile watermark for recovery and alter detection of medical images. LSBs of ROI and hash of ROI are organized to construct the fragile watermark in medical image.

Lai et al. [41] has proposed a transform domain-based hybrid watermarking scheme where watermark has been embedded in the element of singular values of the DWT subbands. This experiment is performed for general grayscale images only. A Sharma et al. [42] projected a watermarking scheme for medical images using DWT-DCT. Hash algorithm and Rivest–Shamir–Adleman is performed on watermark data before embedding it into host image, however the computational complexity cost to some extent.

Support Vector Machine: In statistical learning theory SVM is a new class of supervised machine learning methods, which can be used for regression or classifi-cation problems, but generally it is used in classification issues, however [43, 44]. In this classification algorithm, each data item is plotted as a point in n-dimensional fea-ture space of items with the value of a coordinate, then a classification is performed to distinguish the two classes by finding the best hyperplane, depicted in Fig. 4.

Margin defines the utmost breadth of two parallel lines to the hyperplane, which does not contain any data points within the parallel slabs; The data points nearby to the separating hyperplane are named as support vectors, support vector points lies on the edge of the slab; + representing data points of category positive, and − representing data points of category negative.

The classification of ROI and NROI region in medical image, is an important procedure for embedding the watermark in non-essential portion, but it can be overcome using SVM as a classifier that avoid any distortion to the diagnosis part of the image [45, 46]. So, such classification technique can ameliorate the schematic watermarking methodology [47]. Ramly et al. [48] have explained a watermarking model using SVM and SS. SVM has been used for classification of ROI and NROI in medical images, whereas Spread Spectrum has been used for embedding and extraction of patient information. However, embedding has been done in spatial domain and no attacks have been performed in the proposed model. Yen et al. [49] proposed a new technique for watermarking using supervised machine learning algorithm. The watermark is inserted by unsymmetrically tuning blue channels of the surrounding pixels and central. Since embedding has been performed in spatial domain, which is not robust against several image processing attacks.

4 Proposed Watermarking Model

Digital Imaging Communications in Medicine (DICOM) standard is basically used to store medical image and examination for database [50]. This DICOM data can be easily transmitted from one place to another. In medical imaging, DICOM is a standard for printing, storing, handling and transmitting information. It contains a network communications protocol and a file format definition [51]. The transmission of such digital data through open channel network involves more secrecy from being disclosed. Therefore, digital image watermarking can be so much helpful for such applications.

Medical images are more critical images, since it contains patient's information for diagnosis of disease, which must not be distorted while embedding the secret data into the image. This important part of the image is known as Region of Interest (ROI) and rest part of the image, which is not so much essential known as Non-Region of Interest (NROI). A small misclassification may cause big trouble in extracting essential information of the patient. In statistical learning theory SVM is a new class of machine learning method to overcome the over-fitting weakness of neural network. SVM as a classifier plays an important role, which is being used widely for classification problems [52–54]. However, in medical image watermarking SVM has not been used in transform domain for medical image classification of region so far.

Watermarking process can be defined as the embedding of a secret data into a carrier or host signal, which later can be extracted for verification as well as validation [55–57]. It is appropriate to several applications such as temper detection, copyright protection, content authentication and broadcast monitoring. In addition, watermarking is also used for confidentiality, reliability and availability of the data. Image watermarking has various advantages over medical images such as it saves

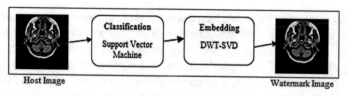

Host Image   Watermark Image

Fig. 5 Watermark classification-embedding process

bandwidth requirement, acquire small storage space, provide confidentiality and protection against unauthenticated access to the secured data [58, 59]. The patient's secret data are generally kept in DICOM file format for any future interrogation. This private information ought to be hold secure from any alteration or tampering [60–62].

There are basically three stages involves for watermarking in medical images, first stage can be described as classification of NROI and ROI, second stage watermark embedding in the host image and last stage is the extraction of watermark information. In classification stage Support Vector Machine (SVM) can plays an important role which is nothing but a machine learning algorithm which is being used widely for classification problems (Burges CJC 1998) as shown at Fig. 5. Whereas, the transform domain methods such as Singular Value Decomposition (SVD), Discrete Cosine Transform (DCT) and Discrete Wavelet Transform (DWT) are robust against various attacks in embedding and extraction process [63, 64]. However, the Discrete Wavelet Transform (DWT) suffers from three major drawbacks i.e. poor directional information, shift sensitive and lack of phase information [65], whereas False positive problem and higher computational cast are the foremost weaknesses of the SVD based image watermarking, in which a false watermark is spotted in watermarked image that was not originally embedded. This false positive problem resolved by using shuffled SVD (SSVD) [66]. It improves the quality of reassembled image by partitioning an image into set of unit images. The permuted original image by SVD with the data independent permutation defines the SSVD [67].

In our proposed model, a double layer security is introduced to ensure the robustness of embedded data. The embedded data is scrambled using a unique key and a transform domain-based hybrid watermarking technique is applied to insert the scrambled data into the coefficients of host image. This proposed scheme exploits the feature of spatio-frequency localization of DWT and intrinsic algebraic properties of SVD.

The proposed watermarking scheme in [68] is implemented for general images by disintegrating the host image into 4 subbands using DWT method and applied the SVD to each band, but in our algorithm, it is extended to the medical images as well as two-layer security. The two-layer security scheme is developed for imperceptibility as well as robustness of medical image. First, the watermark data is scrambled using a unique key then second, this scrambled data is inserted into the transform coefficients of host image.

The soft computing technique has been applied in numerous watermarking model. However, in medical image watermarking, SVM has not been used in transform domain for classification so far. Similarly, owing to the spatial frequency localization properties of DWT, the embedding area in a host image can be easily identified to embed the watermark without much distortion. In SVD technique, the visual perception of the host image is not much affected by slight variations in singular values, which motivate the embedding technique to attain better robustness and imperceptibility. Therefore, a hybrid method i.e. DWT-SVD is applied in our proposed model for data hiding, which is found to more secure against various alteration and attacks. In our watermarking system, SVM involves in classification stage, whereas hybrid watermarking technique (DWT-SVD) is applied in the remaining stages.

Firstly, original host image (H) is pre-processed before embedding a watermark image (W). This process involves classification of host image into two regions; ROI and NROI, afterward watermark embedding in NROI region has taken place. The grayscale image of size 128×128 is divided into two regions and these two parts are scrambled separately, using a secure key.

4.1 Classification Stage

In this classification stage of proposed model, the pixels of medical host image are characterized into ROI and NROI for performing embedding. In SVM classification model, the labelled predictor data is used to build a trained model. To find adequate predictive precision, various SVM kernel functions are available, which requires to be set its parameters for better accuracy. After training the model, it is cross-validated using the test data set. For instance, the host medical image is partitioned into two parts ROI and NROI. The trained SVM model utilizes the available dataset, to manage its related parameters i.e. weight vector and bias, which are mainly used to reduce the misclassification between ROI and NROI. Since ROI is basically use for diagnosis purpose whereas NROI does not have more significance. Therefore, a small misclassification in selection of ROI & NROI can cause a big trouble in diagnosis.

4.2 DWT-SVD Watermarking Stage

The projected hybrid watermarking (DWT-SVD) scheme is performed between the NROI pixels of host image H and the watermark image W, embedding and extraction algorithm are discussed below.

(a) Watermark embedding:

1. The extracted NROI pixels of host image H is disintegrated into 4 subband coefficients (LL, LH, HL and HH) using 1-DWT.

2. The SVD is applied to lower-high & high-low (LH, HL) subband coefficients of NROI image i.e.,

$$H^i = U^i S^i V^{i\prime} \qquad (2)$$

where i = 1, 2.

3. Now, the watermark image W is decomposed into two parts and then scrambled using a secure key.

$$W = P^1 + P^2 \qquad (3)$$

4. Embedding process took place by modification in the singular values of LH and HL subbands with P^1 and P^2, then SVD is performed to the changed singular values.

$$S^i + \alpha W^i = U_W^i S_W^i V_W^{i\prime} \qquad (4)$$

To measure the robustness of inserted watermark a factor is used, known as scale factor (α).

5. The modified DWT coefficient of NROI image is obtained, i.e.,

$$H^{*i} = U^i S_W^i V^{i\prime} \qquad (5)$$

6. A watermarked image is achieved by performing one level inverse DWT on two non-modified (LL, HH) and two modified (HL and LH) coefficients.

Now this watermarked image is added to the ROI pixels of image H to yield the concluding Watermarked image W_D.

(b) Watermark Extraction

1. The watermarked image W_D is disintegrated into 4 subband coefficients (LL$_2$, LH$_2$, HL$_2$ & HH$_2$) using 1-DWT (Fig. 6).
2. The SVD is applied to LH$_2$ and HL$_2$ subbands of watermarked image i.e.,

$$H_W^{*i} = U^{*i} S_W^{*i} V^{*i\prime} \qquad (6)$$

3. Obtain

$$E^{*i} = U_W^i S_W^{*i} V_W^{i\prime} \qquad (7)$$

4. Extraction of both scrambled part of watermark i.e.

$$E_w^{*i} = (E^{*i} - S^i)/\alpha \qquad (8)$$

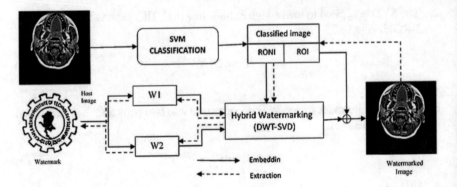

Fig. 6 Flow diagram of proposed watermarking model

5. The obtained scramble part from step 4 is descramble using the same secret key and then combined to form the complete extracted watermark image i.e.,

$$E_w^* = E_w^{*1} + E_w^{*2}. \tag{9}$$

5 Experimental Results and Discussion

5.1 Watermarking

The experiment is performed using the Intel Core i5 3.40 GHz processor, 1 TB HDD, 4 GB DDR3 Memory and Microsoft Windows 10. The proposed watermarking model is experimented with a JPEG grayscale image of size 256×256. The host image and watermarked image is shown in Fig. 7. Database used in this experiment is taken from MATLAB Central library. The grayscale image of size 128×128 is used as a watermark.

In the experiment, each jpeg image is of size 256×256, thus it is consisting of 65,536 predictor data points. This data points are further classified into two regions generally known as training data and testing data which are used for SVM model. For the training purpose, there are 6553 or 10% data points of total available data while the remain 5893 or 90% data points are used to cross validate the trained classification model. The feature values used here for the classification are intensity values and pixel positions of the input image, which are selected by the appropriate threshold value and result is determined by the naked eyes.

To find adequate predictive precision, there are various SVM kernel functions available such as Gaussian kernel, sigmoid kernel, radial basis function kernel. Among these kernels Radial Basis Function (RBF) kernel shown better accuracy for our model than other kernels. The classification mainly depends on two param-

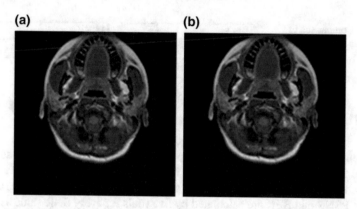

Fig. 7 **a** Host image. **b** Watermarked image

Table 2 SSIM value for different size of watermark

Watermark image (different size)	SSIM
120 Bytes	0.9820
100 Bytes	0.9835
80 Bytes	0.9778
60 Bytes	0.9950
40 Bytes	0.9980
Average	0.9872

Table 3 Comparison of imperceptibility for G&E [68] (PSNR)

S.F. value	0.01	0.03	0.05	0.07	0.09
G&E	37.80	36.79	35.29	33.72	32.26
Ours	52.18	43.92	39.49	36.58	34.40

eters known weight vector and bias, its value is generally determined by using grid search because a small misclassification can cause big trouble in diagnosis. Therefore, the value of w and b are obtained by using grid search, are 1000 and 1.0 respectively. The trained model classified the image into ROI and NROI based on above discussed parameters. After training the model it is cross-validated using test data set. The result received in this experiment reveals precision for proposed classification model is 99.80%. In embedding stage, only those pixels are selected which are labelled as NROI.

The SSIM value is calculated in Table 2 to measure the robustness of proposed model with different size of watermark. The corresponding SSIM value increases when size of watermark decreased. Therefore, this outcome displays that the robustness of the image depends upon the size of watermark, robustness increases as size of watermark decreases. This result displays the trade-off between robustness and capacity (Tables 3 and 4).

Table 4 Comparison of robustness for method [48]

Works	PSNR
Method [48]	48.558
Proposed model	52.18

5.2 Transmission

The simulation of image transmission is performed over an OFDM based downlink physical layer. QPSK modulation is used in physical layer. The BER and PSNR are measured for several values of SNR in the simulation at the receiver end. The average value of PSNR and BER are calculated by five trials for every SNR value as shown in Table 5. The perceptual quality is compared between original and received image at receive end in Figs. 7a and 8. For different SNR values, watermark is obtained in Fig. 9. Figure 9 reveals that, the true recovery of watermark data is obtained at minimum value of SNR of 10.2 dB. The result achieved in this experiment reveal that 10^{-6} BER value is realizable for greater value of SNR i.e. more than 10.4 dB of SNR. The PSNR of received cover image is more than 35 dB, which is acceptable for clinical applications [69].

Several attacks are performed on watermarked image. SSIM [70] is used for determining the resemblance between two images. The SSIM is calculated to determine the robustness of presented model against various attacks, such as pepper and salt noise addition, Gaussian Noise and JPEG compression, as shown in Table 6.

Table 5 BER & PSNR of received host image

S. no.	SNR (dB)	BER (dB)	PSNR (dB)
1	0	0.3240	20.52
2	2.0	0.4248	23.39
3	4.6	0.4327	24.15
4	5.0	0.2140	25.45
5	6.5	0.1140	30.68
6	7.0	0.0431	31.50
7	7.7	0.0284	32.87
8	8.0	0.0212	33.20
9	8.5	0.0184	38.19
10	9.0	0.0172	40.10
11	9.5	0.0163	42.52
12	10.0	0.0048	45.35
13	10.2	0.3201×10^{-3}	52.74
14	10.3	5.0113×10^{-4}	53.19
15	10.4	0.9110×10^{-6}	54.78

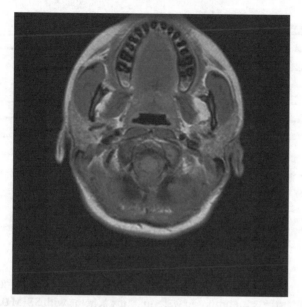

Fig. 8 Extracted watermarked image

SNR 7.1 dB   SNR 7.7   SNR 10.2

Fig. 9 Extracted watermark at different SNR

6 Conclusion

In healthcare industry, medical images or information are transmitted from one place to another using wired or wireless medium. Therefore, the transmission of such information requires more security. The proposed model ensures perceptibility and robustness of medical images during salt and paper noise, Gaussian noise and jpeg compression attacks. The application of SVM as well as double layer security empowers our proposed model against several image processing attacks. Since patient's information is very important, so that in our model SVM involved in classification stage while hybrid watermarking technique (DWT-SVD) is applied in the

Table 6 SSIM values of extracted watermark for several attacks

Sr. no.	Attacks/noise	SSIM
1	Salt and paper noise	0.584
2	Gaussian noise	0.601
3	JPEG attack (90%)	0.487
4	JPEG attack (75%)	0.531
5	JPEG attack (55%)	0.498
6	JPEG attack (45%)	0.601
7	Speckle noise	0.611
8	Histogram equalization	0.588
9	Median filtering	0.583
10	Crop (5%)	0.614
11	Average	0.569

remaining stages. The outcomes from this experiment reveals high quality imperceptibility and robustness for proposed watermarking scheme with SSIM 0.9872. In this
paper, a secured transmission of watermarked image over a simulated LTE downlink
physical layer is performed. The outcomes from this experiment reveals high quality
imperceptibility of received image with PSNR value 52.74 dB at SNR of 10.2 dB,
which is more than the adequate value of 35 dB for visual clinical investigation. The
heartening simulation outcomes can have possible applications in telemedicine.

However, the proposed model may have increased the computational complexity,
which need to be improved. In future work, it can be extended to the color images as
well as CT images or X-ray images for watermarking with other existing techniques.
We would like to further improve the performance of proposed model, which will
be testified in forthcoming communication.

References

1. Emad K, Rajan P, uzair Q, Awab F (2013) Long term evolution. IOSR J Electron Commun Eng
2. State of the Mobile Web (2013) Opera tech report
3. Elhoseny M, Elminir H, Riad A, Yuan X (2016) A secure data routing schema for WSN using
   elliptic curve cryptography and homomorphic encryption. J King Saud Univ Comput Inf Sci
   28(3):262–275. https://doi.org/10.1016/j.jksuci.2015.11.001 Elsevier
4. Elhoseny M, Ramírez-González G, Abu-Elnasr O, Shawkat S, Arunkumar N, Farouk A (2018)
   Secure medical data transmission model for IoT-based healthcare systems. IEEE Access
   6:20596–20608. https://doi.org/10.1109/ACCESS.2018.2817615
5. Shehab A, Ismail A, Osman L, Elhoseny M, El-Henawy IM (2018a) Quantified self using IoT
   wearable devices. In: Hassanien A, Shaalan K, Gaber T, Tolba M (eds) Proceedings of the
   international conference on advanced intelligent systems and informatics 2017. AISI 2017.
   Advances in intelligent systems and computing, vol 639. Springer, Cham. https://doi.org/10.
   1007/978-3-319-64861-3_77

6. Shehab A, Elhoseny M, Muhammad K, Sangaiah A, Yang P, Huang H, Hou G (2018b) Secure and Robust Fragile watermarking scheme for medical images. IEEE Access 6(1):10269–10278. https://doi.org/10.1109/access.2018.2799240

7. Darwish A, Hassanien A, Elhoseny M, Sangaiah A, Muhammad K (2017) The impact of the hybrid platform of internet of things and cloud computing on healthcare systems: opportunities, challenges, and open problems. J Ambient Intell Humanized Comput. https://doi.org/10.1007/s12652-017-0659-1

8. GPP TS 36.300 (2010c) Evolved universal terrestrial radio access (EUTRA); Overall description (Release 8)

9. GPP TS 23.402 (2010a) UTRAN- and E-UTRAN-based systems, Architecture enhancement for non-3GPP access (Release 8)

10. GPP TS 36.211 (2010b) Evolved universal terrestrial radio access (EUTRA), physical channels and modulations (Release 8)

11. Astely D, Dahlman E, Furuskar A, Jading Y, Lindström M, Parkvall S (2009) LTE: the evolution of mobile broadband. IEEE communication magazine

12. Larmo A, Lindström M, Meyer M, Pelletier G, Torsner J, Wiemann H (2009) The LTE link-layer design‖. IEEE Commun Mag 47(4):52–59

13. Jim Z, Wes M (2007) White paper on overview of the 3GPP long term evolution physical layer, freescale semiconductor

14. Lee J, Han J, Zhang J (2009) MIMO technologies in 3GPP LTE and LTE-Advanced. J Wireless Commun Netw 2009

15. Furuskar A, Jonsson T, Lundevall M (2008) Ericsson Research, Sweden, The LTE radio interface- key characteristics and performance.‖ In: IEEE International conference on personal, indoor and mobile radio communications

16. Morelli M, Kuo CCJ, Pun MO (2007) Synchronization techniques for orthogonal frequency division multiple access (OFDMA): a tutorial review. Proc IEEE 95(7):1394–1427

17. YangSun L, Heau JK, Yoon HK (2005) Copyright authentication enhancement of digital watermarking based on intelligent human visual system scheme. LNAI 3682:567–572. Springer, Berlin, Heidelberg KES 2005

18. Speth M, Fechtel S, Fock G, Meyr H (1999) Optimum receiver design for wireless broadband systems using OFDM. Part I. IEEE Trans Commun 47(11):1668–1677

19. Parah A, Shabir SA, Javaid HMA, Bhat GM (2014) Data hiding in scrambled images: A new double layer security data hiding technique. Sci Dir Comput Electr Eng 40:70–82

20. Lu CS, Liao HYM, Huang SK, Sze CJ (1999) Cocktail watermarking on images. In: 3rd international workshop on information hiding, Dresden, Germany

21. Wang RZ, Lin CF, Lin JC (2003) Image hiding by optimal LSB substitution and genetic algorithm. Pattern Recogn 34:671–683

22. Ali HA (2007) Qualitative spatial image data hiding for secure data transmission. GVIP J 7(2):35–37

23. Ghosh S, Talapatra S, Chatterjee N, Maity SP, Rahaman H (2012) FPGA based implementation of embedding and decoding architecture for binary watermark by spread spectrum scheme in spatial domain. Bonfring Int J Adva Image Process 2(4)

24. Barni BF, Piva A (1998) A DCT domain system for robust image watermarking. IEEE Trans Signal Process 66:357–372

25. Hernandez JR, Amado M, Gonzalez PF (2000) DCT-domain watermarking techniques for still images: detector performance analysis and a new structure. IEEE Trans Image Process 9:55–68

26. Tiwari N, Ramaiya MK, Sharma M (2013) Digital watermarking using DWT and DES. In: IEEE 3rd international on advance computing conference (IACC)

27. Gonge SS, Bakal JW (2013) Robust digital watermarking techniques by using DCT and spread spectrum. Int J Electr Electron Data Commun 1. ISSN: 2320-2084

28. Singh AP, Mishra A (2011) Wavelet based watermarking on digital image. Indian J Comput Sci Eng 1

29. Dey N, Roy AB, Dey S (2013) A novel approach of color image hiding using RGB color planes and DWT. Int J Comput Appl 36(5):0975–8887

30. Chakraborty S, Chatterjee S, Dey N, Ashour AS, Hassanien AE (2016) Comparative approach between singular value decomposition and randomized singular value decomposition-based watermarking. In: Intelligent techniques in signal processing for multimedia security vol. 660 of the series Studies in computational intelligence, p 133–149 https://doi.org/10.1007/978-3-319-44790-2_7

31. Dey N, Ashour AS., Chakraborty S, Banerjee S, Gospodinova E, Gospodinov M, Hassanien AE (2016) Watermarking in biomedical signal processing. In: Intelligent techniques in signal processing for multimedia security, vol 660 of the series Studies in computational intelligence, p 345–369 https://doi.org/10.1007/978-3-319-44790-2_16

32. Zhou Z, Tang B, Liu X (2006) A block svd based image watermarking method. In: Proceedings of the 6th world congress on intelligent control and automation, Dalian, China

33. Divecha NH, Jani NN (2012) Image watermarking algorithm using DCT, DWT and SVD. In: IJCA proceedings on national conference on innovative paradigms in engineering and technology

34. Singh AK, Dave M, Mohan A (2015) Robust and secure multiple watermarking in wavelet domain. J Med Imaging Health Informatics 5:406–414

35. Colin RR, Claudia FU, Blas GJT (2007) Data hiding scheme for medical images. In: 17th international conference on electronics. Comput Commun, pp 32–32. https://doi.org/10.1109/conielecomp.2007.14

36. Giakoumaki A, Pavlopoulos S, Koutsouris D (2006) Multiple image watermarking applied to health information management. IEEE Trans Inf Technol Biomed 10(4):722–732. https://doi.org/10.1109/TITB.2006.875655

37. Hyung KL, Hee JK, Ki RK, Jong KL (2005) ROI medical image watermarking using DWT and bit-plane. In: Asia-Pacific conference, communications, pp 512–515

38. Singh AK, Kumar B, Dave M, Mohan A (2015) Multiple watermarking on medical images using selective discrete wavelet transform coefficients. J Med Imaging Health Inf 5:607–614

39. Chiang KH, Chien KCC, Chang RF, Yen HY (2008) Tamper detection and restoring system for medical images using wavelet-based reversible data embedding. J Digital Imaging 21:77–90

40. Eswaraiah R, Reddy ES (2014) A fragile ROI-based medical image watermarking technique with tamper detection and recovery. In: Fourth international conference on communication systems and network technologies

41. Lai CC, Tsai CC (2010) Digital image watermarking using discrete wavelet transform and singular value decomposition. IEEE Trans Instrum Meas 59(11):3060–3063

42. Sharma A, Singh AK, Ghrera SP (2016) Robust and Secure Multiple Watermarking for Medical Images. Wirel Pers Commun

43. Seetha M, MuraliKrishna IV, Deekshatulu BL (2007) Comparison of advanced techniques of image classification. Map World Forum

44. Virmani KJ, Dey N, Kumar V (2015) PCA-PNN and PCA-SVM based CAD systems for breast density classification. In: Applications of intelligent optimization in biology and medicine, vol 96 of the series Intelligent Systems pp 159–180 https://doi.org/10.1007/978-3-319-21212-8_7

45. Bhattacherjee A, Roy S, Paul S, Roy P, Kausar N, Dey N (2016) Classification approach for breast cancer detection using back propagation neural network: a study. In: IGI GLOBAL biomedical image analysis and mining techniques for improved health outcomes https://doi.org/10.4018/978-1-4666-8811-7.ch010

46. Zemmal N, Azizi N, Sellami M, Dey N (2015) Automated classification of mammographic abnormalities using transductive semi supervised learning algorithm. In: Proceedings of the mediterranean conference on information & communication technologies, pp 657–662 https://doi.org/10.1007/978-3-319-30298-0_73

47. Shao Y, Chen W, Liu C (2008) Multiwavelet-based digital watermarking with support vector machine technique. In: CCDC

48. Ramly S, Aljunid SA, Hussain HS (2011) SVM-SS watermarking model for medical images. CCIS 194:372–386

49. Yen SH, Wang CJ (2006) SVM based watermarking technique. Tamkang J Sci Eng 9(2):141–150

50. Rathi SC, Inamdar VS (2012) Medical images authentication through watermarking preserving ROI. Health Inform Int J (HIIJ) 1(1):27–42
51. Coatrieux G, Maitre H, Sankur B, Rolland Y, Collorec R (2000) Relevance of watermarking in medical imaging. In: Proceedings 2000 I.E. EMBS international conference on information technology applications in biomedicine. ITAB-ITIS 2000. Joint meeting third IEEE EMBS international conference on information technology, pp 250–255. https://doi.org/10.1109/itab.2000.892396
52. Burges CJC (1998) A tutorial on support vector machines for pattern recognition. Data Min Knowl Disc 2:121–167
53. Cao XB, Xu YW, Chen D, Qiao H (2009) Associated evolution of a support vector machine-based classifier for pedestrian detection. Inf Sci 179(8):1070–1077
54. Tsai HH, Sun DW (2007) Color image watermark extraction based on support vector machines. Inf Sci 177(2):550–569
55. Lin TC, Lin CM (2009) Wavelet-based copyright-protection scheme for digital images based on local features. Inf Sci 179(19, 9):3349–3358
56. Ruanaidh JJKO, Pun T (1997) Rotation, scale and translation invariant digital image watermarking. Signal Process 66(3):303–317, 28 May 1998
57. Zhou Y, Jin W (2012) A robust digital image multi-watermarking scheme in the dwt domain. In: International conference on techniques and informatics (ICSAI 2012)
58. Singh AK, Kumar B, Dave M, Mohan A (2014) Robust and imperceptible dual watermarking for telemedicine applications. Wirel Pers Commun 80:1415–1433
59. Zain J, Clarke M (2005) Security in telemedicine: issues in watermarking medical images. In: 3rd International conference: sciences of electronic, technologies of information and telecommunications, Tunisia
60. Memon NA, Gilani SAM, Ali A (2009a) Watermarking of chest CT scan medical images for content authentication. In: ICICT, pp 175–180
61. Memon NA, Gilani SAM, Qayoom S (2009b) Multiple watermarking of medical images for content authentication and recovery. In: IEEE: 13th International, INMIC, pp 1–6, 14–15
62. Planitz BM, Maeder AJ (2005) A study of block-based medical image watermarking using a perceptual similarity metric. In: Proceedings in DICTA
63. Singh HV, Singh AK, Yadav S, Mohan A (2014) DCT based secure data hiding for intellectual property right protection. CSI Trans ICT 2:163–168
64. Singh HV, Yadav S, Mohan A (2013) Intellectual property right protection of image data using DCT and spread Spectrum-based watermarking. Int J Electron Secur Digital Forensics 5:218–228
65. Fernandes FCA, Spaendonck RLV, Burrus CS (2002) Shiftable, projection-based complex wavelet transforms. In: Acoustics, speech, and signal processing (ICASSP) on IEEE international conference
66. Zear A, Singh A K, Kumar P (2016a) A proposed secure multiple watermarking technique based on DWT, DCT and SVD for application in medicine. Multimedia tools and applications
67. Zear A, Singh AK, Kumar P (2016b) Multiple Watermarking for Healthcare Applications. J Intell Syst
68. Ganic E, Eskicioglu AM (2004) Robust DWT-SVD domain image watermarking: embedding data in all frequencies. In: Proceedings of the 2004 workshop on multimedia security, Magdeburg, Germany, pp 166–174
69. Basant K, Harsh VS, Singh SP, Anand M (2008) Novel efficient and secure medical data transmission on WiMAX. Telemed e-Health 14(10) https://doi.org/10.1089/tmj.2008.0033
70. Wang Z, Bovik AC, Sheikh HR, Simoncelli EP (2004) Image quality assessment: from error visibility to structural similarity. IEEE Trans Image Process 13(4):600–612

4, 8, 32, 64 Bit Substitution Box Generation Using Irreducible or Reducible Polynomials Over Galois Field GF(P^q) for Smart Applications

Sankhanil Dey and Ranjan Ghosh

Abstract Substitution Box or S-Box had been generated using 4-bit Boolean Functions (BFs) for Encryption and Decryption Algorithm of Lucifer and Data Encryption Standard (DES) in late sixties and late seventies respectively. The S-Box of Advance Encryption Standard have also been generated using Irreducible Polynomials over Galois field GF(2^8) adding an additive constant in early twenty first century. In this chapter Substitution Boxes have been generated from Irreducible or Reducible Polynomials over Galois field GF(p^q). Binary Galois fields have been used to generate Substitution Boxes. Since the Galois Field Number or the Number generated from coefficients of a polynomial over a particular Binary Galois field (2^q) is similar to $\log_2^{q+1}$ bit BFs. So generation of $\log_2^{q+1}$ bit S-Boxes is possible. Now if p = prime or non-prime number then generation of S-Boxes is possible using Galois field GF (p^q), where, q = p − 1.

Keywords Substitution boxes · Irreducible polynomials · Galois fields Finite fields

1 Introduction

Polynomials over Finite field or Galois field GF(p^q) have been of utmost importance in Public Key Cryptography [1]. The polynomials over Finite field or Galois field GF(p^q) that cannot be factored into polynomials with less degree of d and q − d where d = {1,2,..., (q − 1)/2} have been termed as Irreducible polynomials over Finite field or Galois field GF(p^q) and the rest have been termed as Reducible polynomials over Finite field or Galois field GF(p^q) [2]. The polynomials over Galois field GF(p^q) with

S. Dey (✉) · R. Ghosh
Institute of Radio Physics and Electronics, University of Calcutta, Kolkata, India
e-mail: sdrpe_rs@caluniv.ac; sankhanil12009@gmail.com

R. Ghosh
e-mail: rghosh47@yahoo.co

© Springer Nature Switzerland AG 2019
A. E. Hassanien et al. (eds.), *Security in Smart Cities: Models, Applications, and Challenges*, Lecture Notes in Intelligent Transportation and Infrastructure,
https://doi.org/10.1007/978-3-030-01560-2_12

279

coefficient of the highest degree term as 1 have been termed as Monic polynomials Galois field GF(p^q) and rest have been termed as Non-monic Polynomials Galois field GF(p^q) [3]. The polynomials Galois field GF(p^q) with degree q have been termed as Basic Polynomials or BPs over Galois field GF(p^q) and Polynomials with degree $q - 1$ have been termed as Elemental Polynomials or EPs over Galois field GF(p^q) [4].

q bit proper Substitution box or S-Box have 2^q elements in an array where each element is unique and distinct and arranged in a random fashion varies from 0 to q. Polynomials over Galois field GF(p^q) have been termed as binary polynomials if p = 2. The binary number constructed with q = 0 at LSB and q = q at MSB has been termed as binary Coefficient Number or BCN of $\log_2^{q+1}$ bits. The Binary Coefficient Number or BCN over Galois field GF(p^q) has been similar with $\log_2^{q+1}$-bit BFs. The $\log_2^{q+1}$-bit S-Boxes have been generated using $\log_2^{q+1}$-bit BCNs. In this chapter proper 4, 5, 6, 7, and 8-bit S-Boxes have been generated using BCNs and the procure has been continued as a future scope to generate 16 and 32-bit S-Boxes. The non-repeated Coefficients of BPs over Galois field GF(p^q), where $P = 2^{(\log_2 q + 1)}$ and $q = p - 1$ have been used to generate $\log_2^{q+1}$ bit S-Boxes. In this chapter proper 4, 5, 6, 7, and 8-bit S-Boxes have been generated using BCNs and the procure has been continued as a future scope to generate 16 and 32-bit S-Boxes.

Main goal of the Smart Object for Intelligent Application (SOFIA) project is to make "information" in the physical world available for smart services—connecting physical world with information world [5, 6, 7]. Full access to information present in the embedded computing devices has a potential for large impact on the daily lives of people living in this environment [8, 9, 10]. Sharing Information safely has been of utmost importance in SOFIA. Modern block ciphers have been of utmost importance in doing so. The substitution boxes are the major part of ancient as well as modern block ciphers [11, 12].

In this chapter Polynomials over Galois field GF(p^q) and Substitution Boxes have been reviewed in Sect. 2. The generation of 4 and 8-bit S-Boxes using BCNs have been elaborated in Sect. 3. The generation of S-Boxes of 4 and 8 bit using Coefficients of Non-binary Galois Field Polynomials have been depicted in Sect. 4. Conclusion has been given in Sect. 5, Acknowledgement and Reference has been given in last sections.

2 Polynomials Over Galois Field GF(P^q) and $\log_2^{q+1}$ Bit S-Boxes

In this section, Sect. 2.1 has been devoted to a small review of Polynomials. Section 2.2 has been of Utmost importance since in it A four bit bijective Crypto or Proper S-Box has been defined in brief. At last in Sect. 2.3 The equation among 2^{15} Galois field Polynomials and a 4-bit Bijective Crypto S-Box has been elaborated in details.

2.1 Polynomials Over Galois Field GF(P^q)

Polynomials over Galois field GF(p^q) have been of utmost importance in Cryptographic Applications. Polynomials with degree q have been termed as Basic Polynomials and Polynomials with degree less than q have been termed as Elemental Polynomials over Galois field GF(p^q). Polynomials with leading coefficient as 1 have been termed as Monic Polynomials irrespective of BPs and EPs. An example, of the said criteria have been described as follows, the Example of Basic Polynomial or BP over Galois field GF(p^q) has been given below,

$$BP(x) = co_q x^q + co_{q-1} x^{q-1} + co_{q-1} x^{q-2} + \cdots + co_2 x^2 + co_1 x^1 + a_0 \qquad (1)$$

In Eq. (1), BP(x) has been represented as Basic Polynomial over Galois field GF(p^q)since the highest degree term of the said Polynomial over Galois field GF(p^q) is $\in$q. The BP has been called as a Monic BP if $co_q = 1$. The number of Terms in a BP over Galois field GF(p^q) has been 0 to q i.e. (q + 1). The number of possible values of a particular coefficient co_q, where $0 \leq q \leq q$ has been from 0 to p i.e. $\in$ (p + 1). If the value of q has been <q then The Polynomial over Galois field GF(p^q) has been termed as Elemental Polynomial over Galois field GF(p^q). If a BP over Galois field GF(p^q) can be factored into two non-constant EPs then the BP can be termed as Reducible Polynomials over Galois field GF(p^q). If the two factor of a BP over Galois field GF(p^q) have been the BP itself and a constant Polynomial then The BP have been said as an Irreducible Polynomial over Galois field GF(p^q).

2.2 4-Bit Crypto S-Boxes

A 4-bit bijective Crypto S-Box can be written as Follows, where each element of the first row of Table 1, entitled as index, are the position of each element of the S-Box within the given S-Box and the elements of the 2nd row, entitled as S-Box, are the elements of the given Substitution Box. It can be concluded that the 1st row is fixed for all possible bijective crypto S-Boxes. The values of each element of the 1st row is distinct, unique and vary between 0 and F. The values of each element of the 2nd row of a bijective crypto S-Box is also distinct and unique and also vary between 0 and F. The values of the elements of the fixed 1st row is sequential and monotonically increasing where for the 2nd row they can be sequential or partly sequential or non-sequential. Here the given Substitution Box is the 1st 4-bit S-Box of the 1st S-Box out of 8 of Data Encryption Standard [13, 14, 15].

Table 1 4-bit bijective crypto S-Box

Row	Col	1	2	3	4	5	6	7	8	9	A	B	C	D	E	F	G
1	Index	0	1	2	3	4	5	6	7	8	9	A	B	C	D	E	F
2	S-Box	E	4	D	1	2	F	B	8	3	A	6	C	5	9	0	7

Table 2 Input and output BCNs of the substitution box

2.Row	Column	1	2	3	4	5	6	7	8	9	A	B	C	D	E	F	G	H. Decimal Equivalent
1	Index	0	1	2	3	4	5	6	7	8	9	A	B	C	D	E	F	
2	IBF4	0	0	0	0	0	0	0	0	1	1	1	1	1	1	1	1	00255
3	IBF3	0	0	0	0	1	1	1	1	0	0	0	0	1	1	1	1	03855
4	IBF2	0	0	1	1	0	0	1	1	0	0	1	1	0	0	1	1	13107
5	IBF1	0	1	0	1	0	1	0	1	0	1	0	1	0	1	0	1	21845
6	S-Box	E	4	D	1	2	F	B	8	3	A	6	C	5	9	0	7	–
7	OBF4	1	0	1	0	0	1	1	1	0	1	0	1	0	1	0	0	42836
8	OBF3	1	1	1	0	0	1	0	0	0	0	1	1	1	0	0	1	58425
9	OBF2	1	0	0	0	1	1	1	0	1	1	1	0	0	0	0	1	36577
A	OBF1	0	0	1	1	0	1	1	0	1	0	0	0	1	1	0	1	13965

2.3 Relation Between 4-Bit S-Boxes and Polynomials Over Galois Field GF(2^{15})

Index of Each element of a 4-bit bijective crypto S-Box and the element itself is a hexadecimal number and that can be converted into a 4-bit bit sequence. From row 2 through 5 and row 7 through A of each column from 1 through G of Table 2 shows the 4-bit bit sequences of the corresponding hexadecimal numbers of the index of each element of the given S-Box and each element of the S-Box itself. Each row from 2 through 5 and 7 through A from column 1 through G constitutes a 16 bit, bit sequence that is a Basic Polynomial over Galois field GF(2^{15}). Column 1–G of Row 2 is termed as 4th IGFP, Row 3 is termed as 3rd IGFP Row 4 is termed as 2nd IGFP and Row 5 is termed as IGFP whereas column 1 through G of Row 7 is termed as 4th OGFP, Row 8 is termed as 3rd OGFP, Row 9 is termed as 2nd OGFP and Row A is termed as 1st OGFP. The decimal equivalent of each IGFP and OGFP are noted at column H of respective rows. Where IGFP stands for Input Galois Field Polynomial and OGFP stands for Output Galois Field Polynomials. The respective Polynomials have been shown in Row 1 through 8 of column 3 of Table 3.

Table 3 Respective polynomials of IGFP4 through IGFP1 and OGFP4 through OGFP1

Col Row	1	2	3
	Index	DCM eqv.	Polynomials over Galois Field GF(2^{15})
1	IGFP4	00255	BP(x) $= x^7 + x^6 + x^5 + x^4 + x^3 + x^2 + x^1 + 1$
2	IGFP3	03855	BP(x) $= x^{11} + x^{10} + x^9 + x^8 + x^3 + x^2 + x^1 + 1$
3	IGFP2	13107	BP(x) $= x^{13} + x^{12} + x^9 + x^8 + x^5 + x^4 + x^1 + 1$
4	IGFP1	21845	BP(x) $= x^{14} + x^{12} + x^{10} + x^8 + x^6 + x^4 + x^2 + 1$
5	OGFP4	42836	BP(x) $= x^{15} + x^{13} + x^{10} + x^9 + x^8 + x^6 + x^4 + x^2$
6	OGFP3	58425	BP(x) $= x^{15} + x^{14} + x^{13} + x^{10} + x^5 + x^4 + x^3 + 1$
7	OGFP2	36577	BP(x) $= x^{15} + x^{11} + x^{10} + x^9 + x^7 + x^6 + x^5 + 1$
8	OGFP1	13965	BP(x) $= x^{13} + x^{12} + x^{10} + x^9 + x^7 + x^3 + x^2 + 1$

3 3 and 8 Bit S-Box Generation by Respective BCNs Over Binary Galois Field GF(2^q) Where Q $\in$ 15 and 255 Respectively

In this chapter 4 and 8-bit Identity S-Boxes have been taken for example for generation of 4 and 8 bit S-Boxes over Binary Galois Fields GF(2^q) where q $\in$ 15 and 255 respectively. The generation of Identity 4-bit S-Box from four BCNs over Binary Galois Field GF(2^{15}) have been elaborated in Sect. 3.1 and The generation of Identity 8-bit S-Box from Eight BCNs over Binary Galois Field GF(2^{255}) have been elaborated in Sect. 3.2. The Algorithm for generation of $\log_2^{q+1}$ bit S-Boxes over Binary Galois Field GF(2^q) has been depicted with Time Complexity of the algorithm in Sect. 3.3.

Table 4 4-bit identity crypto S-box

Row	Column	1	2	3	4	5	6	7	8	9	A	B	C	D	E	F	G
1	Index	0	1	2	3	4	5	6	7	8	9	A	B	C	D	E	F
2	S-Box	0	1	2	3	4	5	6	7	8	9	A	B	C	D	E	F

3.1 Generation of 4-Bit Identity Crypto S-Box from Four Polynomials Over Binary Galois Field $GF(2^{15})$

The Concerned 4-bit Identity S-Box has been shown in Table 4 where each element of the first row of Table 4, entitled as index, are the position of each element of the S-Box within the given S-Box and the elements of the 2nd row, entitled as S-Box, are the elements of the given Identity Substitution Box. It can be concluded that the 1st row is fixed for all possible bijective crypto S-Boxes. The values of each element of the 1st row is distinct, unique and vary between 0 and F. The values of each element of the 2nd row of the Identity crypto S-Box are also distinct and unique and also vary between 0 and F. The values of the elements of the fixed 1st row is sequential and monotonically increasing where for the 2nd row, they can be sequential or partly sequential or non- sequential. Here the given Substitution Box is the 4-bit Identity Crypto S-Box.

Index of each element of a 4-bit bijective crypto S-Box and the element itself is a hexadecimal number and that can be converted into a 4-bit bit sequence. From row 2 through 5 and row 7 through A of each column from 1 through G of Table 5 shows the 4-bit bit sequences of the corresponding hexadecimal numbers of the index of each element of the given S-Box and each element of the S-Box itself. Each row from 2 through 5 and 7 through A from column 1 through G constitutes a 16 bit, bit sequence that is a Basic Polynomial over Galois field $GF(2^{15})$. Column 1 through G of Row 2 is termed as 4th IGFP, Row 3 is termed as 3rd IGFP Row 4 is termed as 2nd IGFP and Row 5 is termed as IGFP whereas column 1 through G of Row 7 is termed as 4th OGFP, Row 8 is termed as 3rd OGFP, Row 9 is termed as 2nd OGFP and Row A is termed as 1st OGFP. The decimal equivalent of each IGFP and OGFP are noted at column H of respective rows. Where IGFP stands for Input Galois Field Polynomial and OGFP stands for Output Galois Field Polynomials. The respective Polynomials have been shown in Row 1 through 8 of column 3 of Table 6.

3.2 Generation of 8-Bit Identity Crypto S-Box from Eight Polynomials Over Binary Galois Field $GF(2^{255})$

The Concerned 8-bit Identity S-Box has been shown in Table 7 where each element of the first row of Table 7, entitled as index, are the position of each element of the S-Box within the given S-Box and the elements of the column 1 through G of 2nd to

Table 5 Input and output BCNs of the identity substitution box

Row	Column	1	2	3	4	5	6	7	8	9	A	B	C	D	E	F	G	H. Decimal Equivalent
1	Index	0	1	2	3	4	5	6	7	8	9	A	B	C	D	E	F	
2	IBCN4	0	0	0	0	0	0	0	0	1	1	1	1	1	1	1	1	00255
3	IBCN3	0	0	0	0	1	1	1	1	0	0	0	0	1	1	1	1	03855
4	IBCN2	0	0	1	1	0	0	1	1	0	0	1	1	0	0	1	1	13107
5	IBCN1	0	1	0	1	0	1	0	1	0	1	0	1	0	1	0	1	21845
6	S-Box	0	1	2	3	4	5	6	7	8	9	A	B	C	D	E	F	–
7	OBCN4	0	0	0	0	0	0	0	0	1	1	1	1	1	1	1	1	00255
8	OBCN3	0	0	0	0	1	1	1	1	0	0	0	0	1	1	1	1	03855
9	OBCN2	0	0	1	1	0	0	1	1	0	0	1	1	0	0	1	1	13107
A	OBCN1	0	1	0	1	0	1	0	1	0	1	0	1	0	1	0	1	21845

Table 6 Respective polynomials of IGFP4 through IGFP1 and OGFP4 through OGFP1

Col Row	1	2	3
	Index	DCM eqv.	Polynomials over Galois field $GF(2^{15})$
1	IGFP4	00255	$BP(x) = x^7 + x^6 + x^5 + x^4 + x^3 + x^2 + x^1 + 1$
2	IGFP3	03855	$BP(x) = x^{11} + x^{10} + x^9 + x^8 + x^3 + x^2 + x^1 + 1$
3	IGFP2	13107	$BP(x) = x^{13} + x^{12} + x^9 + x^8 + x^5 + x^4 + x^1 + 1$
4	IGFP1	21845	$BP(x) = x^{14} + x^{12} + x^{10} + x^8 + x^6 + x^4 + x^2 + 1$
5	OGFP4	00255	$BP(x) = x^7 + x^6 + x^5 + x^4 + x^3 + x^2 + x^1 + 1$
6	OGFP3	03855	$BP(x) = x^{11} + x^{10} + x^9 + x^8 + x^3 + x^2 + x^1 + 1$
7	OGFP2	13107	$BP(x) = x^{13} + x^{12} + x^9 + x^8 + x^5 + x^4 + x^1 + 1$
8	OGFP1	21845	$BP(x) = x^{14} + x^{12} + x^{10} + x^8 + x^6 + x^4 + x^2 + 1$

17th row, entitled as S-Box, are the elements of the given 8-bit Identity Substitution Box sequentially. It can be concluded that the 1st row is fixed for all possible 8-bit bijective crypto S-Boxes. The values of each element of the 1st row is distinct, unique and vary between 0 and F. The values of each element of the column 1 through G of 2nd row through 17th row of the 8-bit Identity crypto S-Box are also distinct and unique and vary between 0 and 256. The values of the elements of the fixed 1st row is sequential and monotonically increasing where for the 2nd to 17th row, they can be sequential or partly sequential or non- sequential. Here the given Substitution Box is the 8-bit Identity Crypto S-Box.

Index of Each element of an 8-bit bijective crypto S-Box and the element itself is a hexadecimal number and that can be converted into a 256-bit long 8-bit bit sequence. From row 2 through 9 and row A through H of column 2 of Table 8 shows the 8-bit bit sequences of the corresponding hexadecimal numbers of the index of each element of the given S-Box and each element of the S-Box itself. Each row from 2 through 9 and A through H of column 2 constitutes a 256 bit, bit sequence that is a Basic Polynomial over Galois field $GF(2^{255})$. Column 2 of Row 2 is termed as 8th IGFP, Row 3 is termed as 7th IGFP, Row 4 is termed as 6th IGFP, Row 5 is termed as 5th IGFP, Row 6 is termed as 4th IGFP, Row 7 is termed as 3rd IGFP, Row 8 is termed as 2nd IGFP and Row 9 is termed as 1st IGFP whereas column 2 of Row A is termed as 8th OGFP, Row B is termed as 7th OGFP, Row C is termed as 6th OGFP, Row D is termed as 5th OGFP, Row E is termed as 4th OGFP, Row F is termed as 3rd OGFP, Row G is termed as 2nd OGFP and Row H is termed as 1st IGFP. The Binary Coefficient Number of each IGFP and OGFP from MSB[256th bit] to LSB[0th bit] have been given in corresponding rows of each IGFP and OGFP. Where IGFP stands for Input Galois Field Polynomial and OGFP for Output Galois Field Polynomials. The respective Polynomial for IGFP8 and OGFP8 has been shown in Table 9.

3.3 Algorithm to Generate S-Box from Polynomials Over Galois Field $GF(2^{15})$ or $GF(2^{255})$

START.
Step OA. Choose 4 Galois field Polynomials over Galois field $GF(2^{15})$ or 8 Galois field Polynomials over Galois field $GF(2^{255})$.
Step. 01. If Number of Terms in BCNs are Half of Number of total terms
Then Go to **Step. 02 Else** Go to Step. 0A.
Step.02. Convert to decimal the 4 or 8 bit binary number generated by bits in same position of 4 BCNs for Galois field Polynomials over Galois field $GF(2^{15})$ or 8 Galois field Polynomials over Galois field $GF(2^{255})$.
STOP.
Time Complexity of the given Algorithm. O(n).

Table 7 8-bit identity crypto S-box

Row	Column	1	2	3	4	5	6	7	8	9	A	B	C	D	E	F	G
1	Index	0	1	2	3	4	5	6	7	8	9	A	B	C	D	E	F
2	S-box	0	1	2	3	4	5	6	7	8	9	10	11	12	13	14	15
3		16	17	18	19	20	21	22	23	24	25	26	27	28	29	30	31
4		32	33	34	35	36	37	38	39	40	41	42	43	44	45	46	47
5		48	49	50	51	52	53	54	55	56	57	58	59	60	61	62	63
6		64	65	66	67	68	69	70	71	72	73	74	75	76	77	78	79
7		80	81	82	83	84	85	86	87	88	89	90	91	92	93	94	95
8		96	97	98	99	100	101	102	103	104	105	106	107	108	109	110	111
9		112	113	114	115	116	117	118	119	120	121	122	123	124	125	126	127
10		128	129	130	131	132	133	134	135	136	137	138	139	140	141	142	143
11		144	145	146	147	148	149	150	151	152	153	154	155	156	157	158	159
12		160	161	162	163	164	165	166	167	168	169	170	171	172	173	174	175
13		176	177	178	179	180	181	182	183	184	185	186	187	188	189	190	191
14		192	193	194	195	196	197	198	199	200	201	202	203	204	205	206	207
15		208	209	210	211	212	213	214	215	216	217	218	219	220	221	222	223
16		224	225	226	227	228	229	230	231	232	233	234	235	236	237	238	239
17		240	241	242	243	244	245	246	247	248	249	250	251	252	253	254	255

Table 8 BCNs for 8 IGFPs and OGFPs

Row	Col	Polynomials (BCNs) [col. 2] MSB → LSB
1	1	
2	IGFP8	0000000000000000000000000000000000000000000000000000000000000000 0000000000000000000000000000000000000000000000000000000000000000 1111111111111111111111111111111111111111111111111111111111111111 1111111111111111111111111111111111111111111111111111111111111111
3	IGFP 7	0000000000000000000000000000000000000000000000000000000000000000 1111111111111111111111111111111111111111111111111111111111111111 0000000000000000000000000000000000000000000000000000000000000000 1111111111111111111111111111111111111111111111111111111111111111
4	IGFP 6	0000000000000000000000000000000011111111111111111111111111111111 0000000000000000000000000000000011111111111111111111111111111111 0000000000000000000000000000000011111111111111111111111111111111 0000000000000000000000000000000011111111111111111111111111111111
5	IGFP 5	0000000000000000111111111111111100000000000000001111111111111111 0000000000000000111111111111111100000000000000001111111111111111 0000000000000000111111111111111100000000000000001111111111111111 0000000000000000111111111111111100000000000000001111111111111111
6	IGFP 4	0000000011111111000000001111111100000000111111110000000011111111 0000000011111111000000001111111100000000111111110000000011111111 0000000011111111000000001111111100000000111111110000000011111111 0000000011111111000000001111111100000000111111110000000011111111
7	IGFP 3	0000111100001111000011110000111100001111000011110000111100001111 0000111100001111000011110000111100001111000011110000111100001111 0000111100001111000011110000111100001111000011110000111100001111 0000111100001111000011110000111100001111000011110000111100001111

(continued)

Table 8 (continued)

8	IGFP 2	0011001100110011001100110011001100110011001100110011001100110011
		0011001100110011001100110011001100110011001100110011001100110011
		0011001100110011001100110011001100110011001100110011001100110011
		0011001100110011001100110011001100110011001100110011001100110011
9	IGFP 1	0101010101010101010101010101010101010101010101010101010101010101
		0101010101010101010101010101010101010101010101010101010101010101
		0101010101010101010101010101010101010101010101010101010101010101
		0101010101010101010101010101010101010101010101010101010101010101
A	OGFP8	0000000000000000000000000000000000000000000000000000000000000000
		0000000000000000000000000000000000000000000000000000000000000000
		1111111111111111111111111111111111111111111111111111111111111111
		1111111111111111111111111111111111111111111111111111111111111111
B	OGFP 7	0000000000000000000000000000000000000000000000000000000000000000
		1111111111111111111111111111111111111111111111111111111111111111
		0000000000000000000000000000000000000000000000000000000000000000
		1111111111111111111111111111111111111111111111111111111111111111
C	OGFP 6	0000000000000000000000000000000011111111111111111111111111111111
		0000000000000000000000000000000011111111111111111111111111111111
		0000000000000000000000000000000011111111111111111111111111111111
		0000000000000000000000000000000011111111111111111111111111111111
D	OGFP 5	0000000000000000111111111111111100000000000000001111111111111111
		0000000000000000111111111111111100000000000000001111111111111111
		0000000000000000111111111111111100000000000000001111111111111111
		0000000000000000111111111111111100000000000000001111111111111111
E	OGFP 4	0000000011111111000000001111111100000000111111110000000011111111
		0000000011111111000000001111111100000000111111110000000011111111
		0000000011111111000000001111111100000000111111110000000011111111
		0000000011111111000000001111111100000000111111110000000011111111

(continued)

Table 8 (continued)

F	OGFP 3	0000111110000111110000111110000111110000111110000111110000111111 0000111110000111110000111110000111110000111110000111110000001111 0000111110000111110000111110000111110000111110000111110000001111 0000111110000111110000111110000111110000111110000111110000001111
G	OGFP 2	0011001100110011001100110011001100110011001100110011001100110011 0011001100110011001100110011001100110011001100110011001100110011 0011001100110011001100110011001100110011001100110011001100110011 0011001100110011001100110011001100110011001100110011001100110011
H	OGFP 1	0101010101010101010101010101010101010101010101010101010101010101 0101010101010101010101010101010101010101010101010101010101010101 0101010101010101010101010101010101010101010101010101010101010101 0101010101010101010101010101010101010101010101010101010101010101

Table 9 Respective polynomial of IGFP8 and OGFP8 of the given 8 bit S-box

BCNs of	Polynomial
IGFP8 & OGFP8	$x^{127} + x^{126} + x^{125} + x^{124} + x^{123} + x^{122} + x^{121} + x^{120} + x^{119} + x^{118} + x^{117} + x^{116} + x^{115} + x^{114} + x^{113} + x^{112} +$ $x^{111} + x^{110} + x^{109} + x^{108} + x^{107} + x^{106} + x^{105} + x^{104} + x^{103} + x^{102} + x^{101} + x^{100} + x^{99} + x^{98} + x^{97} + x^{96} +$ $x^{95} + x^{94} + x^{93} + x^{92} + x^{91} + x^{90} + x^{89} + x^{88} + x^{87} + x^{86} + x^{85} + x^{84} + x^{83} + x^{82} + x^{81} + x^{80} +$ $x^{79} + x^{78} + x^{77} + x^{76} + x^{75} + x^{74} + x^{73} + x^{72} + x^{71} + x^{70} + x^{69} + x^{68} + x^{67} + x^{66} + x^{65} + x^{64} +$ $x^{63} + x^{62} + x^{61} + x^{60} + x^{59} + x^{58} + x^{57} + x^{56} + x^{55} + x^{54} + x^{53} + x^{52} + x^{51} + x^{50} + x^{49} + x^{48} +$ $x^{47} + x^{46} + x^{45} + x^{44} + x^{43} + x^{42} + x^{41} + x^{40} + x^{39} + x^{38} + x^{37} + x^{36} + x^{35} + x^{34} + x^{33} + x^{32} +$ $x^{31} + x^{30} + x^{29} + x^{28} + x^{27} + x^{26} + x^{25} + x^{24} + x^{23} + x^{22} + x^{21} + x^{20} + x^{19} + x^{18} + x^{17} + x^{16} +$ $x^{15} + x^{14} + x^{13} + x^{12} + x^{11} + x^{10} + x^9 + x^8 + x^7 + x^6 + x^5 + x^4 + x^3 + x^2 + x + 1$

Table 10 Constituted 4-bit crypto S-box

Row	Column	1	2	3	4	5	6	7	8	9	A	B	C	D	E	F	G
1	Index	0	1	2	3	4	5	6	7	8	9	A	B	C	D	E	F
2	S-Box	0	1	2	3	4	5	6	7	8	9	A	B	C	D	E	F

4   4 and 8 Bit S-Box Generation by Respective BCNs Over Non Binary Galois Field GF(16^{15}) and Galois Field GF(256^{255}) Respectively

The coefficients of each polynomial over Non Binary Galois Field GF(16^{15}) forms a 4-bit S-Box. The Coefficient of highest or lowest degree term must be the 1st element in 4-bit S-box, the value of other elements is the value of coefficients with immediate degree less than or greater than the previous one. Let the Polynomial be,

$$\mathbf{BP(x)} = 0x^{15} + 1x^{14} + 2x^{13} + 3x^{12} + 4x^{11} + 5x^{10} + 6x^9 + 7x^8$$
$$+ 8x^7 + 9x^6 + 10x^5 + 11x^4 + 12x^3 + 13x^2 + 14x + 15 \qquad (2)$$

For the above Polynomial, the Constituted 4-bit S-Box have been given in Table 10. The Polynomial with coefficients in reverse order,

$$\mathbf{BP(x)} = 15x^{15} + 14x^{14} + 13x^{13} + 12x^{12} + 11x^{11} + 10x^{10} + 9x^9$$
$$+ 8x^8 + 7x^7 + 6x^6 + 5x^5 + 4x^4 + 3x^3 + 2x^2 + 1x + 0 \qquad (3)$$

For the above Polynomial, the Constituted 4-bit S-Box have been given in Table 11.

Table 11 Constituted 4-bit crypto S-box

Row	Column	1	2	3	4	5	6	7	8	9	A	B	C	D	E	F	G
1	Index	0	1	2	3	4	5	6	7	8	9	A	B	C	D	E	F
2	S-Box	15	14	13	12	11	10	9	8	7	6	5	4	3	2	1	0

Table 12 Polynomial to construct 8-bit identity S-box

Polynomial BP(x)=

$0.x^{255} + 1.x^{254} + 2.x^{253} + 3.x^{252} + 4.x^{251} + 5.x^{250} + 6.x^{249} + 7.x^{248} + 8.x^{247} + 9.x^{246}$
$+10.x^{245} + 11.x^{244} + 12.x^{243} + 13.x^{242} + 14.x^{241} + 15.x^{240} + 16.x^{239} + 17.x^{238}$
$+18.x^{237} + 19.x^{236} + 20.x^{235} + 21.x^{234} + 22.x^{233} + 23.x^{232} + 24.x^{231} + 25.x^{230}$
$+26.x^{229} + 27.x^{228} + 28x^{227} + 29.x^{226} + 30.x^{225} + 31.x^{224} + 32.x^{223} + 33.x^{222}$
$+34.x^{221} + 35.x^{220} + 36.x^{219} + 37.x^{218} + 38.x^{217} + 39.x^{216} + 40.x^{215} + 41.x^{214} + 42.x^{213}$
$+ 43.x^{212} + 44.x^{211} + 45.x^{210} + 46.x^{209} + 47.x^{208} + 48.x^{207} + 49.x^{206} + 50.x^{205}$
$+51.x^{204} + 52.x^{203} + 53.x^{202} + 54.x^{201} + 55.x^{200} + 56.x^{199} + 57.x^{198} + 58.x^{197}$
$+59.x^{196} + 60.x^{195} + 61.x^{194} + 62.x^{193} + 63.x^{192} + 64.x^{191} + 65.x^{190} + 66.x^{189}$
$+67.x^{188} + 68.x^{187} + 69.x^{186} + 70.x^{185} + 71.x^{184} + 72.x^{183} + 73.x^{182} + 74.x^{181}$
$+75.x^{180} + 76.x^{179} + 77.x^{178} + 78.x^{177} + 79.x^{176} + 80.X^{175} + 81.x^{174} + 82.x^{173}$
$+83.x^{172} + 84.x^{171} + 85.x^{170} + 86.x^{169} + 87.x^{168} + 88.x^{167} + 89.x^{166} + 90.x^{165}$
$+91.x^{164} + 92.x^{163} + 93.x^{162} + 94.x^{161} + 95.x^{160} + 96.x^{159} + 97.x^{158} + 98.x^{157}$
$+99.x^{156} + 100.x^{155} + 101.x^{154} + 102.x^{153} + 103.x^{152} + 104.x^{151} + 105.x^{150}$
$+106.x^{149} + 107.x^{148} + 108.x^{147} + 109.x^{146} + 110.x^{145} + 111.x^{144} + 112.X^{143}$
$+113.x^{142} + 114.x^{141} + 115.x^{140} + 116.x^{139} + 117.x^{138} + 118.x^{137} + 119.x^{136}$
$+120.x^{135} + 121.x^{134} + 122.x^{133} + 123.x^{132} + 124.x^{131} + 125.x^{130} + 126.x^{129}$
$+127.x^{128} + 128.x^{127} + 129.x^{126} + 130.x^{125} + 131.x^{124} + 132.x^{123} + 133.x^{122}$
$+134.x^{121} + 135.x^{120} + 136.x^{119} + 137.x^{118} + 138.x^{117} + 139.x^{116} + 140.x^{115}$
$+141.x^{114} + 142.x^{113} + 143.x^{112} + 144.x^{111} + 145.x^{110} + 146.x^{109} + 147.x^{108}$
$+148.x^{107} + 149.x^{106} + 150.x^{105} + 151.x^{104} + 152.x^{103} + 153.x^{102} + 154.x^{101}$
$+155.x^{100} + 156.x^{99} + 157.x^{98} + 158.x^{97} + 159.x^{96} + 160.x^{95} + 161.x^{94} + 162.x^{93}$
$+163.x^{92} + 164.x^{91} + 165.x^{90} + 166.x^{89} + 167.x^{88} + 168.x^{87} + 169.x^{86} + 170.x^{85}$
$+171.x^{84} + 172.x^{83} + 173.x^{82} + 174.x^{81} + 175.x^{80} + 176.x^{79} + 177.x^{78} + 178.x^{77}$
$+179.x^{76} + 180.x^{75} + 181.x^{74} + 182.x^{73} + 183.x^{72} + 184.x^{71} + 185.x^{70} + 186.x^{69}$
$+187.x^{68} + 188.x^{67} + 189.x^{66} + 190.x^{65} + 191.x^{64} + 192.x^{63} + 193.x^{62} + 194.x^{61}$
$+195.x^{60} + 196.x^{59} + 197.x^{58} + 198.x^{57} + 199.x^{56} + 200.x^{55} + 201.x^{54} + 202.x^{53}$
$+203.x^{52} + 204.x^{51} + 205.x^{50} + 206.x^{49} + 207.x^{48} + 208.x^{47} + 209.x^{46} + 210.x^{45}$
$+211.x^{44} + 212.x^{43} + 213.x^{42} + 214.x^{41} + 215.x^{40} + 216.x^{39} + 217.x^{38} + 218.x^{37}$
$+219.x^{36} + 220.x^{35} + 221.x^{34} + 222.x^{33} + 223.x^{32} + 224.x^{31} + 225.x^{30} + 226.x^{29}$
$+227.x^{28} + 228.x^{27} + 229.x^{26} + 230.x^{25} + 231.x^{24} + 232.x^{23} + 233.x^{22} + 234.x^{21}$
$+235.x^{20} + 236.x^{19} + 237.x^{18} + 238.x^{17} + 239.x^{16} + 240.x^{15} + 241.x^{14} + 242.x^{13}$
$+243.x^{12} + 244.x^{11} + 245.x^{10} + 246.x^9 + 247.x^8 + 248.x^7 + 249.x^6 + 250.x^5$
$+251.x^4 + 252.x^3 + 253.x^2 + 254x + 255$

The coefficients of each polynomial over Non Binary Galois Field GF(256^{255}) forms a 8-bit S-Box. The Coefficient of highest or lowest degree term must be the 1st element in 4-bit S-Box, the value of other elements is the value of coefficients with immediate degree less than or greater than the previous one. Let the Polynomial be, Let the Polynomial be given in Table 12.

For the above Polynomial, the Constituted 8-bit S-Box have been given in Table 13.

Table 13 Constituted identity 8-bit S-box

Row	Column	1	2	3	4	5	6	7	8	9	A	B	C	D	E	F	G
1	Index	0	1	2	3	4	5	6	7	8	9	A	B	C	D	E	F
2	s-box	0	1	2	3	4	5	6	7	8	9	10	11	12	13	14	15
3		16	17	18	19	20	21	22	23	24	25	26	27	28	29	30	31
4		32	33	34	35	36	37	38	39	40	41	42	43	44	45	46	47
5		48	49	50	51	52	53	54	55	56	57	58	59	60	61	62	63
6		64	65	66	67	68	69	70	71	72	73	74	75	76	77	78	79
7		80	81	82	83	84	85	86	87	88	89	90	91	92	93	94	95
8		96	97	98	99	100	101	102	103	104	105	106	107	108	109	110	111
9		112	113	114	115	116	117	118	119	120	121	122	123	124	125	126	127
10		128	129	130	131	132	133	134	135	136	137	138	139	140	141	142	143
11		144	145	146	147	148	149	150	151	152	153	154	155	156	157	158	159
12		160	161	162	163	164	165	166	167	168	169	170	171	172	173	174	175
13		176	177	178	179	180	181	182	183	184	185	186	187	188	189	190	191
14		192	193	194	195	196	197	198	199	200	201	202	203	204	205	206	207
15		208	209	210	211	212	213	214	215	216	217	218	219	220	221	222	223
16		224	225	226	227	228	229	230	231	232	233	234	235	236	237	238	239
17		240	241	242	243	244	245	246	247	248	249	250	251	252	253	254	255

Note The 32-bit S-boxes can be constituted by Polynomials over Galois field $GF[(2^{32})(2^{2 \cdot 32} - 1)]$ and the 64-bit S-boxes can be constituted by Polynomials over Galois field $GF[(2^{64})(2^{2 \cdot 64} - 1)]$

5 Conclusion

From this Research Article in can be concluded that 4, 8, 32, 64 bit Substitution boxes can be constituted using Basic Polynomials or BPs over Galois Fields GF(16^{15}), GF(256^{255}), GF[$(2^{32})^{(2^{\wedge}32 - 1)}$] and GF[$(2^{64})^{(2^{\wedge}64 - 1)}$] respectively. For this reason, Generation of 4, 8, 32, 64-bit S-Boxes have been generated very easily with very less complexity. It is a very important work in modern cryptography since the work has been dedicated to crypto community for upgradation of complexity of crypto algorithms and ease to develop smart and intelligent devices for future. In the future work, we are planning to apply the proposed model in different secure applications [16–24].

Acknowledgements For this work I would like to acknowledge my Supervisor Dr. Ranjan Ghsoh for his continuous encouragement and support. I would also like to acknowledge Prof. Debatosh Guha for providing me the infrastructure to do this work very elegantly.

References

1. Church R (1935) Tables of irreducible polynomials for the first four Prime Moduli. Ann Math 36(1):198–209
2. Das S, Zaman JKMS Uz, Ghosh R (2013) Generation of AES S-boxes with various modulus and additive constant polynomials and testing their randomization. In: Proceedings (CIMTA) 2013, Procedia Technology vol 10, pp 957–962
3. C'almet J, Loos R (1980) An improvement of Rabin's probabilistic algorithm for generating irreducible polynomials over Gf(P). Inf Process Lett 11(2)
4. Sadique Uz Zaman JKM, Sankhanil D, Ghosh R (2015) An algorithm to find the irreducible polynomials over Galois Field GF(pm). Int J Comput Appl (0975 – 8887) 109(15)
5. Keller PE, Kouzes RT, Kangas LJ, Hashem S (1995) Transmission of olfactory information for telemedicine. Studies in health technology and informatics, vol 18: Interactive Technology and the New Paradigm for Healthcare, pp 168–172
6. Haghpanah V, Saeedi M (2013) Smart article: a scientific crosstalk. Front Physiol 4:161
7. Jusoh S, Alfawareh HM (2012) Techniques, applications and challenging issue in text mining. Int J Comput Sci 9(6), No.2:431–436
8. Akhras G (2000) Smart materials and smart systems for the future. Can Mil J 1
9. Varadan VK (2005) Handbook of smart systems and materials. Institute of Physics Pub, London
10. Meyer G et al (2009) Advanced microsystems for automotive applications 2009 - smart systems for safety. Springer, Sustainability and Comfort
11. Shannon Claude (1949) Communication theory of secrecy systems *(PDF)*. Bell Syst Tech J 28(4):656–715
12. Vaudenay S (2002) Security flaws induced by CBC padding applications to SSL, IPSEC, WTLS…. In: Advances in Cryptology – EUROCRYPT 2002, Proceedings of international conference on the theory and applications of cryptographic techniques, vol 2332, pp 534–545. Springer
13. Data Encryption Standard (1977) Federal information processing standards Publication No. 46, National Bureau of Standards
14. Feistel H (1971) Block Cipher cryptographic system. US Patent 3798359
15. Diffe W (1982) Cryptographyic technology: fiften yera forecast. In: Gersho A (ed) Advances in cryptography - a report on Crypto81, Report 82-04. Dept. CSE, University of California, Santa Barbara

16. Elhoseny M, Hosny A, Hassanien AE, Muhammad K, Sangaiah AK (2017) Secure automated forensic investigation for sustainable critical infrastructures compliant with green computing requirements. IEEE Trans Sustain Comput Issue: 99. (https://doi.org/10.1109/tsusc.2017. 2782737)

17. FarouK A, Batle J, Elhoseny M, Naseri M, Lone M, Fedorov A, Alkhambashi M, Ahmed SH, Abdel-Aty M (2018) Robust general N user authentication scheme in a centralized quantum communication network via generalized GHZ states. Front Phys 13:130306. Springer. (https:// doi.org/10.1007/s11467-017-0717-3)

18. Wang MM, Qu ZG, Elhoseny M (2017) Quantum secret sharing in noisy environment. In: Sun X, Chao HC, You X, Bertino E (eds) Cloud computing and security. ICCCS 2017. Lecture Notes in Computer Science, vol 10603. Springer, Cham (https://doi.org/10.1007/978-3-319-68542-7_9)

19. Elsayed W, Elhoseny M, Sabbeh S, Riad A (2017) Self-maintenance model for wireless sensor networks. Comput Electr Eng, In Press, Available Online (https://doi.org/10.1016/j. compeleceng.2017.12.022)

20. Elhoseny M, Farouk A, Zhou N, Wang M-M, Abdalla S, Batle J (2017) Dynamic multi-hop clustering in a wireless sensor network: performance improvement. Wirel Pers Commun 95(4):3733–3753. Springer US (https://doi.org/10.1007/s11277-017-4023-8)

21. Yuan X, Elhoseny M, El-Minir HK, Riad AM (2017) A genetic algorithm-based, dynamic clustering method towards improved wsn longevity. J Netw Syst Manage 25(1):21–46. https:// doi.org/10.1007/s10922-016-9379-7 Springer, US

22. Elhoseny M, Tharwat A, Farouk A, Hassanien AE (2017) K-coverage model based on genetic algorithm to extend WSN lifetime. IEEE Sens Lett 1(4):1–4. IEEE. (https://doi.org/10.1109/ lsens.2017.2724846)

23. Elhoseny M, Yuan X, ElMinir HK, Riad AM (2016) An energy efficient encryption method for secure dynamic WSN. Secur Commun Netw 9(13):2024–2031. https://doi.org/10.1002/sec. 1459 Wiley

24. Elhoseny M, Elminir H, Riad A, Yuan X (2016) A secure data routing schema for WSN using elliptic curve cryptography and homomorphic encryption. J King Saud Univ Comput Inf Sci 28(3):262–275. https://doi.org/10.1016/j.jksuci.2015.11.001 Elsevier

Local Binary Patterns Based Facial Expression Recognition for Efficient Smart Applications

Swati Nigam, Rajiv Singh and A. K. Misra

Abstract Facial expressions are direct means of communication of human's emotional state. Hence facial expression recognition (FER) has always been a topic of great interest of researchers specially for smart applications. Numerous approaches have been proposed for FER using Local binary patterns (LBP). This chapter presents an analysis of LBP feature descriptor for FER. State of the art approaches have been discussed that use LBP as their main component. Here, basic LBP operator along with several variants and their main properties are described that have been proved useful for FER. A general framework for FER is described which includes four consecutive modules. These modules are preprocessing, feature extraction, dimensionality reduction and classification. LBP based FER results have been reported on three benchmark datasets JAFFE, CK+ and Yale. Experimentation demonstrates usefulness of LBP in FER.

Keywords Emotion recognition · Facial expression recognition
Behavior detection · Feature learning · Visual surveillance and security
Smart applications

1 Introduction

Facial expression recognition is an active area of research in machine learning especially for security and privacy. It provides important cues for human emotions and behavior analysis. Video analytics involving automatic recognition of facial expres-

S. Nigam · A. K. Misra
Computer Science and Engineering Department, S. P. Memorial Institute of Technology,
Kaushambi, Uttar Pradesh, India
e-mail: swatinigam.au@gmail.com

A. K. Misra
e-mail: akmishra@spmit.edu.in

R. Singh (✉)
Department of Computer Science, Banasthali Vidyapith, Banasthali, Rajasthan, India
e-mail: jkrajivsingh@gmail.com

sions is an interesting topic of research that has attracted much attention in past few years. It has potential applications in several areas such as human-computer interaction, biometrics, data-driven animation, and customized applications for consumer products, etc. [1, 2]. A successful facial expression recognition system must possess the properties that represent features efficiently and effectively. These features should be capable of minimizing within class variability and maximizing between class variability [3]. However, the inherent variability of facial expressions may be affected by several factors such as different illumination conditions, pose, alignment, and occlusion problems. These factors make facial expression recognition much more difficult [4].

To deal with aforesaid challenges, several techniques have been proposed in last decade. Few surveys on facial expression analysis have dealt with those challenges and provided possible solutions [5–7]. Depending on the reviews provided, the facial expression analysis approaches can be divided in two categories: geometric features based methods (GFBM) and appearance based methods (ABM). Initial attempts for facial expression recognition were mostly based on geometric features likewise position, distance and angles. One of the most popular geometric feature based approaches is Facial Action Coding System (FACS) proposed by Ekman and Friesen [8]. It recognizes facial expressions using a set of Action Units (AU) that correspond to the physical motion of a specific facial muscle. A facial feature representation approach based on the geometric positions of 34 manually selected fiducial points was presented in [9]. A similar approach in [10] used linear programming for feature extraction and classifier training. In other similar approaches [11, 12], facial expression recognition systems based on tracked fiducial point data are presented. Geometric features based approaches provided good results in recognition of action units. However, they can be applied only after accurate detection of facial components.

Appearance based methods (ABM) remove the shortcomings of geometric features based methods (GFBM). In order to retrieve changes in facial appearance, ABM apply an image filter or filter bank on whole face or a specific regions of face. The most common techniques of ABM are Principal Component Analysis (PCA) [13] and Independent Component Analysis (ICA) [14]. PCA uses only holistic information of an image whereas ICA uses local information also. Few other ABM are Gabor-wavelets [15, 16] and local feature analysis [17]. A hybrid approach of local feature analysis and image vector analysis is proposed in [18].

Recently, Local binary patterns (LBP) and their variants based facial expression analysis methods have gained much attention due to their effectiveness towards facial features [19]. The LBP based methods are theoretically simple yet efficient. LBP is originally a texture operator. It was initially used for face recognition and later on proved very significant for facial expression recognition [20]. LBP retrieves local information by thresholding neighborhood values of center pixel. It provides a method which is computationally efficient and robust towards monotonic changes in illumination. For the smart applications, this process is more complicated [21–23]. Therefore, in this chapter, we have focused on comparative analysis of FER methods based on LBP. Several variants of LBP and their useful properties have been used in different algorithms developed in this chapter. Three publicly available benchmark datasets are used for this purpose viz. JAFFE, CK+ and Yale. Experimental results demonstrate usefulness of LBP in FER.

The rest of the chapter is organized as follows: basic local binary pattern (LBP) operator is discussed in Sect. 2. In Sect. 3, brief details of different variants of LBP are given. Section 4 defines major properties of LBP and its variants. Section 5 explains the proposed FER framework. Experimentation and results are discussed in Sect. 6. Finally, Sect. 7 concludes this study.

2 Local Binary Pattern (LBP)

Local binary pattern (LBP) labels the pixels of an image by thresholding the neighborhood of each pixel with the value of center pixel and outputs a binary number [24–28]. A simple LBP feature is represented by using the relative gray values of an eight connected region as shown in Fig. 1.

An LBP feature can be constructed for a specific circular pixel neighborhood of radius R. Intensities of the P sample pixel points are compared in the circular neighborhood with center pixel in clockwise or anticlockwise direction. This comparison determines the value of the sample point that should be either one (1) or zero (0). A value one (1) is assigned for the sample point if gray value of sample point is greater than or equal to gray value of center pixel, and a value zero (0) is assigned for the sample point if its gray value is less than gray value of center pixel. A popular choice for number of sample points is 8 (i.e. $P = 8$) and for radius is 1 (i.e. $R = 1$). Although, we can use other combinations of P and R also. If a sample point is located between pixels, the intensity value for this point is determined by using bilinear interpolation. After extracting value of each sample point, the value of center pixel is replaced by a binary pattern. This procedure is followed for every pixel in an image except border pixels as they do not have neighboring pixels. The procedure for computation of LBP code is shown in Fig. 2.

Fig. 1 Eight connected neighborhood of a pixel

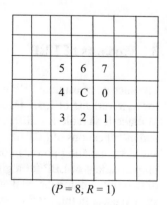

$(P = 8, R = 1)$

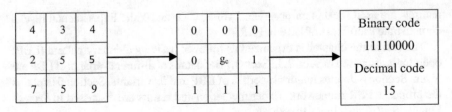

Fig. 2 An example local binary pattern

With the help of these considerations, feature vector of a single pixel of an image, denoted by $LBP_{P,R}$, is given below

$$LBP_{P,R}(x, y) = \sum_{h=0}^{P-1} s(g_h - g_c)2^h \tag{1}$$

where (x, y) is the location of the center pixel, g_c represent gray value of the center pixel, g_h represent gray value of neighboring pixel and sign function $s(z)$ of a sequence z is defined as

$$s(z) = \begin{cases} 1, & z \geq 0 \\ 0, & z < 0 \end{cases} \tag{2}$$

Equation (1) can be interpreted as sign function of differences in surrounding region represented as a P bit binary number that results in 2^P different values for binary patterns. The feature vector $LBP_{P,R}$ is a histogram of 2^P LBP values of image pixels. For a neighborhood of $P = 8$ and $R = 1$, the histogram size is $2^8 = 256$. For M regions in an image, all histograms can be merged into one histogram of size $M \cdot 2^P$. The name 'Local Binary Pattern' reflects nature of the operator $LBP_{P,R}$, i.e. a local neighborhood is thresholded at gray value of center pixel into a binary pattern. The $LBP_{P,R}$ feature vector computation is shown in Fig. 3.

3   Variants of LBP

The success of original LBP in different vision applications has inspired researches for new variants of LBP. Due to flexible nature of LBP, it can be easily modified to make it suitable for different vision applications. Therefore, several extensions and modifications of LBP have been proposed to increase and explore its robustness and discriminating power in different types of applications [29–36]. Selection of a proper variant of LBP for a given application is based on the discriminating power of descriptor, computational efficiency, robustness to illumination, and the imaging system used. In this section, we have described different variants of LBP to provide a basic understanding of their role in FER.

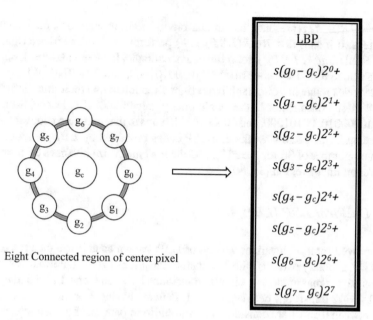

Fig. 3 LBP in an eight connected region

3.1 Rotation Invariant LBP (LBP_RI)

In many applications, it is desirable to have robust features that are invariant against rotations of input image. As the LBP features are obtained by a circular sampling around the center pixel, rotation of input image has two effects [37, 38]

i. Each local neighborhood is rotated into other pixel location, and
ii. Within each neighborhood the sampling points on the circle surrounding the center pixel are rotated into a different orientation.

Rotation of an input image causes the LBP pattern to translate into a different location and then rotate about the origin. When an image is rotated, the gray value g_h will correspondingly move along the perimeter of the circle, so different $LBP_{P,R}$ may be computed. A modified version of LBP has been proposed to achieve rotation invariance and to reduce the histogram dimension of the LBP [27]. Computing the histogram of LBP codes using rotation invariant mapping reduces the effect of rotation. In rotation invariant mapping, each LBP binary code is circularly rotated to achieve its minimum value.

The modified version of $LBP_{P,R}$ with rotation invariance, denoted as $LBP_{P,R}^{ri}$, is defined as follows

$$LBP_{P,R}^{ri}(x, y) = \min\{ROR(LBP_{P,R}, h)|h = 0, 1, \ldots, P - 1\} \qquad (3)$$

where $ROR(u, h)$ represents the circular bitwise right rotation of a bit sequence u by h steps. It means that $ROR(LBP_{P,R}, h)$ performs a circular bitwise right shift on the P-bit number $LBP_{P,R}$ for h times. For example, for $P = 8$ and $R = 1$, suppose there are three 8-bit LBP codes 10000010, 00101000, and 00000101. When 0-1 bit of these codes rotate in a circular bitwise right direction to find their minimum value, then all these codes result in the single binary value 00000101. Hence, 8-bit LBP codes 10000010, 00101000, and 00000101 all maps to the minimum code 00000101. Proceeding in the same way, the 256 LBP codes produced by $LBP_{P,R}$ reduce to a minimum number of 36 for $LBP_{P,R}^{ri}$, and the histogram size becomes 36 instead of 256 over an image region [39, 40].

3.2   Uniform LBP (LBP_U2)

An improved rotation invariant version of LBP, known as uniform patterns or uniform LBP, is binary patterns with few spatial transitions [41, 42]. Formally, uniform patterns have a maximum two circular transitions between 0 and 1 bit i.e. they may have 0, 1 or 2 transitions between 0 and 1. Patterns having 3 or more than 3 transitions between 0 and 1 are considered as non-uniform patterns. For example, pattern 00000001 have only one and pattern 11111011 have two transitions between 0 and 1, therefore they are uniform patterns. On the other hand, pattern 11110100 have 3 and pattern 00001010 have 4 transitions between 0 and 1, therefore they are non-uniform patterns. Figure 4 shows all 0-1 transitions of uniform patterns for $P = 8$ and $R = 1$.

Fig. 4 All uniform LBP for $P = 8$ and $R = 1$. Red circles represent bit value 0 and blue circles represent bit value 1

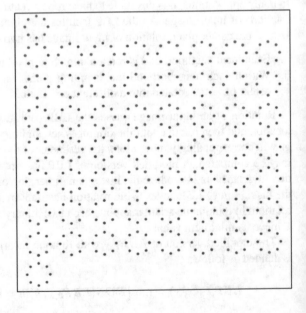

The gray scale and rotation invariant uniform LBP, denoted as $LBP_{P,R}^{riu2}$, is defined as

$$LBP_{P,R}^{riu2} = \begin{cases} \sum_{h=0}^{P-1} s(g_h - g_c), & if \ U(LBP_{P,R}) \leq 2 \\ P+1, & otherwise \end{cases} \quad (4)$$

and

$$U(LBP_{P,R}) = |s(g_{P-1} - g_c) - s(g_0 - g_c)| + \sum_{h=1}^{P-1} |s(g_h - g_c) - s(g_{h-1} - g_c)| \quad (5)$$

where the uniformity measure U corresponds to the number of spatial transitions (bitwise changes between 0 and 1) in a pattern and superscript *riu2* corresponds to a rotation invariant operator with uniform patterns having at most 2 transitions between 0 and 1 bit. In uniform LBP mapping, there is a separate output label for each uniform pattern and all non-uniform patterns are assigned to a single label.

3.3 Multiscale LBP

One of the shortcomings of original LBP operator is that it is computed in a small spatial neighborhood area. A general choice of size of the neighborhood is 3×3 pixels. However, the features computed in 3×3 local neighborhoods cannot capture information about large structures. These large structures may be dominant features in some images [43, 44]. Also, adjacent LBP codes are not totally independent of each other as shown in Fig. 5.

Figure 5 shows three adjacent 4-bit LBP codes. Each LBP code limits the set of possible codes adjacent to it and creates a neighborhood of a single code slightly larger than 3×3 pixels. Therefore, actual local support area of the operator changes and it does not become robust against local variations in the pattern. In order to remove this shortcoming, an operator with a larger spatial support area is often needed that is robust to handle local changes within its area.

Fig. 5 Adjacent LBP neighborhoods in a four connected region

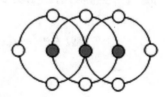

An explicit way of increasing the neighborhood area is to merge the information retrieved by different LBP operators by varying (P, R) so that each pixel in an image gets different LBP values. This multiscale operation could be performed by varying values of P and R as follows. For example, LBP values for pairs $(P = 4, R = 1)$, $(P = 8, R = 1)$, $(P = 12, R = 2)$, and $(P = 16, R = 2)$ could be merged together to form an operator of larger spatial support area. A multiscale LBP feature histogram obtained by merging the rotation invariant uniform LBP operator $LBP_{P,R}^{riu2}$ for the pairs $(P = 4, R = 1)$, $(P = 8, R = 1)$, $(P = 12, R = 2)$, and $(P = 16, R = 2)$ is denoted by $LBP_{(4,1)+(8,1)+(12,2)+(16,2)}^{riu2}$. More accurate information can be achieved by using the joint distribution of these values [45].

It is important to note that such distributions become sparse with increasing image size. For example, let us take the joint distribution of four rotation invariant uniform LBP codes for the pairs $(P = 4, R = 1)$, $(P = 8, R = 1)$, $(P = 12, R = 2)$, and $(P = 16, R = 2)$. The number of histogram bins are 15, 59, 135 and 243 according to $P \times (P-1)+3$. Therefore, their joint distribution consists of number of bins $15 \times 59 \times 135 \times 243 \approx 3 \times 10^7$. Due to this large size, only marginal distributions of different LBP descriptors have been considered even though the output of different LBP descriptors on a pixel may not be statistically independent. Although the statistical independence may not be achieved by LBP operators of different radii for a few typical cases, use of multiscale analysis often increases the discriminating power of the resulting LBP codes. In many applications, this explicit way of building a multiscale LBP operator has resulted in better accuracy [43–45].

3.4 Center Symmetric LBP (CSLBP)

The center symmetric LBP (CSLBP) has been developed for representation of region of interest. It results into lesser number of LBP codes and hence provides a lesser number of values in histogram to be used as region descriptors. Moreover, CSLBP is developed to provide better representation of flat regions in an image [46].

CSLBP is different from LBP in the sense that, instead of comparing pixel values with center pixel, CSLBP compares pixel values with the opposite pixel symmetrically about the center pixel. An illustration of CSLBP for 8 neighbors has been shown in Fig. 6.

To increase the robustness of CSLBP operator in flat regions, a non-zero threshold Th has been imposed on the differences of pixel gray values. Now the CSLBP operator, denoted as $CSLBP_{P,R,Th}$, can be defined as

$$CSLBP_{P,R,Th}(x, y) = \sum_{h=0}^{(P/2)-1} s(g_h - g_{h+(P/2)} - Th) 2^h \qquad (6)$$

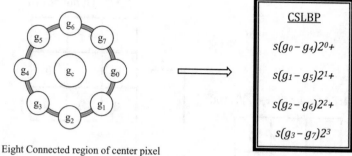

Eight Connected region of center pixel

Fig. 6 CSLBP in an eight connected region

where,

$$s(z) = \begin{cases} 1, z \geq 0 \\ 0, otherwise \end{cases} \tag{7}$$

where g_h and $g_{h+(P/2)}$ correspond to the gray values of the center symmetric pairs of pixels of P equally spaced pixels on a circle of radius R. It should be noted that the CSLBP is closely related to the gradient operator as it considers gray level differences between pairs of opposite pixels in a neighborhood.

3.5 Local Ternary Pattern (LTP)

In order to make LBP more robust against small changes in pixel values of flat image regions where gray values of pixels do not change significantly, the thresholding scheme of descriptor is modified and a three-level operator called local ternary pattern (LTP) is proposed [47]. The LTP descriptor is suitable to deal with the problems in near constant gray value areas. In binary encoding the difference between the center pixel and a neighboring pixel is encoded by two values zero (0) and one (1) whereas in ternary encoding the difference between the center pixel and a neighboring pixel is encoded by three values (1, 0 and -1). The ternary pattern has been divided into two binary patterns taking into account its positive and negative components. The histograms from these positive and negative components computed over a region of interest are then concatenated. The LTP operator with a threshold value t, denoted as $LTP_{P,R}$, can be defined as

$$LTP_{P,R}(x, y) = \sum_{h=0}^{P-1} s(g_h - g_c)2^h \tag{8}$$

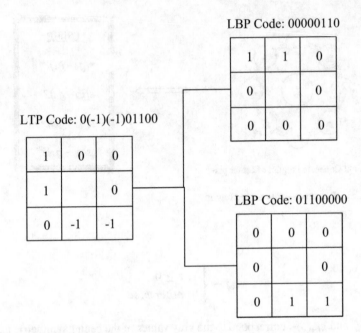

Fig. 7 Illustration of an LTP code

where,

$$s(z) = \begin{cases} 1, & z \geq g_c + t \\ 0, & |z - g_c| < t \\ -1, & z \leq g_c - t \end{cases} \qquad (9)$$

An example of splitting a ternary code into positive and negative binary codes is shown in Fig. 7. Here it can be noted that LTP operator is similar to the texture spectrum operator as both use a three-valued output instead of two [48].

4 Properties of LBP and Its Variants

In this section, we have explored the important properties of LBP and its variants. These properties are gray scale invariance, rotation invariance and multiresolution analysis. The mentioned properties have been used in many vision applications to achieve better performance and robustness.

4.1 Gray Scale Invariance

The gray scale invariance in LBP can be achieved in following way. Consider an image $f(x, y)$ and if g_c represent the gray value of a pixel at location (x, y) and g_h represent gray level of sample points in an equispaced circular neighborhood where number of sample points are P and radius is R around the pixel at location (x, y). Then gray value g_h of sample point at neighboring location (x_h, y_h) of center pixel can be denoted as

$$g_h = f(x_h, y_h), \quad h = 0, \ldots, P - 1 \tag{10}$$

LBP achieves gray scale invariance by assuming that the gray scale transformation is a linear function. It realizes gray scale invariance by global normalization of the input image using histogram equalization. In order to achieve gray scale invariance, let the joint distribution of the gray values of number of pixels $P + 1$ with $P > 0$ in an image $f(x, y)$ is given by

$$T = t(g_c, g_0, g_1, \ldots, g_{P-1}) \tag{11}$$

Without loss of information, by subtracting the gray value of the center pixel (g_c) from the gray values of the neighboring pixels, we get

$$T = t(g_c, g_0 - g_c, g_1 - g_c, \ldots, g_{P-1} - g_c) \tag{12}$$

Now it is assumed that the center pixel does not depend on neighboring pixels i.e. differences $g_h - g_c$ are independent of g_c, therefore Eq. (12) may be factored in

$$T \approx t(g_c) t(g_0 - g_c, g_1 - g_c, \ldots, g_{P-1} - g_c) \tag{13}$$

Practically, differences $g_h - g_c$ are not perfectly independent of g_c and Eq. (13) represents only an approximation of the joint distribution. However, the possible loss of information is within acceptable range, in achieving the invariance against the changes in gray values [38]. Here, the distribution $t(g_c)$ is the overall intensity distribution of image $f(x, y)$ which can be omitted because it does not contain any significant information. Therefore, only the joint distribution of differences

$$t(g_0 - g_c, g_1 - g_c, \ldots, g_{P-1} - g_c)$$

represent the local distribution of gray values of pixels. For suitable estimation of the joint distribution of differences, sign function of differences is taken as

$$t(s(g_0 - g_c), s(g_1 - g_c), \ldots, s(g_{P-1} - g_c))$$

where sign function $s(z)$ is defined as

$$s(z) = \begin{cases} 1, \ z \geq 0 \\ 0, \ otherwise \end{cases} \tag{14}$$

With help of the above differences, the general local binary pattern operator $LBP_{P,R}$ can be obtained as follows:

$$LBP_{P,R}(x, y) = \sum_{h=0}^{P-1} s(g_h - g_c)2^h \tag{15}$$

Equation (15) has achieved invariance with respect to the scaling of gray scale by considering the signs of the differences only instead of their exact values. Signed differences $g_h - g_c$ are not affected by changes in mean luminance, hence the joint difference distribution is invariant against gray scale changes. Thus, the operator $LBP_{P,R}$ is invariant against any monotonic transformation of gray values i.e. as long as the order of the gray values stays the same, the output of the operator remains constant.

4.2   Rotation Invariance

The $LBP_{P,R}$ operator produces 2^P different output values, corresponding to 2^P different binary patterns formed by P pixels in the neighbor set. When an image is rotated, the gray values g_h will correspondingly move along the perimeter of the circle around g_c. Since, g_0 is always assigned to be the gray value of element $(0, R)$ to the right of g_c, rotating a particular binary pattern naturally results in a different $LBP_{P,R}$ value. This does not apply to patterns comprising of only 0 or 1 e.g. patterns 00000000 and 11111111, as gray values of these patterns remain constant at all rotation angles. To remove the effect of rotation, i.e. to assign a unique identifier to each rotation invariant LBP, Eq. (3) has been defined based on following assumptions.

Let $U(m, r)$ represents a particular uniform LBP pattern (refer Sect. 3.2). The pair (m, r) denotes a uniform pair where m is the number of bits having value 1 in this uniform pattern (m corresponds to row number in Fig. 4 for $P=8$) and r is the rotation of the pattern (r corresponds to column number in Fig. 4 for $P=8$). Let the neighborhood contains P sampling points then m obtains values from 0 to $P + 1$, where $m = P + 1$ is the special label marking all non-uniform patterns. For $m=0$, the pattern is 00000000 and for $m=P$, the pattern is 11111111 and both of these are not affected by any rotation. For all other values of m i.e. if $1 \leq m \leq P - 1$, the rotation of the pattern lies in the range $1 \leq r \leq P - 1$.

Let $f^\beta(x, y)$ represents the rotation of image $f(x, y)$ by β degrees. Within this rotation, the point (x, y) is rotated to a new position (x', y'). A circular sampling neighborhood values on above points $f(x, y)$ and $f^\beta(x', y')$ also rotates by β degrees as shown in Fig. 8.

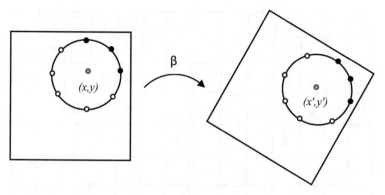

Fig. 8 Points in circular neighborhood after image rotation

Here, rotations are limited to integer multiples of the angle between two sampling points, i.e. $\beta = h\frac{360°}{P}$, $h = 0, 1, \ldots, P - 1$. This rotates the sampling neighborhood in exactly h discrete steps. Therefore, the uniform pattern $U(m, r)$ at point (x, y) is replaced by uniform pattern $U(m, \ r + h \ \text{mod} \ P)$ at the point (x', y') of the rotated image. In this way, to remove the effect of rotation, the neighboring h bits around a pixel are clockwise rotated h times so that a maximum number of the most significant bits are used to express this pixel.

4.3 Multiresolution Analysis

In previous sections, we have overviewed a general gray scale and rotation invariant operator for recognizing the spatial patterns and the contrast of local image objects using a circularly symmetric neighbor set of P pixels placed on a circle of radius R. By changing the values of P and R, the operators can be realized for different quantization of the angular space and spatial resolutions. The general formulation of original LBP operator puts no limitation to the size of neighborhood or to number of sampling points. Therefore, the multiresolution analysis can be accomplished by analyzing and representing the information provided by multiple operators of varying (P, R). Figure 9 illustrates circularly symmetric neighbor sets for different values of P and R. These different (P, R) pairs provide information for multiresolution analysis.

5 Facial Expression Recognition Framework

A facial expression recognition system consists of four consecutive modules. These are

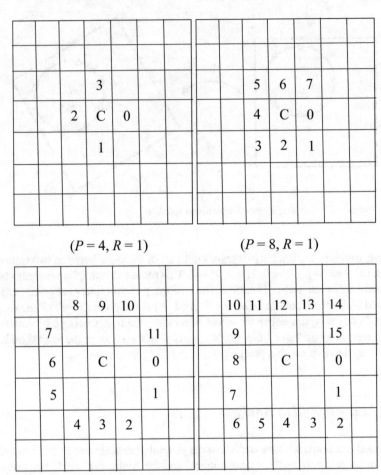

Fig. 9 Circularly symmetric neighbor sets for different (P, R)

(1) Pre-processing
(2) Feature extraction
(3) Dimensionality reduction
(4) Classification

A general block diagram of facial expression recognition system is shown in Fig. 10.

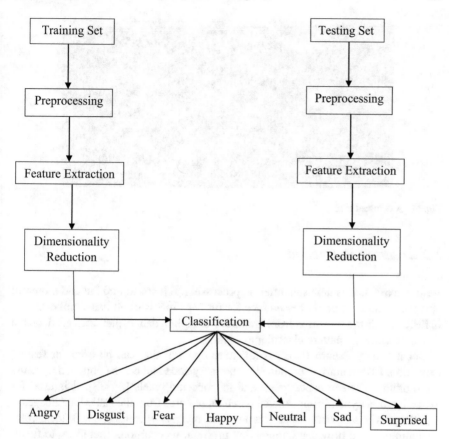

Fig. 10 Block diagram of 7-class facial expression recognition framework

5.1 Preprocessing

Preprocessing is used before feature extraction to improve performance of facial expression recognition system. It includes different processes viz. face detection, scaling, cropping, normalization, contrast enhancement and additional enhancement processes to improve recognition rate.

Face detection is a preprocessing method for which Viola-Jones presented an algorithm based on Adaboost learning algorithm and haar like features to detect the location of face in an image [49] as shown in Fig. 11.

Face cropping is performed over cropped face to extract detected face from whole image. Image scaling [50, 51] is performed on face image to increase or decrease size of face. Interpolation methods is used for this purpose. Normalization step removes effect of face variations.

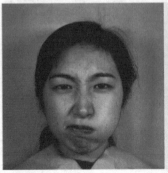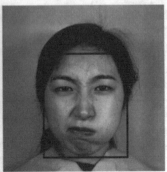

Fig. 11 A detected face

5.2 Feature Extraction

Feature extraction is next step after preprocessing. It helps to find out and represent positive features of preprocessed face for further processing. It has significant role in FER as it shows transition from graphical to implicit data depiction. This depicted data is fed into the pattern classifier as input.

Local Binary Pattern (LBP) is a texture operator that can be used for feature extraction. LBP feature is produced with binary code and can be obtained by using thresholding between center pixel and neighboring pixels [24–33]. It is used for extracting non dynamic appearance based on features from the static face images.

LBP provides accurate, sparse and simple analysis. It can work with local color configuration and provides illumination invariant recognition. Therefore, to facilitate more accurate and simple FER, we have explored and exploited LBP. LBP is originally a simple and efficient texture descriptor that labels the pixels of an image by thresholding the neighborhood of each pixel with the value of the center pixel and outputs the result as a binary number [24, 25]. Owing to its discriminating power and computational simplicity, it has later been used for several vision applications. The two main advantages of LBP are:

i. The most important advantage of the LBP operator in real-time applications is its robustness towards monotonic gray scale changes caused by illumination variations.
ii. The second important advantage of LBP is its computational simplicity that makes it suitable for analysis of images as well as videos in real time.

5.3 Dimensionality Reduction

Several extracted features are high dimensional vectors. Due to the large size of feature vectors, it is not practically possible to use all the patterns as input to our classifier, for fear of misclassification and possible system crash. The principal component analysis (PCA) feature reduction algorithm has special speed advantage in increasing classification process [52, 53]. The higher dimensional feature vectors can be reduced into lower dimensional feature vectors with the use of PCA. It can be used as a tool for finding the best projection directions that represent the original data on the condition of the least mean-square [52, 53]. Therefore, we have formulated a PCA-based algorithm to select a few deserving portions of the LBP and its variants. The selected features represent samples of the facial deformation patterns of the expressive face. A brief overview of PCA is given below.

Let number of classes of facial images in training set is C and each class i ($i = 1, 2, ..., C$) comprises of a number of face images M. These images are $F_{i1}, F_{i2}, ..., F_{iM}$ where each image F is a $m \times n$ size 2D array of intensity values and each image can be converted into a vector of $m \times n$ pixels. If training set of $N = C \times M$ images is defined by $X_k = [F_{i1}, F_{i2}, ..., F_{ij}]$, then the covariance matrix is defined as

$$\Phi = \frac{1}{N} \sum_{k=1}^{N} X X^T \qquad (16)$$

where $X = [F_{i1} - \bar{F}, F_{i2} - \bar{F}, ..., F_{ij} - \bar{F}]$ and $\bar{F} = \sum_{i=1}^{C} \sum_{j=1}^{M} F_{ij}$ is mean of training set. The eigenvalues and eigenvectors are then calculated from the covariance matrix. Let $P = (p_1, p_2, ..., p_r)$ $(r < N)$ are the r normalized eigenvectors corresponding to r largest eigenvalues. Each of the r eigenvectors is called an eigenface. Now, each face image of the training set is projected into the eigenfaces space to obtain its corresponding eigenface based features Y_s, which is defined as

$$Y_s = P^T A_k, \quad k = 1, 2, ..., N \qquad (17)$$

where A_k is the mean subtracted image of X_k. For an image of size 256×256 pixels, we get a multiscale LBP feature vector of size 1×3584 pixels. This large size of feature vector is reduced into a smaller feature vector of size 64×2 pixels using PCA.

5.4 Classification

Classification is the final stage of FER system in which the classifier categorizes the expression such as smile, sad, surprise, anger, fear, disgust and neutral. Support Vector Machine (SVM) is one of the classification techniques in which two types of approaches are involved. They are one against one and one against all approaches. One against all classification means it constructs one sample for each class [12]. One against one classification means it constructs one class for each pair of classes

[54]. SVM is one of the strongest classification methods for advanced dimensionality troubles [55].

SVM is the supervised machine learning technique and it uses four types of kernels for its better performance [56]. They are linear, polynomial, Radial Basis Function (RBF) and sigmoid. The linear kernel maps the high dimensional data and it is linearly separable. The RBF kernel uses the function that maps the single feature into the high dimensional data. The polynomial kernel learns the nonlinear models and also resolves their similarity.

The facial expressions for a given video sequence have been classified using multiclass SVM classifier in following manner:

$$V \in \begin{cases} expression_1 \ iff \ V \in class_1 \\ expression_2 \ iff \ V \in class_2 \\ \qquad \cdots\cdots\cdots \\ expression_n \ iff \ V \in class_n \end{cases} \tag{18}$$

6   Experiments and Results

Benchmark facial expression recognition datasets are JAFFE dataset [57], Cohn-Kanade dataset [58], Extended Cohn-Kanade dataset [59], Yale dataset [60] and MMI dataset [61], etc.

We have implemented and tested our proposed facial expression recognition method over several datasets of facial expression recognition. Here, we present results for three most popularly used datasets that are—the Japanese female facial expression (JAFFE) recognition dataset [57], Extended Cohn-Kanade dataset (CK+) [59], and the Yale facial expression recognition dataset [60]. The experiments have been performed in Matlab 8.1 environment on an Intel® Core™ i3 2.27 GHz machine.

For evaluation, we have separated the subjects randomly into n groups of roughly equal size and applied a leave one group out cross validation test scheme. These groups are subject-independent. The same subjects did not appear in both training and testing. Therefore, the testing is done with novel faces and person-independent.

6.1   Japanese Female Facial Expression (JAFFE) Dataset Results

Japanese Female Facial Expression (JAFFE) Dataset [57] is a popular dataset for facial expression recognition. It consists of 213 image sequences involving seven expressions of happy, sad, angry, disgust, fear, surprise and neutral. It is generated with the help of 10 Japanese female objects. It is the first dataset provided for downloading which made it very popular for expression recognition task. A few samples

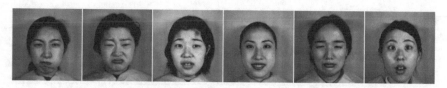

Fig. 12 Sample expressions of JAFFE dataset

Table 1 Confusion matrix for LBP feature

Expressions	Angry	Disgust	Fear	Happy	Neutral	Sad	Surprise
Angry	*96.7*	3.3	0	0	0	0	0
Disgust	10	*0*	26.6	20	6.7	6.7	30
Fear	0	0	*100*	0	0	0	0
Happy	0	0	0	*100*	0	0	0
Neutral	0	0	0	0	*100*	0	0
Sad	0	0	0	0	0	*100*	0
Surprise	0	0	0	0	0	0	*100*

Italics texts show the highest accuracy in the corresponding columns

Table 2 Confusion matrix for LBP_RI feature

Expressions	Angry	Disgust	Fear	Happy	Neutral	Sad	Surprise
Angry	*86.8*	3.3	3.3	0	3.3	3.3	0
Disgust	50	*0*	16.6	6.7	16.6	3.3	6.8
Fear	56.6	0	*43.4*	0	0	0	0
Happy	23.3	0	13.3	*63.4*	0	0	0
Neutral	36.6	0	6.6	26.6	*30.2*	0	0
Sad	16.6	0	16.6	26.6	10	*30.2*	0
Surprise	30	0	13.3	13.3	16.7	13.3	*13.4*

Italics texts show the highest accuracy in the corresponding columns

of JAFFE dataset are shown in Fig. 12. Confusion matrices for LBP and its few variants are shown in Tables 1, 2 and 3. A comparison of recognition accuracies of different LBP variants is shown in Fig. 13. From Tables 1, 2 and 3 and Fig. 13, it is clear that LBP and its variants show high recognition rate for different expressions of JAFFE dataset.

6.2 Extended Cohn–Kanade (CK+) Dataset Results

Extended Cohn–Kanade (CK+) AU-coded facial expression dataset [59] is used in an abundance of approaches for facial expression recognition. This dataset is generated

Table 3 Confusion matrix for LBP_U2 feature

Expressions	Angry	Disgust	Fear	Happy	Neutral	Sad	Surprise
Angry	*96.7*	3.3	0	0	0	0	0
Disgust	30	*0*	26.7	20	10	3.3	10
Fear	0	0	*100*	0	0	0	0
Happy	3.3	0	3.3	*93.4*	0	0	0
Neutral	3.3	0	6.6	10	*76.8*	3.3	0
Sad	3.3	0	16.6	3.3	20	*56.8*	0
Surprise	6.6	0	0	3.3	16.6	10	*63.5*

Italics texts show the highest accuracy in the corresponding columns

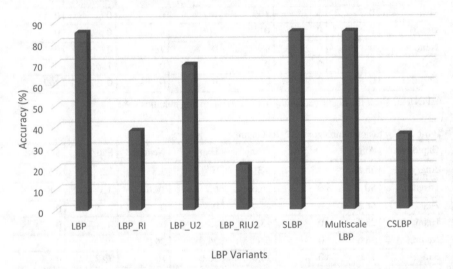

Fig. 13 Recognition accuracy of different LBP variants on JAFFE dataset

by 123 humans of ages lying between 18 and 50. It can be divided into different groups of people. For example (1) 31% men and 69% women and (2) 81% Euro-American people, 13% Afro-American people and 6% others. The images are captured from front side as well as with 30° angle. 593 sequences are generated for front side view. A few samples of CK+ dataset are shown in Fig. 14. Confusion matrices for LBP and its few variants are shown in Tables 4, 5 and 6. Comparison of recognition accuracy of different LBP variants is shown in Fig. 15. From Tables 4, 5, 6 and Fig. 15, it can be observed that LBP and its variants demonstrate high accuracy for CK+ dataset also.

Fig. 14 Sample expressions of CK+ dataset

Table 4 Confusion matrix for LBP feature

Expressions	Angry	Disgust	Fear	Happy	Sad	Surprise
Angry	*89.7*	2.7	0	0	7.6	0
Disgust	0	*97.5*	2.5	0	0	0
Fear	0	2	*73.0*	22.0	3.0	0
Happy	0	0.4	0.7	*97.9*	1.0	0
Sad	10.3	0	0.8	0.8	*83.5*	4.6
Surprise	0	0	1.3	0	0	*98.7*

Italics texts show the highest accuracy in the corresponding columns

Table 5 Confusion matrix for Boosted LBP feature

Expressions	Angry	Disgust	Fear	Happy	Sad	Surprise
Angry	*96.9*	0	0	0	3.1	0
Disgust	2.6	*94.8*	0	0	2.6	0
Fear	0	0	*94.6*	3.6	1.8	0
Happy	0	0.9	2.9	*96.2*	0	0
Sad	2.7	0	0	0	*95.9*	1.4
Surprise	0	0	0	0	1.3	*98.7*

Italics texts show the highest accuracy in the corresponding columns

Table 6 Confusion matrix for LBP-TOP feature

Expressions	Angry	Disgust	Fear	Happy	Sad	Surprise
Angry	*96.9*	0	0	0	3.1	0
Disgust	2.6	*94.8*	0	0	2.6	0
Fear	0	0	*94.6*	3.6	1.8	0
Happy	0	0.9	2.9	*96.2*	0	0
Sad	2.7	0	0	0	*95.9*	1.4
Surprise	0	0	0	0	1.3	*98.7*

Italics texts show the highest accuracy in the corresponding columns

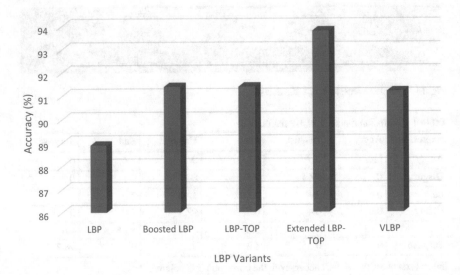

Fig. 15 Recognition accuracy of different LBP variants on CK+ dataset

Fig. 16 Sample expressions of Yale dataset

6.3 Yale Dataset Results

The Yale facial expression dataset [60] contains 165 grayscale images of 15 individuals. There are 11 images per subject, one per different facial expression or configuration: center-light, with glasses, happy, left-light, without glasses, normal, right-light, sad, sleepy, surprised, and wink. A few samples of Yale dataset are shown in Fig. 16. Confusion matrix for multiscale LBP feature is shown in Table 7. A comparison of recognition accuracies of different LBP variants is shown in Fig. 17. It can be observed from Table 7 and Fig. 17, that LBP and its variants show high accuracy for different facial expressions of Yale dataset.

Table 7 Confusion matrix for Multiscale LBP feature

Expressions	Happy	Normal	Sad	Sleepy	Surprised	Wink
Happy	*100*	0	0	0	0	0
Normal	13.4	*86.6*	0	0	0	0
Sad	6.7	6.7	*86.6*	0	0	0
Sleepy	6.7	0	13.3	*80*	0	0
Surprised	0	0	0	6.7	*93.3*	0
Wink	0	0	0	6.7	0	*93.3*

Italics texts show the highest accuracy in the corresponding columns

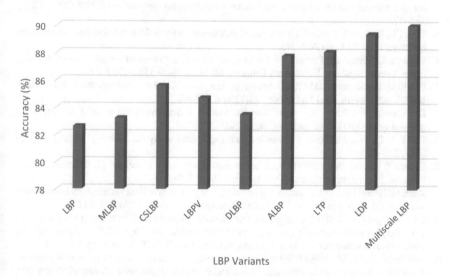

Fig. 17 Recognition accuracy of different LBP variants on Yale dataset

7 Conclusions and Future Work

This chapter presents an analysis of facial expression recognition techniques. This analysis is based on local binary patterns and its several variants such as multiscale LBP, center symmetric LBP, etc. The framework is divided into four steps. These are Preprocessing, Feature extraction, Dimensionality reduction and classification. Preprocessing step includes face detection, cropping and normalization. Feature extraction includes retrieval of LBP feature and its variants. Dimensionality reduction step uses PCA for reduction in size of extracted features. Finally, multiclass SVM is used for expression recognition. Analysis demonstrates high recognition accuracy of LBP and its variants in facial expression recognition. This analysis may further be improved in future by using different combinations of LBP.

Acknowledgements This work is supported by Science and Engineering Research Board, Department of Science and Technology, Government of India under grant number PDF/2016/003644.

References

1. Bousmalis K, Mehu M, Pantic M (2013) Towards the automatic detection of spontaneous agreement and disagreement based on nonverbal behaviour: a survey of related cues, databases, and tools. Image Vis Comput 31(2):203–221
2. Bartlett MS, Littlewort G, Frank M, Lainscsek C, Fasel I, Movellan J (2005) Recognizing facial expression: machine learning and application to spontaneous behavior. In: 2005 IEEE computer society conference on computer vision and pattern recognition (CVPR'05), vol 2, pp 568–573
3. Tian YL, Kanade T, Cohn JF (2005) Facial expression analysis. Handbook of face recognition. Springer, New York, pp 247–275
4. Rudovic O, Pantic M, Patras I (2013) Coupled Gaussian processes for pose-invariant facial expression recognition. IEEE Trans Pattern Anal Mach Intell 35(6):1357–1369
5. Pantic M, Rothkrantz LJM (2000) Automatic analysis of facial expressions: the state of the art. IEEE Trans Pattern Anal Mach Intell 22(12):1424–1445
6. Bettadapura V (2012) Face expression recognition and analysis: the state of the art. arXiv preprint arXiv:1203.6722, http://arxiv.org/pdf/1203.6722
7. Fasel B, Luettin J (2003) Automatic facial expression analysis: a survey. Pattern Recogn 36(1):259–275
8. Ekman P, Friesen WV (1978) Facial action coding system: a technique for the measurement of facial movement. Consulting Psychologists Press, USA
9. Zhang Z (1999) Feature based facial expression recognition: sensitivity analysis and experiment with a multi-layer perceptron. Int J Pattern Recognit Artif Intell 13(6):893–911
10. Guo G, Dyer CR (2003) Simultaneous feature selection and classifier training via linear programming: a case study for face expression recognition. In: 2003 IEEE computer society conference on computer vision and pattern recognition (CVPR'03), vol 1, pp 346–352
11. Valstar M, Pantic M (2006) Fully automatic facial action unit detection and temporal analysis. In: IEEE computer society computer vision and pattern recognition workshop (CVPRW'06), pp 149–149
12. Valstar MF, Patras I, Pantic M (2005) Facial action unit detection using probabilistic actively learned support vector machines on tracked facial point data. In: IEEE computer society computer vision and pattern recognition workshop (CVPRW'05), vol 3, p 76
13. Franco L, Treves A (2001) A neural network facial expression recognition system using unsupervised local processing. In: 2nd international symposium on image and signal processing and analysis (ISPA'01), pp 628–632
14. Uddin MZ, Lee JJ, Kim TS (2009) An enhanced independent component-based human facial expression recognition from video. IEEE Trans Consum Electron 55(4):2216–2224
15. Lekshmi VP, Sasikumar M (2009) Analysis of facial expression using Gabor and SVM. Int J Recent Trends Eng 1(2):47–50
16. Owusu E, Zhan Y, Mao QR (2014) A neural-AdaBoost based facial expression recognition system. Expert Syst Appl 41(7):3383–3390
17. Kumbhar M, Jadhav A, Patil M (2012) Facial expression recognition based on image feature. Int J Comput Commun Eng 1(2):117–119
18. Shinohara Y, Otsu N (2004) Facial expression recognition using fisher weight maps. In: Sixth IEEE international conference on automatic face and gesture recognition, pp 499–504
19. Ahmed F, Bari H, Hossain E (2014) Person-independent facial expression recognition based on compound local binary pattern (CLBP). Int Arab J Inf Technol 11(2):195–203

20. Nigam S, Khare A (2015) Multiscale local binary patterns for facial expression-based human emotion recognition. In: Computational vision and robotics. Springer, New Delhi, pp 71–77
21. Darwish A, Hassanien AE, Elhoseny M, Sangaiah AK, Muhammad K (2017) The impact of the hybrid platform of internet of things and cloud computing on healthcare systems: opportunities, challenges, and open problems. J Ambient Intell Humanized Comput. First Online: 29 Dec 2017. https://doi.org/10.1007/s12652-017-0659-1
22. Elhoseny M, Abdelaziz A, Salama A, Riad AM, Sangaiah AK, Muhammad K (2018) A hybrid model of internet of things and cloud computing to manage big data in health services applications. Future Gen Comput Syst. Available online 15 Mar 2018 (In Press). https://doi.org/10.1016/j.future.2018.03.005
23. Abdelaziza A, Elhoseny M, Salama AS, Riad AM (2018) A machine learning model for improving healthcare services on cloud computing environment. Measurement 119:117–128. https://doi.org/10.1016/j.measurement.2018.01.022
24. Zhang J, Tan T (2002) Brief review of invariant texture analysis methods. Pattern Recogn 35(3):735–747
25. Pietikäinen M (2005) Image analysis with local binary patterns. In: Image analysis. Springer, Berlin, pp 115–118
26. Mäenpää T, Pietikäinen M (2005) Texture analysis with local binary patterns. In: Handbook of pattern recognition and computer vision, vol 3, pp 197–216
27. Mäenpää T (2003) The Local binary pattern approach to texture analysis: extensions and applications. Doctoral dissertation, University of Oulu
28. Turtinen M (2007) Learning and recognizing texture characteristics using local binary patterns. Doctoral dissertation, University of Oulu
29. Ilea DE, Whelan PF (2011) Image segmentation based on the integration of color–texture descriptors—a review. Pattern Recogn 44(10):2479–2501
30. Nigam S, Khare A (2016) Integration of moment invariants and uniform local binary patterns for human activity recognition in video sequences. Multimedia Tools Appl 75(24):17303–17332
31. Wang A, Wang S, Lucieer A (2010) Segmentation of multispectral high-resolution satellite imagery based on integrated feature distributions. Int J Remote Sens 31(6):1471–1483
32. Burçin K, Vasif NV (2011) Down syndrome recognition using local binary patterns and statistical evaluation of the system. Expert Syst Appl 38(7):8690–8695
33. Nigam S, Khare A (2015) Multiresolution approach for multiple human detection using moments and local binary patterns. Multimedia Tools Appl 74(17):7037–7062
34. Mattivi R, Shao L (2010) Spatio-temporal dynamic texture descriptors for human motion recognition. In: Intelligent video event analysis and understanding. Springer, Berlin, pp 69–91
35. Nanni L, Brahnam S, Lumini A (2011) Combining different local binary pattern variants to boost performance. Expert Syst Appl 38(5):6209–6216
36. Nanni L, Lumini A, Brahnam S (2012) Survey on LBP based texture descriptors for image classification. Expert Syst Appl 39(3):3634–3641
37. Pietikäinen M, Ojala T, Xu Z (2000) Rotation-invariant texture classification using feature distributions. Pattern Recogn 33(1):43–52
38. Ojala T, Pietikäinen M, Mäenpää T (2000) Gray scale and rotation invariant texture classification with local binary patterns. In: European conference on computer vision (ECCV'2000). Springer, Berlin, pp 404–420
39. Zhao G, Ahonen T, Matas J, Pietikainen M (2012) Rotation-invariant image and video description with local binary pattern features. IEEE Trans Image Process 21(4):1465–1477
40. Guo Z, Zhang L, Zhang D (2010) Rotation invariant texture classification using LBP variance (LBPV) with global matching. Pattern Recogn 43(3):706–719
41. Ojala T, Pietikainen M, Maenpaa T (2002) Multiresolution gray-scale and rotation invariant texture classification with local binary patterns. IEEE Trans Pattern Anal Mach Intell 24(7):971–987
42. Lahdenoja O, Poikonen J, Laiho M (2013) Towards understanding the formation of uniform local binary patterns, vol 2013, Article ID 429347, 20 p. https://doi.org/10.1155/2013/429347

43. Chan CH, Kittler J, Messer K (2007) Multi-scale local binary pattern histograms for face recognition. Springer, Berlin, pp 809–818
44. Liao S, Zhu X, Lei Z, Zhang L, Li SZ (2007) Learning multi-scale block local binary patterns for face recognition. In: Advances in biometrics. Springer, Berlin, pp 828–837
45. Guo Z, Zhang D (2010) A completed modeling of local binary pattern operator for texture classification. IEEE Trans Image Process 19(6):1657–1663
46. Heikkilä M, Pietikäinen M, Schmid C (2009) Description of interest regions with local binary patterns. Pattern Recogn 42(3):425–436
47. Tan X, Triggs B (2010) Enhanced local texture feature sets for face recognition under difficult lighting conditions. IEEE Trans Image Process 19(6):1635–1650
48. Wang L, He DC (1990) Texture classification using texture spectrum. Pattern Recogn 23(8):905–910
49. Viola P, Jones MJ (2004) Robust real-time face detection. Int J Comput Vision 57(2):137–154
50. Elhoseny M, Ramírez-González G, Abu-Elnasr OM, Shawkat SA, Arunkumar N, Farouk A (2018) Secure medical data transmission model for IoT-based healthcare systems. IEEE Access 6:20596–20608. https://doi.org/10.1109/access.2018.2817615
51. Shehab A, Elhoseny M, Muhammad K, Sangaiah AK, Yang P, Huang H, Hou G (2018) Secure and robust fragile watermarking scheme for medical images. IEEE Access 6:10269–10278. https://doi.org/10.1109/access.2018.2799240
52. Jolliffe I (2005) Principal component analysis. Wiley, New York
53. Bengio Y, Courville A, Vincent P (2013) Representation learning: a review and new perspectives. IEEE Trans Pattern Anal Mach Intell 35(8):1798–1828
54. Gang L, Xiao-hua L, Ji-Liu Z, Xiao-gang G (2009) Geometric feature based facial expression recognition using multiclass support vector machines. In: 2009 IEEE international conference on granular computing (GRC'09), pp 318–321
55. Burges CJ (1998) A tutorial on support vector machines for pattern recognition. Data Min Knowl Disc 2(2):121–167
56. Cortes C, Vapnik V (1995) Support-vector networks. Mach Learn 20(3):273–297
57. Dailey MN, Joyce C, Lyons MJ, Kamachi M, Ishi H, Gyoba J, Cottrell GW (2010) Evidence and a computational explanation of cultural differences in facial expression recognition. Emotion 10(6):874
58. Kanade T, Cohn JF, Tian Y (2000) Comprehensive database for facial expression analysis. In: Proceedings of the fourth IEEE international conference on automatic face and gesture recognition, 2000. IEEE, pp 46–53
59. Lucey P, Cohn JF, Kanade T, Saragih J, Ambadar Z, Matthews I (2010) The extended cohn-kanade dataset (CK+): a complete dataset for action unit and emotion-specified expression. In: 2010 IEEE computer society conference on computer vision and pattern recognition workshops (CVPRW). IEEE, pp 94–101
60. Yale Facial Expression Database. http://vision.ucsd.edu/content/yale-face-database
61. Pantic M, Valstar M, Rademaker R, Maat L (2005) Web-based database for facial expression analysis. In: ICME 2005. IEEE international conference on multimedia and expo, 2005. IEEE, 5 pp

Efficient Generation of Association Rules from Numeric Data Using Genetic Algorithm for Smart Cities

Pardeep Kumar and Amit Kumar Singh

Abstract Machine learning plays an important role to develop smart cities by gathering real time information using several state of the art algorithms. In the recent past, association rule mining plays an important role in the discovery of accurate information from databases to satisfy the need of real time applications in smart cities ranging from healthcare to intelligent transport systems. It is used in various applications for decision making, detection and prediction etc. because of its robustness to derive associations among various attributes of datasets. This technique seems to be simple in case of categorical data but becomes quite complex in case of numeric data. In this work, we have mainly concentrated on the problem of generating association rules from numeric data in an efficient way. For accomplishing this task we have taken genetic algorithm as the base of the solution to this problem. Genetic algorithm is selected for this task because of its nature of self-improving and ability to handle large solution set. Here we have proposed genetic algorithm based association rule mining algorithm which generates random association rules on the basis of general property of datasets. The generated rule set is improved at each run of algorithm and filtered for more and more interesting and accurate rules.

Keywords Association rule mining · Decision making · Categorical data
Numeric data · Genetic algorithm · Smart cities

1 Introduction

Large amounts of data is being collected routinely in the task of day-to-day management and decision making in banking, business, administration, the delivery

P. Kumar (✉)
Department of Computer Science and Engineering, Jaypee University
of Information Technology, Solan 173234, Himachal Pradesh, India
e-mail: pardeepkumarkhokhar@gmail.com

A. K. Singh
Department of Computer Science and Engineering, National Institute of Technology Patna, Patna
University Campus, Patna 800005, Bihar, India
e-mail: amit_245singh@yahoo.com

© Springer Nature Switzerland AG 2019 323
A. E. Hassanien et al. (eds.), *Security in Smart Cities: Models, Applications,
and Challenges*, Lecture Notes in Intelligent Transportation and Infrastructure,
https://doi.org/10.1007/978-3-030-01560-2_14

of social and health services, works related to environmental protection, handling security and in politics. Such data are primarily used for accounting service such as decision building, prediction, etc. to users of the system, later it can be used for maintenance in long term. Typically, data sets to be considered for this task are very large and constantly growing and contain a large number of complex dimensionality. While these data sets reflect properties of the managed subjects and relations between them, and are thus potentially of some use to their owner, but they often have relatively less useful information as compared to its volume. Hence, robust, efficient, simple and computationally efficient algorithm is required to extract information from such datasets.

One of the important data mining tasks is generating association rules, which deals with finding IF-THEN rules by analyzing similarity between combinations of attributes of data. As it can be seen that IF-THEN rules are easy to understand and can be used in multiple applications for deriving inferences for relevant tasks, therefore generating such kind of rules is considered as an important task in data mining.

To generate such algorithm in an efficient way, one of the important subjects of computer science also helps data mining, i.e. machine learning. Machine learning is a subject which keeps efficient solution of various general problems which can be used in any field of study such as data mining, optimization, information retrieval etc. We will also be using one of the machine learning techniques in our work i.e. genetic algorithm.

2 Basic Terminology

2.1 Numeric Association Rule Mining

Association rule mining is a well-known data mining technique that is used for extracting information from raw data in form of IF-THEN statement. These statements can be used in many fields for prediction and decision making. First introduced in Agrawal et al. [1], the process of extracting such rules was quite cumbersome where all the possible combinations are checked for co-relations. The co-relation between attributes was checked on basis of interestingness. Each rule is checked for interestingness using various interestingness parameters such as support, confidence, lift, conviction etc. Various such parameters are given in Tew et al. [2]. Later same author improved the procedure of extracting rules in Agrawal and Srikant [3] where the authors proposed a well-known Apriori algorithm in which only frequent occurring pairs are checked for further co-relation. After the introduction of Apriori, lots of other algorithms were proposed for extracting association rules. Like in Houtsma and Swami [4], association rule mining using sql query was explained but it requires multiple scans of database. In Hidber [5], an online technique for generation of association rule was proposed where the rules are generated at real time. A comparison between two well-known algorithms Apriori–Eclat was discussed in Borgelt [6]. But all these algorithms were designed for categorical data only. None of these algorithm works well for numeric dataset. The rules needed for numeric dataset are of form:

$$\text{attribute1}[lb1, ub1], \text{attribute2}[lb2, ub2] \Rightarrow \text{attribute3}[lb3, ub3]$$

where there is a lower and upper bound attached with every attribute. These kinds of rules are more general in nature as well as contains more information as compared to categorical association rules. By extracting numeric association rule, various inter disciplinary fields can be benefited such as medication, structural engineering, pollution analysis, image processing etc.

2.2 Genetic Algorithm

Genetic Algorithm (GA) explained in Goldberg [7] is a powerful algorithmic technique that can be used for generating solution for a search and optimization problem [8, 9]. It basically follows one of the nature's principle i.e. survival of the fittest as stated in Darwin's theory of evolution. It tells that the one that is stronger survives for longer. This algorithm finds better solution in each generation and keeps on improving the solution in each generation till it finds a near optimal solution.

The process of general genetic algorithm starts with the description of problem; on the basis of problem and the solution needed an encoding scheme is selected. Chromosomes are generated as initial population by using that encoding scheme. Fitness value is calculated for each chromosome and on the basis of that it is decided whether the solution is appropriate or not. Further the population is changed by applying crossover and mutation operators which is meant for introducing diversity in the algorithm. By implementing these operators, a new generation of population is evolved then again on that population fitness is checked. This is an iterative self-improving process which improves its own solution iteration by iteration. The iterations are stopped when the desirable results are generated. The population in the final iteration is considered to be the best solution set. As per the requirement of the problem results can be formulated from the population of last generation. A general work flow of genetic algorithm is illustrated in Fig. 1 given in Obitko [10].

3 Literature Survey

To derive association rules of form IF-THEN statement, we need to get an antecedent and a related consequent part which are co-related to each other. For applying GA for finding such rules as mentioned in Freitas [11], four main designing components are needed i.e. Chromosome encoding for representing rules in form of GA understandable chromosomes, Crossover operator for generating new child chromosomes, Mutation operators for introducing diversity in the population and fitness function

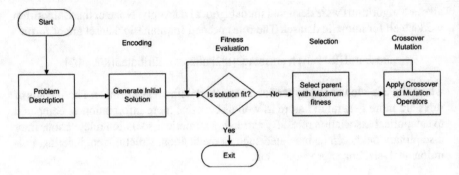

Fig. 1 Flowchart of basic genetic algorithm

which will evaluate fitness measure of each rule. After deciding these factors only thing left is to apply basic genetic algorithm which is defined in introduction part above, using all these components.

3.1 Chromosome Representation

Broadly there are two chromosome representation techniques are present in literature for ARM. In the first approach (Pittsburgh approach), a batch of possible association rules are encoded in a chromosome. This encoding approach is well suitable for classification rule mining, where the goal is to generate a good quality set of rules. However, in ARM, the main goal is to find a set of rules where each rule is good in itself. Therefore, for ARM case, the Michigan approach explained in Goldberg [7] is mostly used. Each chromosome represents exactly one rule in itself, is more suitable in Michigan approach. Most of the ARM techniques use Michigan chromosome representation.

In an early work by Ghosh and Nath [12], the authors use the Michigan approach as follows: each chromosome has length $2k$, where k is the number of items. The chromosomes in this case were binary strings where each attribute is given two bits. If these two bits are 00 or 11, then the attribute is present in the antecedent part or consequent part of the rule, respectively, else the attribute is not present in the rule.

In paper by Anand et al. [13], each chromosome has two parts. The first part indicates the location of the attribute in the rule and the other part indicates the categorical value it has. The prior part consists of two bits where the attribute appears in the antecedent part or the consequent part of the rule, if the bits are 10 and 11, respectively; else, it is meant to be absent from the rule. The other part represents the categorical values carried by attributes in binary form. However, the authors in the paper did not given any justification of how the binary value of the attribute in the second part will appear and how categorical state will be managed if the number of states for an attribute is not an exact power of two. The main demerit of choosing a binary encoding scheme is that the length of the chromosome is large when the number of attributes increases, because at least two bits are required for each attribute representation. An integer encoding can be used as a solution to this problem.

An integer encoding scheme has been used in association rule mining using multi objective genetic algorithm by (ARMGA) Yan et al. [14]. In this work the chromosomes encode the index of the attributes. A chromosome represented as encoding a k-rule, k is the total number of attributes in the antecedent part and the consequent part of the rule, and this chromosome will have $k+1$ genes. The first gene position represents the differentiating position of the chromosome where the attributes of antecedent and the consequent are separated. For example, if Ai represents the ith item, then the chromosome {3 | 5 4 2 1 3} represents the rule $A2, A5, A4 \Rightarrow A1, A3$. This kind of representation reduces length of the chromosome significantly.

In Alatas et al. [15], encoding scheme used for chromosomes representation where each attributes has three parts. The first part tells whether the attribute is present or not in the rule, and if present, in which part of the rule (antecedent or consequent) it is present. The second and third part tells the lower and upper bounds of the value of attribute. The first part can have integer numbers such as 0, 1, or 2, which represents the occurrence of the attribute in the antecedent part of the rule, the occurrence of the attribute in the consequent of the rule, and the absence of the attribute from the given rule, respectively. Next, the second and the third part can take real numbers from the related attribute range. Further it is to be noted that as MODENAR uses differential evolution as an optimization technique and works on real-valued chromosomes, the authors considered a round off operator for handling the integer value part of the chromosome. In Martin et al. [16], authors use positional encoding scheme. Metawa et al. [17] used binary chromosome encoding scheme in their genetic algorithm based proposed approach for optimizing bank lending decisions to specify the selected customer in the lending decision.

3.2   Objective Functions

Even though support count and confidence are two well-known objectives that are generally to be maximized yet there are several other parameters available to measure the interestingness or strength of association rules as mentioned in Berzal et al. [18]. Some of the parameters, which can be used by different algorithms for optimization in a multi-objective scenario, are comprehensibility, conviction, interestingness, performance, lift, coverage, precision etc. Each function has its own significance and combination of best of these functions gives best result in the end. In Alatas et al. [15], objective function tries to optimize four criteria of the rules: support, confidence, amplitude of the intervals and comprehensibility, which make up the subparts of the rule. In Martin et al. [16], authors use comprehensibility, interestingness, and performance for fitness calculation. Anand and Vinodchandra [19] proposed association rule mining algorithm using treap data structure where they focused on running time rather than rule set quality. Umit and Bilal [20] proposed exploration of association rules within numerical databases with Gravitational Search Algorithm (GSA) with flexible fitness function. In Uroš et al. [21], authors proposed a modified single-objective binary cuckoo search for association rule mining (MBCS-ARM) where objective function composed of support and confidence only. The quality of rule sets was still an issue in literature.

3.3 Evolutionary Operators

Mostly, when binary encoding scheme is used, some of the standard crossover and mutation operators are used. For example, in Ghosh and Nath [12], bit-flip mutation and multipoint crossover have been used. In Anand et al. [13], mutation operator, bit flip has been adopted. However, the authors did not specifically mention which crossover operator should be used.

In Qodmanan et al. [22], where integer encoding for the chromosomes is used, an order-1 crossover technique is taken into consideration. In this technique, first a segment is selected from any two parent chromosomes and these are copied to the two child chromosomes. Next, starting from the right side of the segment, the values of the genes that didn't appear in the selected segment of the first parent, are copied to the first child. The same procedure is repeated for the second child as well. The mutation operator introduces diversity by replacing a selected attribute value from the chromosome with a random attribute value not present in the chromosome currently. Alatas et al. [15] used differential evolution based crossover and mutation operators.

4 Proposed Approach

In the literature, till date a lot of work has been done in the field of association rule mining but most of the work is done for extracting rules from categorical data only. A very few algorithm can be seen for dealing with numeric data to get the relevant and interesting association rules. We have kept our focus mainly on mining association rule from numeric data. To handle numeric data efficiently, it is required that the data related to each attribute should be checked individually which increases the search space exponentially. Therefore, to deal with such problem, we have used genetic algorithm which finds solution with the least possible domain knowledge. In this work we have analyzed the basic structure of previously available work on same domain and proposed an algorithm which generates better results from the previously known works of similar kind.

In the proposed work we will be generating the rules of the form:

attribute1[lb1, ub1], attribute2[lb2, ub2] $\rightarrow$ attribute3[lb3, ub3]

Chromosome Representation

A	B	C	D	E	F
$1, lb_1, ub_1$	$0, lb_2, ub_2$	$2, lb_3, ub_3$	$3, lb_4, ub_4$	$0, lb_5, ub_5$	$1, lb_6, ub_6$

Fig. 2 Chromosome representation for proposed approach

which signifies if the value of attribute1 is between lb1 and ub1, attribute2 is between lb2 and ub2 then the value of attribute3 will be between lb3 and ub3. In this approach there is no limitation on length of either of antecedent or consequent part.

4.1 Technical Details of Implementation

As discussed earlier, for using genetic algorithm for any problem, the basic components of genetic algorithm (i.e. encoding scheme, initial population, crossover and mutation operator and fitness function) have to be decided. Therefore, in our case for generating rules using genetic algorithm, the part of algorithm that we have used is as follows:

Chromosome Representation

We have used Michigan encoding scheme in this implementation where one chromosome represent one rule in the population. The chromosome is represented as shown in Fig. 2 where column under A represents a gene corresponding to attribute A which consists of 3 parts. First part represents the location of attribute whether it will be in antecedent part or it will be in consequent part. If first bit of gene is 0 indicates this attribute is in antecedent part, 1 indicates it to be in consequent part, whereas 2 indicate that the attribute is not present in the rule. Other two parts of the gene are lower bound and upper bound value of the attribute. Chromosome in Fig. 2 shows a rule of form:

$$B[lb_2, ub_2], E[lb_5, ub_5] => A[lb_1, ub_1], F[lb_6, ub_6]$$

which represents rule as IF B has a value between lb_2 and ub_2 and value of E is between lb_5 and ub_5 THEN the value of A will be in lb_1 and ub_1 and the value of F will be in lb_6 and ub_6.

Population Generation

Initial population generation in implementation is done by keeping in mind the frequency of value of attribute that occurs in the dataset i.e. a value that is more frequently coming into dataset are selected for generating chromosome. This is because the initial population is an important factor for better output of genetic algorithm as the whole flow of genetic algorithm is broadly based on the first generation. It is the solution of the first generation which is improved further in other generation. If we keep a check on initial population then there is chance of getting better results as the output of the genetic algorithm.

Fitness Calculation

For every chromosome in the population, fitness is calculated. This fitness is the indicator of interestingness and importance of the rule. Various important interestingness parameters have been compared in Tan et al. [23]. It is to be decided prior that which interestingness parameters to be selected for better results. We have selected support, confidence, lift and conviction for our proposed work. If chromosome has fitness greater than the threshold fitness then it is considered for next generation else it is restricted to enter in the next generation. The threshold is calculated as the average of the fitness of rules of last generation. In this implementation we have taken a power fitness function. For calculating fitness function, initially support, confidence, lift and conviction of the chromosome is calculated.

For a sample rule like 'if A then B':

Interestingness Measures used in proposed algorithm are:

1. Support [1]—indicates how frequently A and B occur in the dataset.

$$\text{Support} = \frac{\text{Count}(A \cup B)}{\# \text{ of Records}}$$

2. Confidence [1]—indicates how frequently A and B occurred when it is known that A was already present in the record.

$$\text{Confidence} = \frac{\text{Support}(A \cup B)}{\text{Support}(A)}$$

3. Lift [24]—indicates how frequently A and B occurred when it is known that B was already present in the record.

$$\text{Lift} = \frac{\text{Confidence}(A \rightarrow B)}{\text{Support}(B)}$$

4. Conviction—indicates the comparison of the probability that A appears without B if they are dependent with the actual frequency of the appearance of A without B.

$$\text{conviction} = \frac{P(A)P(\sim B)}{P(A(\sim B))}$$

First fitness filter is created on the basis of support. A minimum Support value is taken and only the chromosomes having fitness value greater than support will be considered for rest of the fitness value calculation process.

Rest of the fitness calculation is done by a multi-objective maximizing function of form

$$Fitness = \left(Support^4\right) + \left(Confidence^3\right) + \left(Lift^2\right) + \left(Conviction^2\right)$$

where constant 4, 3, 2, 2 in the power are chosen on the basis of importance of these interestingness parameter. The main motive of this function is to maximize the effect of interestingness parameter. The chromosome with the fitness value less than the average fitness value is not considered for next generation. This chromosome is sent to the mutation repository for generating new chromosome out of it with few improvements.

Crossover Operator

Crossover is done among two fit parent chromosome to generate two new off springs. Various crossover operators and their performance have been discussed in Magalhães-Mendes and Cidem [25]. For this implementation, we have taken a hybrid of two crossover operator i.e. single point and every point crossover operator. In one generation single point crossover is implemented and in next generation every point crossover is applied. This is repeated in the same way for all iteration. This is done to introduce better structure to generate new population. The idea behind combining the two crossover operator is that, at certain iteration we crossover at one point and on certain other iterations we do the crossover at every point in which swapping of gene of each attribute takes place. First operator preserves the quality of fit chromosome whereas every point crossover will introduce more chromosomes with better quality of previous chromosome. By keeping the balance among both of them will generate better chromosomes at different generation.

Mutation Operator

Mutation is applied to bring diversity in population. In mutation some of the bits of previous chromosome are changed to some other random value. For our implementation, we have implemented a selective bit mutation where a bit is selected to change in the chromosome and then generate a new chromosome. And while applying the mutation operator we also have kept a provision of generating some new random chromosome similar to the way it was generated in initial population phase. We apply mutation operator over chromosomes which are not fit i.e. the chromosome that have not come in the threshold fitness range in previous generation. There we select any of the gene (attribute) of the chromosome and increase or reduces the lower bound and upper bound or change the first bit which is representing location of attribute in the rule. By changing lower bound and upper bound there is chance of improvement in chromosome. Once the chromosome become fit by changing range of some attribute next time it will enter the population and may be selected for crossover in next few iterations.

Selection of Chromosome

We have used the probability based procedure for selecting two chromosomes for crossover in which the support count is used as a probability parameter. A parent chromosome is selected having value higher than the average support count of entire population.

4.2 Flow Chart for Implementation

Flow chart of our proposed approach is depicted in Fig. 3.

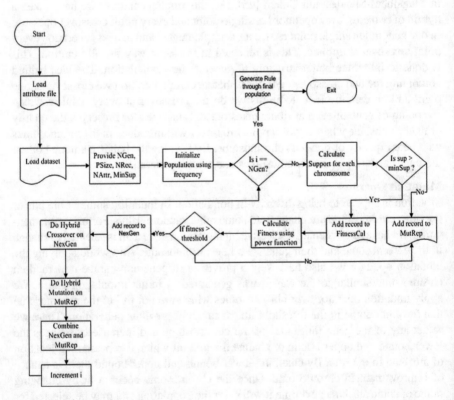

Fig. 3 Flowchart for proposed approach

4.3 Pseudo Code for Implementation

Algorithm: **Implementation**

Input: Dataset D

PSize - Population Size

NGen - Number of generations

RMut- Mutation Rate

RCross- Crossover Rate

Output: Association Rules

1.Begin
2. fori=1 to PSize //No. of Chromosome in one generation
3.   for each attribute
4.   Check upper bound and lower bound.
5.   Calculate frequency of values in dataset.
6.   Generate chromosome from step 4 – 5 information
7.   End for
8. End for   //Initial population generated
9. for each Psize // for each chromosome
10.   Calculate Support
11.   If Support >MinSupport
12.   Calculate Fitness
13. End for
14. for each number of generated rules of Step 5-9
15.   If (Fitness >AvgFitness) //AvgFitness of last generation
16.   Add rule to nexGen
17.   Else
18.   Add rule to MutRep   //MutRep:Rules for mutation
19. End for

20.   for i=1 to NGen // No. of iteration

21.   do Crossover on nextGen
22.   do Mutation on MutRep
23.   Repeat Step 9 – Step 19
24. End for
25. End

5   Experimental Setup

We have implemented the algorithm in JAVA and tested results on various datasets of keel repository to analyze the proposed algorithm. The program is implemented on NetBeans 8.0.2 and executed on a DELL Inspiron Laptop with Intel® Core™ i5-4210U CPU @ 1.70 GHZ 2.50 GHz and 8 GB RAM.

5.1   Datasets

The above proposed algorithms are implemented in such a way that it can be executed for any dataset of real numbers. For presenting the result, we have implemented and tested our algorithm on 4 datasets of keel repository as mentioned in Alcala-Fdez et al. [26]. These datasets are shown in Table 1.

The implementation done in this work is a generalized implementation which can be applied to any dataset. As the dataset and the name of the attributes have to be given in form of a .txt file in the beginning of the execution of the process. Here we have discussed the performance of the proposed algorithm on the measures of number of rules generated, execution time of algorithm and trends in the average fitness value of the rules at different generation of the execution.

6   Results of Proposed Algorithm

We have first illustrated different results that we got on implementation of proposed algorithms on Quake and Stocks dataset. Later we have illustrated the comparison between algorithms proposed by different authors and our proposed algorithm. We have taken crossover rate as 0.2 and mutation rate as 0.02 carry out results. Results of proposed algorithm on Quake dataset are shown in Table 2.

Figure 4 illustrate the number of rules that are generated by implementation on different support counts. As discussed earlier support count is the first filter of selecting chromosomes for calculating fitness. Minimum support has to be given manually at the starting on the basis of need of application. It can be shown that with increase

Table 1 Datasets taken for analysis

Dataset name	No. of attributes	No. of records	Type of dataset
Quake	4	2178	Real valued
Stocks	10	950	Real valued
Pollution	16	60	Real and integer
Stulong	5	1419	Integer

Table 2 Parameter values for implementation on Quake dataset

Parameters	Values
Dataset: Quake dataset	(4/2178) (attributes/records)
Population size	100
Number of generation	10

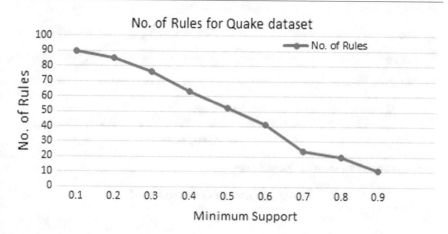

Fig. 4 No. of rules generated for Quake dataset

in minimum support count, the numbers of rules that are generating are decreasing which signify that selecting min. support for filter purpose is a correct decision.

Figure 5 illustrates the execution time taken for quake dataset for single generation. It can be seen that with increasing the minimum support count, execution time is decreasing and it can also be concluded from this graph that for minimum support of 0.1, 0.2, 0.3, the execution time is much higher because number of rules generated are much higher at such a low min. support value. Therefore, execution time of simulation gets increased whereas from minimum count 0.4 onwards, the number of rules that get filtered out for further processing is low. Hence, execution time also decreases to a relevant extent.

Figure 6 illustrates the average fitness of rules with minimum support count. As the min. support count is increasing, the average of fitness value of the generated rules is also increases. It signifies that with increase in minimum support, the relevancy of the rules increases i.e. more interesting rules get generated.

Further we have illustrated some more plots for different generations to analyze the improving nature of rules at each generation. It is a property of genetic algorithm that all iterations of execution lead to improve the solution set. To see the effect of this property in our proposed implementation, we have plotted the results of different generation which are illustrated here.

Figure 7 illustrates the fitness of rules generated for Stocks dataset for generation 1. The fitness of those chromosomes is shown as dots in the figure.

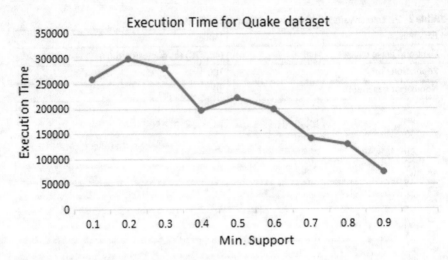

Fig. 5 Execution time taken for Quake dataset

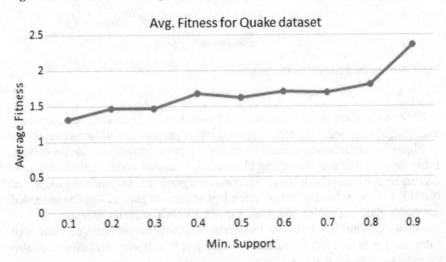

Fig. 6 Average fitness of rules generated for Quake dataset

Figure 8 illustrates the fitness of rules that are being generated at different generation i.e. for generation 1, generation 4, generation 6, generation 8 and generation 10 of the implementation for Stocks dataset. It can be shown in the figure that fitness of rules gets better and better with increasing generation, i.e. at all iteration rules are getting more interesting in terms of fitness. Also the numbers of rules are decreasing at each generation because of filtering of unfit or not-interesting rules at all iteration.

Figure 9 illustrates the fitness of rules at different generation for quake dataset. It shows the similar kind of results as of Fig. 8 i.e. with increase in generation, the rules are getting more refined due to increase in the fitness value of rules (shown by

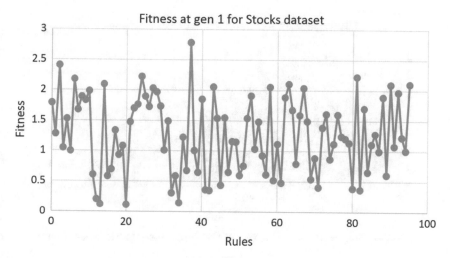

Fig. 7 Fitness of rules generated for Stocks dataset

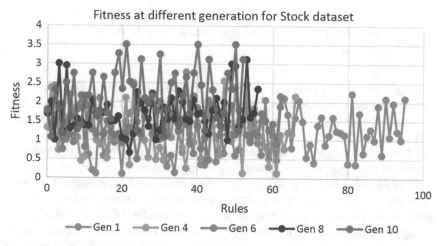

Fig. 8 Fitness of rules generated for Stocks dataset at different generation

peaks in the graph) and decreasing the number of rules (shown by tail of the plot for each generation).

We can see that our proposed algorithm is following the basic property of genetic algorithm i.e. the evolving nature of the solution at all iterations with the results of both the above graphs.

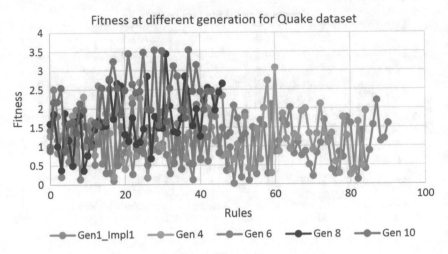

Fig. 9 Fitness of rules generated for Quake dataset at different generation

6.1 Comparison of Proposed Approach with Existing Approach

Next, to compare our proposed algorithm with previously known algorithms, we have plotted the comparison bar graphs. We have compared our results with four previously known algorithms that are EARMGA [14], GENAR [27], MODENAR [15] and MOPNAR [16]. Details of these algorithms have been discussed in the literature survey section. We have compared our algorithm on the basis of number of rules these algorithms have generated at min. support count of 0.5, average of support count of the rules that are generated by these algorithms. For comparison we have taken four datasets from the Keel dataset repository i.e. Quake, Pollution, Stocks and Stulong. The comparison values of previous approaches i.e. number of rules and average support for previously known algorithm have been taken from the literature Martin et al. [16] and that is compared with the results generated by implementing proposed algorithms.

Figure 10 is the graphical bar plot of data presented in Table 3 taken from Minaei-Bidgoli et al. [28], Al-Maqaleh [29], Martin et al. [16] and the results found on implementation of our proposed approach. This bar plot illustrate the comparison between various previously known algorithms with the proposed algorithm of our work on the basis of number of rules generated by these algorithms with different datasets. From this plot, we can conclude that our proposed algorithm have generated less number of rules as compared to EARMGA, MODENAR and MOPNAR (see Fig. 11 and Table 4).

This bar plot illustrates the average support of rules generated by different algorithms on different datasets. Higher average support of rules is a desirable parameter, as higher value of support count indicates the more interesting rules. As per the plot

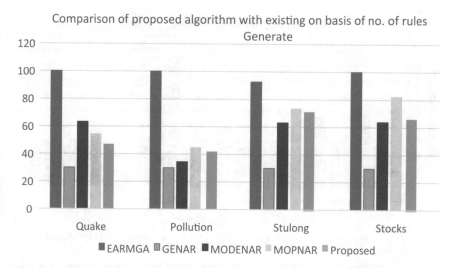

Fig. 10 Comparison of proposed algorithm with existing on basis of no. of rules generated

Table 3 Number of rules generated by different algorithms

	Quake	Pollution	Stulong	Stocks
EARMGA	100	100	92.6	100
GENAR	30	30	30	30
MODENAR	63.3	34.4	63.2	63.8
MOPNAR	54.6	45	73.6	82.4
Proposed approach	45	38	68	64

Table 4 Average support of different algorithms

Avg. support	Quake	Pollution	Stulong	Stocks
EARMGA	0.27	0.25	0.27	0.37
GENAR	0.55	0.22	0.88	0.29
MODENAR	0.36	0.27	0.52	0.48
MOPNAR	0.27	0.26	0.27	0.31
Proposed approach	0.47	0.41	0.35	0.42

shown above it can be concluded that our proposed algorithm results in better average support count than EARMGA, MODENAR, MOPNAR which signify that rules generated by our approach are more interesting than the rules generated by previous algorithms.

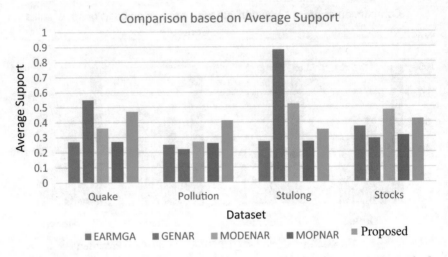

Fig. 11 Comparison of proposed algorithm with existing on basis of average support of rules generated

7 Conclusion and Future Scope

We have studied and analyzed the problem of association rule mining for numeric datasets. Processing of numeric dataset for getting information is a tedious and computationally complex task because of the nature of the data. Association rule mining for numeric dataset was a challenging task as the huge number of expected rules will be generated. Therefore, we needed a technique which runs for many solutions in parallel so that we can process huge number of rules and also have the tendency of improving its solution by itself for generation of more interesting rules. The pre-existing technique that follows these criteria is genetic algorithm which has the property of running for solution in parallel and improving their own solution in all iteration i.e. generation in terms of genetic algorithm. Therefore we have decided to solve the problem of association rule mining with genetic algorithm.

In this work, we have proposed an approach for mining numeric association rules using genetic algorithm and both the algorithms are implemented and analyzed using keel repository datasets. After implementation and analysis, we can conclude that genetic algorithm is a good approach to be used for mining numeric association rules as it analyzes multiple rules together at a time which decreases the probability of getting stuck in a local optimal solution. Further, in all iterations, best rules are more refined and new random rules joined the population which makes the scenario continuous improving i.e. even if we have started from a random initial population of rules but with all iterations rules get improved. Rather than using traditional methods for optimization i.e. derivative function etc., fitness functions are used for optimization which makes it more efficient.

In the current work we have focused only on generating association rules from testing datasets. It is still a long way to go for using this solution in real life scenario. To improve the current solution and embedding that solution to real world problems, various research opportunities could be:

1. Genetic algorithm is a robust technique for handling NP Hard problems and as association rule mining is also an NP Hard problem therefore we can say that we are going in correct direction. But it is still a long way to go as the size of the dataset increases more refine operators and fitness function have to be used. The future research direction for this work would be to develop more refined operators that could be used for better results.
2. Further Genetic algorithm can be applied to many real world problem some of which are discussed in Moin et al. [30], Barak and Modarres [31].We would like to test our proposed work on similar kind of real world problem.
3. Also one of the issues with the proposed algorithm is that with the increase in the size of dataset, the efficiency of the algorithm degrades i.e. proposed algorithms are not scalable enough. Therefore study regarding scalability of such algorithms can also be the future research direction of this work.

The proposed algorithm is efficient one and may be one of the possible candidates for the research community to get real time information from databases in smart city applications ranging from healthcare applications to intelligent transportation system [32–34].

References

1. Agrawal R, Imielinski T, Swami AN (1993) Mining association rules between set of items in large databases. In: ACM SIGMOD international conference on management of data, Washington D.C.
2. Tew C, Giraud-Carrier C, Tanner K (2013) Behaviour-based clustering and analysis of interestingness measures for association rule mining. Springer Data Min Knowl Discov J 28(4):1004–1045
3. Agrawal R, Srikant R (1994) Fast algorithms for mining association rules in large databases. In: Proceedings of the twentieth international conference on very large databases, Santiago de Chile, Chile
4. Houtsma M, Swami A (1995) Set-oriented mining for association rules in relational databases. In: Proceedings of the 11th IEEE international conference on data engineering, Taipei, Taiwan
5. Hidber C (1999) Online association rule mining. In: Proceedings of the ACM SIGMOD international conference on management of data, Philadelphia, PA, USA
6. Borgelt C (2003) Efficient implementations of Apriori and Eclat. In: Proceedings of the workshop frequent itemset mining implementations, Melbourne, FL
7. Goldberg DE (1989) Genetic algorithms in search, optimization and machine learning. Addison-Wesley Publishing Co., Inc., Boston
8. Elhoseny M, Yuan X, Yu Z, Mao C, El-Minir H, Riad A (2015) Balancing energy consumption in heterogeneous wireless sensor networks using genetic algorithm. IEEE Commun Lett 19(12):2194–2197. https://doi.org/10.1109/LCOMM.2014.2381226

9. Yuan X, Elhoseny M, El-Minir H, Riad A (2017) A genetic algorithm-based, dynamic clustering method towards improved WSN longevity. J Netw Syst Manage Springer, US 25(1):21–46. https://doi.org/10.1007/s10922-016-9379-7

10. Obitko M (n.d.) Genetic algorithm tutorial. Retrieved from https://courses.cs.washington.edu/courses/cse473/06sp/GeneticAlgDemo/index.html

11. Freitas AA (2003) A survey of evolutionary algorithms for data mining and knowledge discovery. Advances in evolutionary computing. Springer, New York, NY, USA, pp 819–845

12. Ghosh A, Nath B (2004) Multi-objective rule mining using genetic algorithms. J Inf Sci Elsevier 163:123–133

13. Anand R, Vaid A, Singh PK (2009) Association rule mining using multiobjective evolutionary algorithms: strengths and challenges. In: Proceedings of the nature & biologically inspired computing, Coimbatore, India

14. Yan X, Zhang C, Zhang S (2009) Genetic algorithm-based strategy for identifying association rules without specifying actual minimum support. J Expert Syst Appl Elsevier 36(2):3066–3076

15. Alatas B, Akin E, Karci A (2008) MODENAR: multi-objective differential evolution algorithm for mining numeric association rules. J Appl Soft Comput Elsevier 8(1):646–656

16. Martin D, Rosete A, Alcala J (2014) A new multiobjective evolutionary algorithm for mining a reduced set of interesting positive and negative quantitative association rules. IEEE Trans Evol Comput 18(1):54–69

17. Metawa N, Hassan MK, Elhoseny M (2017) Genetic algorithm based model for optimizing bank lending decisions. Expert Syst Appl 80:75–82

18. Berzal F, Blanco I, Sanchez D, Vila M (2002) Measuring the accuracy and interest of association rules: a new framework. J Intell Data Anal 6(3):221–235

19. Anand HS, Vinodchandra SS (2016) Association rule mining using treap. Int J Mach Learn Cybern. https://doi.org/10.1007/s13042-016-0546-7

20. Umit C, Bilal A (2017) Automatic mining of quantitative association rules with gravitational search algorithm. Int J Softw Eng Knowl Eng 27(3):343–372

21. Uroš M, Milan Z, Iztok F Jr, Iztok F (2017) Modified binary cuckoo search for association rule mining. J Intell Fuzzy Syst 32(6):4319–4330

22. Qodmanan HR, Nasiri M, Minaei-Bidgoli B (2011) Multiobjective association rule mining with genetic algorithm without specifying minimum support and minimum confidence. J Expert Syst Appl Elsevier 38(1):288–298

23. Tan P, Kumar V, Srivastava J (2002) Selecting the right interestingness measure for association patterns. In: Proceedings of the 8th international conference of KDD, Edmonton, AB, Canada

24. Brin S, Motwani R, Ullman JD, Tsu S (1997) Dynamic itemset counting and implication rules for market basket data. In: Proceedings of the ACM SIGMOD international conference on management of data, New York, NY, USA

25. Magalhães-Mendes J, Cidem (2013) A comparative study of crossover operators for genetic algorithms to solve the job shop scheduling problem. WSEAS Trans Comput 12(4):164–173

26. Alcala-Fdez J, Fernandez A, Luengo J, Derrac J, Garcia S, Sanchez L, Herrera F (2011) KEEL data-mining software tool: data set repository, integration of algorithms and experimental analysis framework. J Multiple Value Logic Soft Comput 17(2–3):255–287

27. Mata J, Alvaez J, Riquelme J (2001) Mining numeric association rules with genetic algorithms. In: Proceedings of the 5th international conference artificial neural network genetic algorithms, pp 264–267

28. Minaei-Bidgoli B, Barmaki R, Nasir M (2013) Mining numerical association rules via multi-objective genetic algorithms. J Inf Sci Elsevier 233:15–24

29. Al-Maqaleh B (2013) Discovering Interesting association rules: a multi-objective Genetic algorithm approach. Int J Appl Inf Sci 5(3):47–52

30. Moin NH, Sin OC, Omar M (2015) Hybrid genetic algorithm with multiparents crossover for job shop scheduling problems. In: Mathematical problems in engineering, Hindawi, 2015

31. Barak S, Modarres M (2015) Developing an approach to evaluate stocks by forecasting effective. Expert Syst Appl Elsevier 42:1325–1339

32. Darwish A, Hassanien A, Elhoseny M, Sangaiah A, Muhammad K (2017) The impact of the hybrid platform of internet of things and cloud computing on healthcare systems: opportunities, challenges, and open problems. J Ambient Intell Humanized Comput. First Online: 29 Dec 2017. https://doi.org/10.1007/s12652-017-0659-1

33. Elhoseny H, Elhoseny M, Riad AM, Hassanien A (2018) A framework for big data analysis in smart cities. In: Hassanien A, Tolba M, Elhoseny M, Mostafa M (eds) The international conference on advanced machine learning technologies and applications (AMLTA2018). AMLTA 2018. Advances in Intelligent Systems and Computing, vol 723. Springer, Cham. https://doi.org/10.1007/978-3-319-74690-6_40

34. Shehab A, Ismail A, Osman L, Elhoseny M, El-Henawy IM (2018) Quantified self using IoT wearable devices. In: Hassanien A, Shaalan K, Gaber T, Tolba M (eds) Proceedings of the international conference on advanced intelligent systems and informatics 2017. AISI 2017. Advances in intelligent systems and computing, vol 639. Springer, Cham. https://doi.org/10.1007/978-3-319-64861-3_77

Online User Authentication System Based on Finger Knuckle Print Using Smartphones

Atrab A. Abd El-aziz, Eman M. EL-daydamony, Alaa. M. Raid
and Hassan H. Soliman

Abstract In recent years, the widespread use of smartphone devices has raised security issues relating to these devices to protect sensitive data. The risk is greater without an adequate method to control access to the applications accessed by smartphones. Knuckle print based biometrics is being used to control access to several different services via smartphones such as payment gateways etc. In this paper, we present an evaluation of a finger knuckles recognition system using HOG descriptor with SVM on the images collected using the smartphones built-in camera. In this work, we used the HK_PolU database of 187 individuals with 3 knuckle images with a different uniform environment. We evaluate the existing performance metrics based on Recognition Rate (RR), False Acceptance Rate (FAR), False Reject Rate (FRR) and Receiver operating characteristic (ROC) curve. The proposed system has been executed using smartphone emulator with the Android operating system. The best performance achieved was with HOG of 16-cell-size = 95.4% at FAR = 3.4%, FRR = 5.5% and AUC-ROC 0.954 with recognition time from (1–7) seconds which indicates the applicability of the proposed FKP based mobile authentication system in a real-life scenario.

A. A. Abd El-aziz (✉)
Information Technology Department, Faculty of Computers and Information,
Kafr El- Sheikh University, Kafr El- Sheikh, Egypt
e-mail: Atrab_Ahmed@fci.kfs.edu.eg

E. M. EL-daydamony · H. H. Soliman
Information Technology Department, Faculty of Computers and Information,
Mansoura University, Mansoura, Egypt
e-mail: eman.8.2000@gmail.com

H. H. Soliman
e-mail: hsoliman@mans.edu.eg

Alaa. M. Raid
Information System Department, Faculty of Computers and Information,
Mansoura University, Mansoura, Egypt
e-mail: amriad2000@mans.edu.eg

© Springer Nature Switzerland AG 2019
A. E. Hassanien et al. (eds.), *Security in Smart Cities: Models, Applications, and Challenges*, Lecture Notes in Intelligent Transportation and Infrastructure,
https://doi.org/10.1007/978-3-030-01560-2_15

Keywords Smartphone biometrics · Finger knuckle
Histogram of oriented gradients · Authentication · Smartphone sensors

1 Introduction

1.1 Smartphone

The number of smartphones has grown rapidly during the first years, which has contributed to the increase of confidential information on these devices. The latest generation of smart devices, specifically tablets and smartphones devices are the commonly used personal devices in human daily life. Presently, 1.3 million Android device and 400,000 Apple devices are activated [1, 2]. The usage of smartphones is quite distinctive than the usage of PCs and laptops. Smartphone users review their smartphones nearly 150 times a day [2]. Some of the reasons behind the popularity of smartphones include its better batteries, powerful processors, advanced hardware including powerful embedded sensors and faster connectivity chips as shown in Fig. 1.

In order to get a better idea of the threats to user data, a security company in the United States (US), conducted a social research in five main cities in North America. They left fifty smartphones in unrestricted and unsecured places [3]. The results revealed that 96% of people who obtained the smartphones actually accessed them, and 43% of them accessed online bank accounts. All smartphones currently available in the market continuously collect the user's location coordinates. Smartphones offer

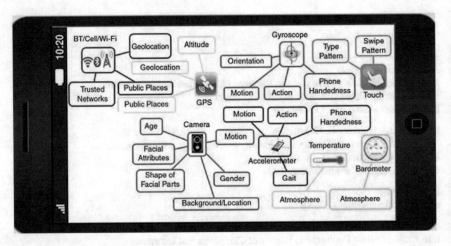

Fig. 1 Smarphone sensors

users the capabilities to access mobile devices, bank accounts, and social networks over the common favorite applications, such as Facebook, etc. [3].

All online mobile applications store confidential data in the user's privacy, which usually becomes simply accessible since the smartphone is accessed. Therefore, any illegal access to these devices could have dangerous consequences and become a pain for the victim.

1.2 Biometrics

Person recognition is an essential process in an extensive diversity of applications including unrestricted security, computer systems and access control. Beside the improvement of stability and accuracy, biometric technologies are rising quickly for finding solutions for this problem [4]. Biometric systems are a pattern recognition systems used to verify a person's identity by analyzing and comparing person's stored patterns against enrolled records of those patterns.

These biometric technologies became the premise of a large assortment of vital secure personal identification and authentication solutions. Therefore, biometrics points to the personal physiological traits (e.g., fingerprint, palm print, face, iris, etc.) or behavioral traits (e.g., gait, voice, keystroke dynamics, etc.) features that recognize a someone else's Therefore, everybody incorporates a distinctive biometric pattern taking under consideration that their biometric signatures cannot be forgotten, lost, or stolen [5]. Biometric systems offer several advantages over traditional methods. Biometric traits can't be duplicated, forged or distributed easily, unlike passwords and tokens. Biometric systems additionally raise user aid by assuaging the requirement to remember passwords. Furthermore, use of biometrics can provide non-repudiation which is not attainable through traditional methods.

Currently, there is an advanced improvement in the field of biometric technologies. The use of the individual's hand biometrics is one of those technologies that offer a reliable, safe, easy-to-use, low-cost solution for a variety of the security applications. Therefore, being a new technology of hand, Finger-Knuckle-Print [6], has attracted an increasing amount of regards. FKP has many benefits that correspond to different available methods.

1.3 Finger Knuckle Print Biometric

The hand of the individual has many biometric characteristics, such as fingerprint, knuckle and palm print, and hand geometry. In the midst of these characteristics, the impression of a finger knuckle is approximately a new biometric feature in contrast to the famous biometric features such as fingerprint, face, palm and iris [7], as shown in Fig. 2.

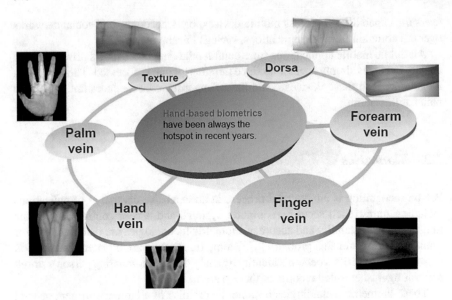

Fig. 2 Hand-based biometrics

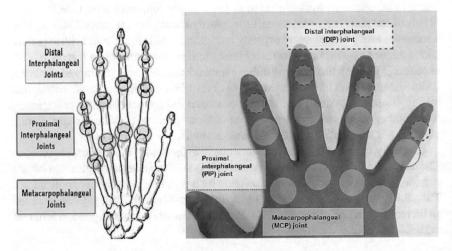

Fig. 3 Taxonomy of the joints of the finger knuckles: the blue circles represent the distal interphalangeal joints (DIP), the circles of green color point to the proximal interphalangeal joints (PIP) and the points of the circles of red color to the metacarpophalangeal joints (MCP)

The outside finger surface has three knuckles: first, a proximal interphalangeal (PIP) joint. Second, a distal interphalangeal (DIP) joint. Finally, a metacarpophalangeal (MCP) joint as shown in Fig. 3.

Kumar et al. [8] classified those joints of three knuckles in minor and major knuckles, where the DIP joint is the fundamental minor, an MCP is the second minor knuckle and a PIP is the major knuckle.

It is simple to obtain such finger knuckle patterns by a smartphone camera. This advantage empowers us to generate a flexible biometric authentication method using the smartphone camera. A finger knuckle is thus assumed to be unique as well as a palm print and fingerprint. This paper concentrates on the use of knuckle patterns with fingers to develop an individual online authentication system that uses the smartphone's camera for the security of online systems.

2 Related Work

The skin patterns of the knuckle surface of the fingers, such as lines, textures, and folds, have a great capacity for differentiation to be universally identified as a biometric identifier. Previously, because of the deficiency of the benchmark datasets and also the lower recognition rates, this field failed to receive decent attention and consequently, a finite range of efforts were declared until 2007.

Following that, in depth efforts are enforced within the FKP based mostly authentication system region. Region of interest (ROI) exposure, PIP majority-wise, feature extraction, and classification play a very important role within the recognition system supported FKP.

There are some further conditions, like guide size and memory storage, however they can't be done at constant time.

In this section, we present a report related to the extraction of ROI, extraction of features, classification techniques.

2.1 ROI Extraction of FKP

Many types of research concentrated on the texture patterns of PIP joints recognition algorithms to evaluate FKP performance. The existing systems obtain ROI based upon local convexity features of PIP.

Many of the existing systems experimented against two publicly accessible datasets i.e., IIT Delhi FKP and PolyU FKP Image dataset. Samples from the PolyU FKP database are taken under controlled conditions since the subject lays his finger on the fixed parts to overcome spatial variations and obtain the clear characteristics of the line of a finger knuckle.

In [9], the authors proposed an FKP ROI extraction algorithm by using improved contrast and better asymmetric samples.

Subsequently, [10] provides an ROI location and exposure method based on central point for FKP images. This system is strong towards the displacement of the finger and the rotation in the horizontal direction.

In [11], the authors declared a modified Gabor filter that combines the curvature parameter to verify the response of the characteristics of the convex curve line around the main joint of the finger. The center line of the knuckle and the center point of the knuckle are measured to extract the ROI from the FKP image. Eventually, the ROI is produced based on the center point of the knuckle and obtained 95% accuracy with the PolyU FKP data set.

2.2 Feature Extraction of FKP

Most researches in [13–16, 18] applied coding procedures to obtain characteristics by implementing spatial filters to images and binarizing their answers, where a variety of prototypes of Gabor filter are usually employed as a spatial filter.

In [12], the authors used the Gabor 2-D filter to extract local orientation information and preserve it in a feature vector called the competence code. In addition, in [13], the authors modified the method of knuckle codes by utilizing the- radon transformation in improved knuckle images. They reported 1.14% EER and RR of 98.6%.

In [14], the authors proposed a multi-algorithmic procedure based on the principal component analysis (PCA), independent component analysis (ICA) and linear discriminant analysis (LDA) matching scores. They achieved EER of 1.39%.

In [15], an important method named a monogenic code that revealed the phase information and orientation of the knuckle images were wrapped.

In further work [16], the authors have applied a weighted add rule fusion technique to realize optimum results.

In [17], the SIFT key points were extracted from the improved FKP pictures supported the Gabor filter.

In [18], the authors executed a known local feature descriptor called LBP on ROI based on FKP images.

In [19], proposed a biometric scheme for permitted access to mobile devices. The major contribution of the paper was to develop a feature extraction methods called Probabilistic Hough Transform dedicated to mobile phones. The knuckle system was experimented using the IIT Delhi database acquired from a digital camera. They obtained EER 1.02%. They didn't provide any method to deal with knuckle data collected using smartphone camera.

In [20], Meraoumia, Abdallah, et al. introduced a biometric authentication system using FKP. The study achieved using PolyU-FKP database. They used HOG descriptor for feature extraction. The main objective of this paper was to measure the identification system using a combination of multiple fingers. They achieved 0.606% EER. Additionally, the results would vary if finger knuckle images were acquired using smartphone camera.

In [21], the texture of the knuckles has been used as an identifier in smartphones applications. This was the initial attempt to develop smartphones practical biometric authentication system using the PIP joints. As a result, the performance of the recognition was not necessarily good. They used 1D Log-Gabor filters for feature extraction. Finally, they were the first to measure time. The user's average authentication time does not exceed 5 s. Accuracy rates of 72% and an EER 9% have been achieved.

2.3 Matching of FKP

In [14], the authors have introduced a recognition algorithm for finger knuckle using multiple patterns obtained from the fingers. They proved that the matching score estimated by combining four PIP joints is effective for the authentication of the person.

In [22], the authors have applied fusion using a weighted sum rule for fusing local and global information to accomplish optimum results.

In [23], the authors have applied a score level fusion to integrate the obtained matching scores caused by fragility masks.

In [24], the authors matched minutia samples retrieved using the cylindrical code, minutiae triangulation, and spectral minutia-based approaches.

There are a few studies for smartphone authentication using finger knuckle recognition. The only attempt to develop a practical FKP authentication system for smartphones was in [21]. The recognition performance was not good. Consequently, this paper focuses on the use of HOG descriptor as feature extraction algorithm with SVM classifier for the first time on public available database acquired from the smartphone camera to improve the recognition rate with time reduction.

The sections of the paper are arranged as follows: Sect. 3 displays in brief a summary of our FKP authentication system. The technique of feature extraction is explained in Sect. 4, which incorporates a summary of HOG technique. Section 5 briefly explained the SVM classification algorithm. In Sect. 6 the matching strategies. Section 7 briefly presents the experimental database. In Sect. 8, we explain the development environment. Section 9 displays the experimental results. Finally, Sect. 10 concludes the conclusions and future work.

3 The Proposed Workflow

Figure 4 shows the block diagram of our FKP based authentication system. The image capture process for FKP samples is explained in [21]. FKP images were acquired using the integrated camera of HTC Desire HD A9191 smartphone camera with autofocus capability from 187 subjects. The knuckle images were acquired from varying uniform background in real environments.

The preprocessing steps as explained in [21] have been done using the histogram equalization to minimize the influence of varying illumination and to adjust the contrast. Then: The median filter was used to overcome the speckle-like noise with

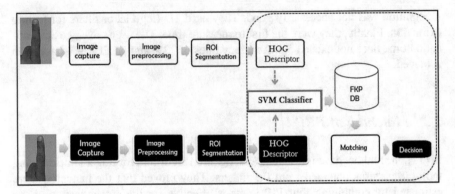

Fig. 4 Block diagram of FKP based authentication system

Canny edge detector. Subsequently, the image rotation and normalization have been done to localize the ROI.

3.1 The Enrollment Phase

In this step, the HOG descriptor has been used to extract the feature vector from each ROI sub-image. After that, the SVM classifier has been trained using all the feature vectors (extracted from all enrolled subjects) and therefore forming the biometric authentication system.

3.2 The Authentication Phase

The HOG feature vector is obtained from the FKP-ROI of the external person. Then the SVM classifier used it as an input vector for make the decision to accept or reject.

4 Feature Extraction

One of the essential steps in any biometric system is to get a decision to accept or reject a person. The primary function of the feature extraction algorithm is to extract the discriminating information existing within the extracted region of interest (ROI) of FKP sub-image. In this model, FKP feature vector created by HOG descriptor which describes each sub-image ROI as a histogram 1-D (one-dimensional vector) vector.

Fig. 5 Process of formation
of HOG feature descriptor

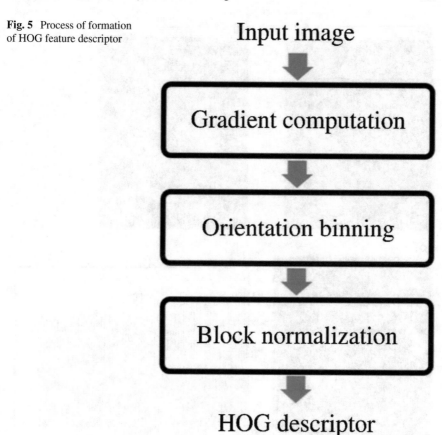

Fig. 5 Process of formation of HOG feature descriptor

4.1 Histogram of Oriented Gradients Descriptor

In the image processing domain, HOG is a descriptor of the characteristics used to detect objects. The HOG descriptor method counts the occurrence of gradient orientation in the localized interest region (ROI).

The HOG algorithm implementation involves the following steps: first, the ROI sub-image is split into small connected blocks. Next, each block is divided into small connected regions named cells. For each cell, it calculates the orientations of the edges or the histogram of the gradient directions for all the pixels within the cell. The groups of normalized histograms describe the block histogram. The set of these blocks represents the descriptor [25].

The HOG is a popular descriptor that was initially proposed for the detection of the pedestrian by Dalal and Triggs [25]. Since that time it has been utilized for various classification problems. HOG features were applied for detection of human cells in [26], and for vehicle detection in [27]. The primary method of generating HOG descriptor for an image is shown in Fig. 5.

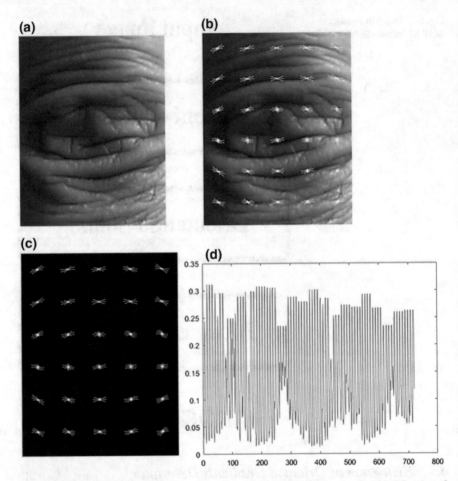

Fig. 6 **a** The ROI of FKP. **b** The HOG visualization. **c** The ROI with its visualized HOG points. **d** The HOG descriptor histogram

This paper contributes to the exploration of feature extraction technique based on HOG descriptor technique. Our study is the first to apply HOG algorithm on FKP images acquired using smartphone camera. The HOG descriptor was obtained using two different cell size ([32 * 32], [16 * 16]), block size as 2 * 2 and bin size as 9. Therefore, it was found that the length of the descriptor was 720 for 32 cells and 3564 for 16 cells for ROI-size of 200×160 pixels. Figure 6 shows an example of the HOG descriptor.

Classification is the process of automatically assigning an object to one of various (often pre-defined) classes based on the attributes of the object. In the classification process, we are ordinarily given a training set of objects and their attributes are named features. Simultaneously with this, we are granted the categories or class labels of each object. A classifier is then created based on the training set.

When a training dataset has class labels, it is known as supervised classification. There are various methods of classification, example support vector machines, nearest neighbour's, decision trees, neural networks, etc.

The classification has a two-step process. The first step is the model construction which describes a group of present categories. Each sample in this is considered to belong to a predefined class, as determined by the class label attribute. The model is described as classification rules, etc. The second step is the model usage which used for classifying future or unknown objects.

4.2 Support Vector Machine Classifier

SVM was initially shown by Boser, etc. in [28] and since then, it is very attractive due to its simplicity and excellent performance. SVM was developed from Statistical Learning Theory.

It is a supervised learning algorithm that is employed for classification and regression and belongs to the family of generalized linear classifiers. Among all existing classifiers, SVM [28] is surely the most popular for the two reasons: (i) It relies on reducing the empirical risk (or maximizing the margin and (ii) It does not execute any assumption about the data (score) distribution.

Statistical learning hypothesis models data as a function estimation process. It represented as a system that receives some data as input and then outputs a function that can be applied to predict some features of future data.

Therefore, a set of training patterns or the training input feature vectors $\{(x_1, y_1), (x_2, y_2) …, (x_n, y_n)\}$ is given, where x_i is the ith training samples and y_i represents the class label associated with ith. Thus, SVM determines an optimal separating hyper plane and the theory that is able to accurately classify the given training data as shown in Fig. 7.

Assume that there is a set of training patterns $\{x_i; i = 1 … n\}$ assigned to one of the two classes $\omega 1$ and $\omega 2$, with labels $y_i = \pm 1$ are given, such that

$$\text{For } y = +1 \quad w_i x_i + b > 0 \tag{1}$$
$$\text{For } y = -1 \quad w_i x_i + b < 0 \tag{2}$$

In order to classify the given dataset, we require finding a linear discriminant function, or decision boundary such that

$$g(x) = w_i x + b = 0 \tag{3}$$

The desired hyper-plane that divides the lines and maximizes the margin between the closest points on the convex sides of the two datasets. The distance from the nearest point in a hyper-plane to the origin can be calculated by increasing x since x is located on the hyper-plane.

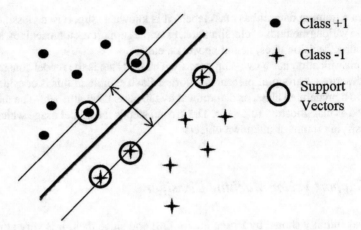

Fig. 7 Example to show support vectors, margin and decision boundary

So, for the points on the other side, we have a similar story. Therefore, when subtracting and solving the two distances, we obtain the sum of distances of the separation hyper-plane to the closest points. Greatest Margin $= M = 2/\|w\|$ Therefore, the decision rule is given as

$$g(x) = w_i + b \begin{Bmatrix} > 0 \\ < 0 \end{Bmatrix} \rightarrow x \in \begin{Bmatrix} w_i; y_i = +1 \\ w_2; y_i = -1 \end{Bmatrix} \tag{4}$$

5 Matching

The matching process selects an unidentified FKP image to one of the predefined two classes (genuine or imposter). Thus, the HOG feature vector is classified using SVM classifier and the predicted output of the unidentified FKP image is either accepted or rejected.

$$g(x) = w_i + b \begin{Bmatrix} > 0 \\ < 0 \ \substack{decision} \end{Bmatrix} \underset{decision}{\Rightarrow} \begin{Bmatrix} Accepted \\ Rejected \end{Bmatrix} \tag{5}$$

6 Database

Our proposed model results were evaluated using the Hong Kong Polytechnic University (HK_PolU) public available database in [29]. The FKP images in HK_PolU

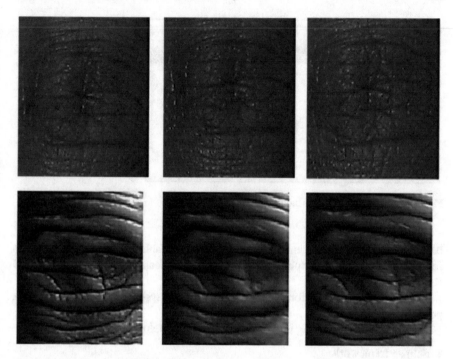

Fig. 8 Example images from HK-PolU database

database acquired using HTC Desire HD A9191 smartphone [21]. The database acquired from 187 subjects with three pictures each (Fig. 8).

7 Development Environment

In this paper, we proposed a web service online FKP user authentication system using the smartphone platform shown in Fig. 9.

Our model is divided into two parts. The first is the client side (smartphone device). The second is the server side. On the client side, the user smartphone used only to send the ROI sub_image of FKP image. On the server side, the remote server used not only to store the system database for the authorized users, but also for the feature extraction and matching process.

This is considered a new and good idea because it provides more security, usability, power, cost and time saving to different online applications that need to authenticate smartphone users.

The system execution speed is fast when a faster CPU is supported, a larger RAM and a faster internet speed. Since most of the Android smartphones products already had more than 1.2 GHz CPU and more than 1 GB of RAM. Then, the runtime would

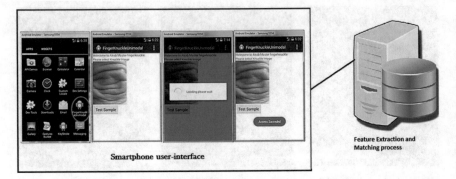

Fig. 9 Smartphone FKP online biometric system

be better than our experiments. The minimum acceptable hardware specifications for the runtime of the smartphone system would be at least 1.2 GHz CPU and 1 GB RAM.

In order to test the performance of our proposed system, we have used JAVA programming language.

As a result, the following tools were used in the development process:

- Eclipse framework
- Java Development Kit (JDK)
- Android SDK
- MCRInstaller
- Apache Tomcat 8
- Apache Axis
- ksoap2-android-assembly.

The smartphone emulator we employed to develop and evaluate our multimodal authentication system is Samsung Galaxy Nexus emulator the details of smartphone specifications are described in the following:

- Platform: Android OS v4.4.2 (Kit Kat)
- CPU: Dual-core 1.2 GHz
- Camera: 5 MP, autofocus, LED flash
- RAM: 1 GB
- Internal Storage (ROM): 1 GB.

8   Results

In this paper, performance is measured using False Acceptance Rate (FAR in %), False Rejection Rate (FRR in %), Recognition Rate, Recognition Time, and corresponding Receiver Operating Characteristic curve (ROC).

Table 1 Confusion matrix for FKP recognition using 32-HOG descriptor

		Actual class	
		Genuine	Imposter
Predicted class	Genuine	283 true accept	17 false accept
	Imposter	26 false reject	235 true reject

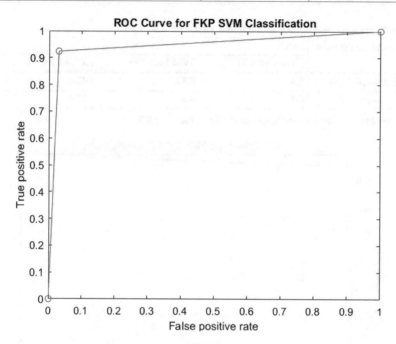

Fig. 10 ROC curve of FKP using HOG with 32 cell size in experiment

Our experiment was performed using 187 subjects of the database. Two samples were selected for training and one for the test. For training, we labeled 103 users as genuine and 80 as imposters. Two images per each person were used for training. For testing, we used four people who have not been used for training as impostors with a sample for each person who has not been used for training. The correct numbers of recognitions were then counted for 561 samples to compute the accuracy.

The HOG descriptor was found using two different cells size of 16 * 16 and 32 * 32, block size of 2 * 2 and bin size as 9. Thus, we have two parts of results.

Firstly, For HOG with cell-size 32 * 32, we obtained an accurate recognition rate of 92.3%, FRR 5.5% and FAR 3.4% as shown in Table 3. Table 1 shows the confusion matrixes of 32-cell size HOG descriptor.

The ROC (receiver operating characteristic) curve of Fig. 10 represents the false positive rate (FPR) versus true positive rate (TPR) for 32-cell of HOG descriptor.

Table 2 Confusion matrix For FKP recognition using HOG descriptor

		Actual class	
		Genuine	Imposter
Predicted class	Genuine	292 true accept	9 false accept
	Imposter	17 false reject	243 true reject

Table 3 Measurement of FKP with different HOG cell size

Performance measure	HOG		
	SVM FAR (%)	SVM FRR (%)	SVM RR (%)
Cell-size (16 * 16)	3.4	5.5	95.4
Cell-size (32 * 32)	6.8	8.3	92.3

For each method, the first row describes the FAR, FRR and RR

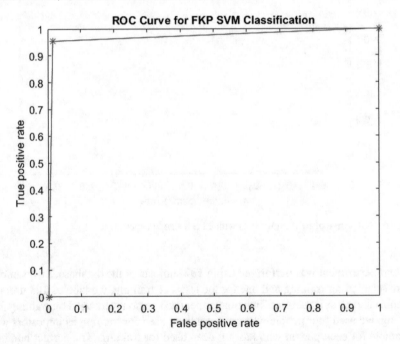

Fig. 11 ROC curve of FKP using HOG with 16 cell size in experiment

Second, For HOG with cell-size 16 * 16, we obtained an accurate recognition rate of 95.4%, FRR 8.3% and FAR 6.8%. Table 2 shows the confusion matrixes of 16-cell size HOG descriptor (Table 3).

The ROC (receiver operating characteristic) curve of Fig. 11 represents the false positive rate (FPR) versus true positive rate (TPR) for 16-cell of HOG descriptor.

It was found from the experiments that HOG feature descriptor with 16 cell-sizes outperforms the HOG with 32 cell-sizes in terms of accuracy as shown in Fig. 12.

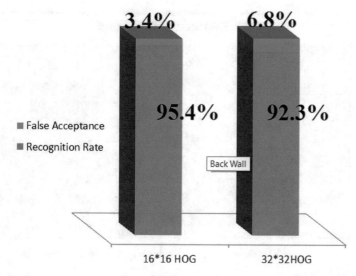

Fig. 12 FKP recognition results

8.1 Time Investigation

The proposed model is implemented in Java language. The smartphone emulator has been equipped with CPU Dual-core 1.2 GHz processors with a RAM of 1 GB and Android as its operating system. The system detailed time analysis is shown in Fig. 13. The proposed model can achieve a single online verification in less than 7 s in case of 16-cell-size and less than 3 s in case of 32-cell size in our prototype model, which is considered fast enough for real-time online applications. Subsequently, the 32 Cell-Size outperforms the 16 cell-size in term of recognition time.

8.2 Comparative Analysis

In this paper, we have only explained FKP over Smartphone papers as the performance comparison. As we mentioned above, there is only a single work that used the smartphone camera for FKP recognition [21].

- **Accuracy Evaluation**

The method in [21] used Log-Gabor filter with Hamming distance for FKP recognition and achieved an accuracy rate 72%, While we achieved 95.4% recognition rate. SO, we conclude that the HOG descriptor with SVM is more useful than the Log-Gabor filter with Hamming distance for FKP recognition as we achieved higher recognition rate. We have empirically observed that a 32-cell-size does not achieve

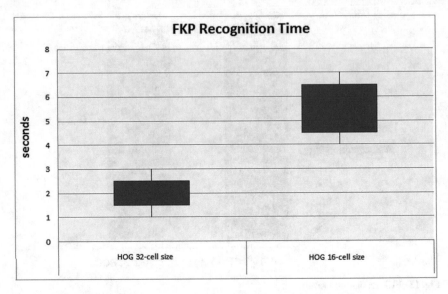

Fig. 13 FKP recognition time

the high performance. Thus, we used the 16-cell-size bit code. Cell size reduction has scaled the number of features of each pixel, So It resulted in a better performance.

- **Usability Evaluation**

The method in [21] ignored the system usability and concentrated on exploiting knuckle images for automated authentication for smartphones. On the other hand, we concentrated on usability by performing a trade-off between the FAR and FRR in our proposed model we achieved FAR 3.4 and 5.5% FRR.

- **Time Evaluation**

We evaluate the recognition time for our schemes. Ideally, we would like the recognition time to be as low as possible to prevent the adversary from accessing confidential data using their devices. For the evaluation of the recognition time, the 32-cell HOG achieved the fastest in the range of 1–3 s. Another scheme generally achieves recognition time in less than 7 s. Empirically, this may be due to the fact that at the feature vector dimensionality increased.

The method in [21] recognition time doesn't exceed 5 s in offline mode. Then, our proposed framework recognition time is fast enough for real-time online applications.

9 Conclusion

This paper proposed a new online finger knuckle authentication system using the smartphone built-in camera. The proposed system has been carried out with the aim to develop a FKP recognition system for applications such as financial, e-business and banking schemes. In this work, the proposed algorithm is based on HOG descriptor with SVM classifier. The major contributions of this paper include an efficient FKP recognition approach using the histogram of oriented gradients as a feature descriptor. It was found that Histogram of Oriented gradients worked best both in accuracy and speed as compared to 1D LOG-Gabor filter with Hamming distance. Initially, the extracted ROI samples are further reconstructed using HOG descriptor to obtain 1D feature vector representations. The Hog 1D feature vector is matched using the SVM classification algorithm. To add merit to the work, the recognition time is evaluated. Our proposed methods are better in terms of accuracy (95.4%), usability (i.e. FAR 3.4%, FRR 5.5%) and computation time don't exceed 7 s (i.e. (1–7) s).

Future challenges in this work are that additional efforts are required to assure precise location and detection of the finger knuckle, from dorsal knuckle images with non-uniform environments and noisy background, since recognition accuracy is significantly influenced by accuracy of limited segmentation.

References

1. Truong A (2012) More iPhones are sold than babies are born each day. http://news.discovery.com/tech/gear-and-gadgets/more-phones-are-sold-than-babies-are-born-each-day.htm
2. How smartphones are on the verge of taking over the world. Homepage, http://www.nydailynews.com/life-style/smartphones. Last accessed 17 Dec 2017
3. Introducing the Symantec smartphone honey stick project. Homepage, http://www.symantec.com/connect/blogs/ introducing-Symantec-smartphone-honey-stick-project. Last accessed 18 Dec 2017
4. Wang L (2012) Some issues of biometrics: technology intelligence, progress and challenges. Int J Inf Technol Manage 11(1-2):72–82
5. Zhang D et al (2011) Online joint palmprint and palmvein verification. Expert Syst Appl 38.3:2621–2631
6. Meraoumia AA, Chitroub S, Bouridane A (2014) Robust human identity identification system by using hand biometric traits. In: 2014 26th international conference on microelectronics (ICM). IEEE
7. Ross A, Nandakumar K, Jain AK (2008) Introduction to multibiometrics. Handbook of biometrics. US, Springer, pp 271–292
8. Kumar A, Xu Z (2014) Can we use second minor finger knuckle patterns to identify humans? In: Proceedings of the IEEE conference on computer vision and pattern recognition workshops
9. Kong T, Yang G, Yang L (2014) A new finger-knuckle-print ROI extraction method based on probabilistic region growing algorithm. Int J Mach Learn Cybernet 5(4):569–578
10. Yu H et al (2015) A new finger-knuckle-print ROI extraction method based on Two-stage center point detection. Int J Sig Process Image Process Pattern Recogn 8.2:185–200
11. Nigam A, Tiwari K, Gupta P (2016) Multiple texture information fusion for finger-knuckle-print authentication system. Neurocomputing 188:190–205
12. Zhang L, Zhang L, Zhang D (2009) Finger-knuckle-print: a new biometric identifier. In: 2009 16th IEEE international conference on image processing (ICIP). IEEE
13. Kumar A, Zhou Y (2009) Human identification using knucklecodes. In: IEEE 3rd international conference on biometrics: theory, applications, and systems, 2009. BTAS'09. IEEE
14. Kumar A, Ravikanth C (2009) Personal authentication using finger knuckle surface. IEEE Trans Inf Forensics Secur 4(1):98–110

15. Zhang L, Zhang L, Zhang D (2010) Monogeniccode: a novel fast feature coding algorithm with applications to finger-knuckle-print recognition. In: 2010 international workshop on emerging techniques and challenges for hand-based biometrics (ETCHB). IEEE
16. Zhang L et al (2011) Ensemble of local and global information for finger–knuckle-print recognition. Pattern Recogn 44.9:1990–1998
17. Morales A et al (2011) Improved finger-knuckle-print authentication based on orientation enhancement. Electron Lett 47(6):380–381
18. Yu PF, Zhou H, Li HY (2014) Personal identification using finger-knuckle-print based on local binary pattern. Appl Mech Mater 441. Trans Tech Publications
19. Choraś M, Kozik R (2012) Contactless palmprint and knuckle biometrics for mobile devices. Pattern Anal Appl 15(1):73–85
20. Meraoumia A et al (2015) Finger-Knuckle-print identification based on histogram of oriented gradients and SVM classifier. In: 2015 first international conference on new technologies of information and communication (NTIC). IEEE
21. Cheng KY, Kumar A (2012) Contactless finger knuckle identification using smartphones. In: 2012 BIOSIG-Proceedings of the international conference of the biometrics special interest group (BIOSIG). IEEE
22. Zhang L et al (1998) Ensemble of local and global information for finger–knuckle-print recognition. Pattern Recogn 44.9:1990–1998
23. Gao G et al (2014) Integration of multiple orientation and texture information for finger-knuckle-print verification. Neurocomputing 135:180–191
24. Kumar A, Wang B (2015) Recovering and matching minutiae patterns from finger knuckle images. Pattern Recogn Lett 68:361–367
25. Dalal N, Triggs B (2005) Histograms of oriented gradients for human detection. In: IEEE computer society conference on computer vision and pattern recognition, 2005. CVPR 2005, vol 1. IEEE
26. Barbu T (2012) SVM-based human cell detection technique using histograms of oriented gradients. Cell 4:11
27. Cao X et al (2011) Linear SVM classification using boosting HOG features for vehicle detection in low-altitude airborne videos. In: 2011 18th IEEE international conference on image processing (ICIP). IEEE
28. Vapnik VN (1998) Statistical learning theory. Springer
29. The Hong Kong Polytechnic University Mobile Phone Finger Knuckle Database, http://www4.comp.polyu.edu.hk/~csajaykr/knuckle.htm 2012

Printed in the United States
By Bookmasters